Environmental Coastal Regions III

THIRD INTERNATIONAL CONFERENCE ON
ENVIRONMENTAL PROBLEMS IN COASTAL REGIONS

COASTAL ENVIRONMENT III

CONFERENCE CHAIRMEN

G.R. Rodríguez
Universidad de Las Palmas de Gran Canaria, Spain

C.A. Brebbia
Wessex Institute of Technology, UK

E. Pérez-Martell
Universidad de Las Palmas de Gran Canaria, Spain

LOCAL ORGANISING COMMITTEE

G.R. Rodríguez
E. Pérez-Martell
M. Pacheco
L. García-Weil
A. Tejera

INTERNATIONAL SCIENTIFIC ADVISORY COMMITTEE

J. Baldasano
C. Borrego
A. H-D. Cheng
N.F. Ebecken
R.A. Falconer

K. Jensen
M. Koch
J.G. Kretzschmar
B. O'Connor
R. Rajar

Organised by
Wessex Institute of Technology, UK
Universidad de Las Palmas de Gran Canaria, Spain

Sponsored by

Ayuntamiento de
Las Palmas de Gran Canaria

Universidad de
Las Palmas de Gran Canaria

Gobierno Autónomo
de Canarias

Environmental Coastal Regions III

EDITORS:

G.R. Rodríguez
Universidad de Las Palmas de Gran Canaria, Spain

C.A. Brebbia
Wessex Institute of Technology, UK

E. Pérez-Martell
Universidad de Las Palmas de Gran Canaria, Spain

G.R. Rodríguez
Universidad de Las Palmas de Gran Canaria, Spain

C.A. Brebbia
Wessex Institute of Technology, UK

E. Pérez-Martell
Universidad de Las Palmas de Gran Canaria, Spain

Published by

WIT Press
Ashurst Lodge, Ashurst, Southampton SO40 7AA, UK
Tel: 44 (0) 238 029 3223; Fax: 44 (0) 238 029 2853
Email: witpress@witpress.com
http://www.witpress.com

For USA, Canada and Mexico

Computational Mechanics Inc
25 Bridge Street, Billerica, MA 01821, USA
Tel: 978 667 5841; Fax: 978 667 7582
Email: cmina@ix.netcom.com
US site: http://www.compmech.com

British Library Cataloguing-in-Publication Data

A Catalogue record for this book is available
from the British Library

ISBN: 1-85312-827-9
ISSN: 1462-6098

The texts of the papers in this volume were set individually by the authors or under their supervision. Only minor corrections to the text may have been carried out by the publisher.

No responsibility is assumed by the Publisher for any injury and/or any damage to persons or property as a matter of products liability, negligence or otherwise, or from any use or operation of any methods, products, instructions or ideas contained in the material herein.

PREFACE

As population growth and industrial and tourist activity development continue in the coastal zone, the environmental stress on nearshore water, air, and ground increases. Coastal waters are considered a convenient dumping ground for waste materials or energy dispersion, because of its large dilution capacity. However, discharges of municipal and industrial wastes, and spills of potentially toxic substances from other sources contribute to a host of water quality problems. Location of conventional or nuclear power plants along the coast can give rise to serious problems of water and air contamination. The increasing demand of recreational areas and the need to mitigate storm hazards and coastal erosion has lead to the construction of coastal structures. These and other activities gives rise to physical and ecological changes in the coastal hydrodynamics, declining biodiversity or populations, and in extreme cases to health problems for human beings.

Due to its high ecological, social, economic and cultural interest, almost all activities or actions in the coastal zone generate conflicts among different interested parts: users, landowners, engineers, ecologists, economists, politicians, etc. Thus, the development of effective strategies for coastal management should consider the coast as a dynamical integrated system in which inputs, outputs and recirculation of matter and energy control the environmental quality.

This conference address the subjects of monitoring, analysing, and modelling coastal water, air and ground phenomena which need to be understood to prevent, alleviate, or minimise environmental problems, allowing a balanced use of the coastal regions as a common resource.

The editors are grateful to all participants and reviewers and most particularly to all the authors who contributed to the high quality of the book. Special recognition is expressed to all the organisations that helped the conference become a reality. The support from the University of Las Palmas de Gran Canaria and the Wessex Institute of Technology are gratefully acknowledged.

The Editors
Las Palmas de Gran Canaria
September 2000

CONTENTS

Section 6: Sediment Transport

Section 7: Remote Sensing

Section 8: Pollution Studies

Section 1
Environmental Management

Modelling the transport of sediment and plutonium from the Mururoa lagoon

R. Rajar & D. Zagar
Faculty of Civil and Geodetic Engineering, University of Ljubljana, Ljubljana, Slovenia

Abstract

Between 1966 and 1996 the French Government carried out a programme of nuclear weapons testing at Mururoa and Fangataufa atolls in the Tuamotu Archipelago in French Polynesia. In 1997/98 an international study was elaborated with the goal to show the impact of the testing on the environment. One of the aspects of this study was also to study the possible future outflow of radioactively contaminated sediments from the Mururoa lagoon into the Pacific Ocean. A 3D hydrodynamic (HD) and sediment transport model was used to calculate the circulation and the sediment outflow from the lagoon. The model is shortly described in the paper. Two scenarios were simulated: (a) Normal conditions, where tidal forcing and easterly trade wind of 8 m/s were taken into account. The outflow of sediment through the pass in the lagoon rim was determined to be about 80 000 tons/year; (b) Storm conditions, with wind from NW, max. speed of 150 km/h were calculated. Frequency of these storms (tropical cyclones) is about 0.1 storm/year. The sediment outflow through the pass in these conditions would be about 720 000 tons/storm. A large amount of sediments would also be outwashed by waves over the south rim of the atoll, i.e. about 3 200 000 tons/storm. Taking into account both normal and storm conditions, the average annual outflow of sediment-bound Plutonium would be about 77 Gbq/year.

1 Problem description

At the Mururoa and Fangataufa Atolls in French Polynesia, the French government conducted 46 atmospheric and 147 underground nuclear experiments between the years 1966 and 1996. Recent measurements in both lagoons have shown that the sediment is partly contaminated, primarily by isotopes of plutonium, which for the most part is the consequence of the atmospheric tests. As part of the study *The Radiological Situation at the Atolls of Mururoa and Fangataufa* under the International Atomic Energy Agency, modelling of the transport of the contaminated bottom sediment of the Mururoa lagoon was undertaken to asses its radiological impact.

The Mururoa atoll is situated in the Tuamotu Archipelago. Its maximum dimensions are approx. 25 x 10 km (see Fig. 2). The maximum depth of the lagoon is 55 m, its average depth being 33 m. The exchange of water with the Pacific Ocean is enabled mainly through the pass in the NW rim, which is 8 km long with average depth of 8 m. The main objective of this part of the study is to provide answers to the following questions:

(a) What amount of bottom sediments in the Mururoa lagoon would be resuspended and transported out of the lagoon into the Pacific Ocean per year according to different scenarios (circulation due to trade wind and tide, tropical cyclones)?

(b) What is the outflow of radionuclides transported with the sediments out of the lagoon per year and by each typical cyclone.

The need for using modelling tools is obvious. It is practically impossible to carry out extensive measurements, which could provide us with the necessary answers, especially in the case of storm conditions. However, measurements are necessary for calibration and verification of the model results.

2 Model description

A three-dimensional (3D) model PCFLOW3D, developed at the Ljubljana University, was used for the simulation of water circulation in the lagoon and for the sediment resuspension and transport inside the lagoon and out of it (Rajar and Cetina [1], Rajar et. al. [2]). The integrated model is composed of several modules. Only hydrodynamic and sediment transport modules were used in this study.

The hydrodynamic module, which simulates water circulation in the lagoon in different conditions is three-dimensional (3D) and non-linear. A one-equation turbulence model of Koutitas is included for the calculation of vertical turbulent viscosity. The finite-volume numerical method is used for the solution of the basic equations. The module is baroclinic, but for the Mururoa case the simulations were carried out with uniform water density.

Sediment transport module is based on the equations of Van Rijn [3]. Resuspension of sediment from the bottom depends on the bottom shear stress, caused by current velocities and waves. The transport and dispersion of sediments inside the lagoon and out of it (through the pass) was simulated depending on the combined action of currents and waves.

It was estimated that a significant amount of sediment can also be washed out of the lagoon over the SE and S rim during cyclones. To determine the water discharge over the rim, all the available data on the observations of the phenomena (wind, waves, storm surge) in the Mururoa and similar lagoons were taken into account [4] together with an empirical procedure for the calculation of water discharge due to wave overtopping of the rim described in Sylvester [5].

The discharge of sediment bound radionuclides from the lagoon for different scenarios was calculated with the following assumptions:

(a) Only $^{239+240}$Pu was taken into account, as the most important element bound to sediment.

(b) As the half-life time of plutonium is very long (for ^{239}Pu it is about 24000 years), the decay rate was neglected.

(c) With the concentration of radionuclides in sediment Csed (Bq/kg) given as input data, the outflow discharge of the sediment bound radionuclides R (Bq/year) was calculated to be **R (Bq/year) = Csed (Bq/kg) * Qsed (kg/year)**, where Qsed is the outflow discharge of sediment.

Data on bottom sediments were taken from the report of Masse and Mussa, [6], based on measurements. The D50 particle diameter is 0.1 mm.

3 Simulated cases and results

3.1 Normal conditions: Effect of trade wind and tide

Normal conditions, governing the circulation during the greater part of the year were taken into account first: trade wind, easterly of 8 m/s, together with the tide, with an amplitude of 0.3 m and a period of 12.4 h (Fig.1a).

Hydrodynamic simulation of current velocities was carried out first (Fig. 2). The wave height caused by east wind of 8 m/s increases from east to west in the lagoon and attains a maximum value of nearly 1 meter near the westernmost coast. The computed hydrodynamic field was in close agreement with the simulation results of Tartinville (7), whose part of the study was to determine the outflow of radionuclides dissolved in water.

Near the pass, the current velocities are of the order of 20 cm/s. Most of the sediment resuspension occurs in the vicinity of the pass, and some in the shallow water along the southern coast. There is a resuspension and transport of suspended sediments out of the lagoon during ebb tide. The simulation results show that during one tidal cycle a mass of 1.1×10^5 kg is washed out of the lagoon. This results in a **rate of sediment outflow due to wind and tide of about 80 000 tons/year,** which corresponds to about 3 kg/s over the whole cross section of the pass.

The outwashed mass of 80 000 tons would mean an erosion of 0.4 mm/year over the whole lagoon. This value is given only for illustrative purposes, because the erosion would not be distributed uniformly over the lagoon, and some sediments are also formed and deposited by various processes. No measurements of this phenomenon were available for the Mururoa lagoon.

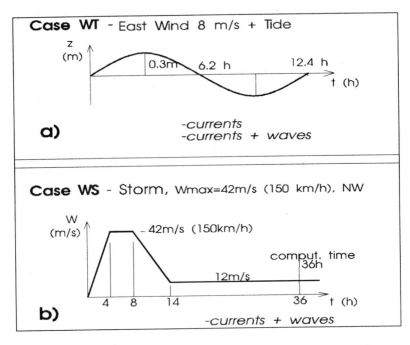

Fig.1. Forcing factors – wind and tide for (a) Normal conditions
(b) Storm conditions

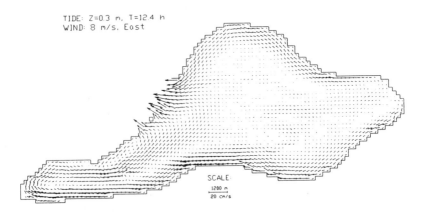

Fig. 2. Simulated depth-averaged velocities at ebb tide: Normal conditions.

3.2 Storm conditions

There is evidence of tropical cyclones in the region of Mururoa and also of their impact on the sediment transport. The forcing conditions for this case were determined mainly from [4]. The *"maximum historically probable cyclone"* in this region has been defined as a NW wind of 80 knots (150 km/h or 42 m/s). Fig. 1b. shows the assumed time development of wind speed for such a storm. The duration of maximum wind speed was estimated to be 4 hours. During the passage of the cyclone the direction of the wind is changing, but the NW direction was taken into account, as it was decided that this has the greatest influence on the phenomenon.

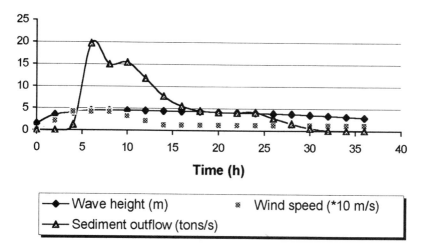

Fig. 3. Wave height, wind speed and sediment outflow through the pass: Storm conditions.

The frequency of such cyclones was estimated from [4] to be about 0.1 per year (one cyclone per 10 years). Although the statistical data show smaller frequency during the last 100 years, there were 5 cyclones in the first half of 1983 in the Tuamotu Archipelago (though not all of them touched Mururoa). Gabrie and Salvat [8] describe the frequency in the region of Mururoa to be "four to eight per century".

Resulting flow of water and sediments through the pass. The 3D hydrodynamic simulations have shown surface velocities of the order of 1.3 m/s and maximum depth-averaged velocities (Fig. 4) to the order of 40-50 cm/s, which, for the most part, is in agreement with the observed values, described in [4]. Wave height is described to be almost 10 m in the ocean and about 5 m in the central part of the lagoon (due to transmission through the pass, and due to wave diffraction, refraction and reflection). This was taken into account in the hydrodynamic and sediment transport simulations.

Fig. 5 shows the distribution of the ration of the bottom shear stress to the critical shear stress. Wherever this ratio is greater than 1.0 resuspension of the sediment will occur. The final results of the modelling for storm case are shown in Fig. 3. The time integration of the sediment discharge over the storm duration shows that about 720 000 tons of sediment are transported out of the lagoon during one storm event.

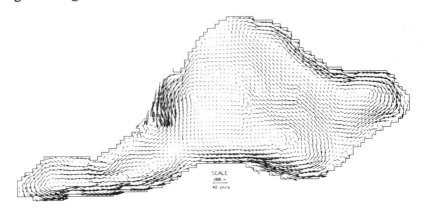

Fig. 4. Depth-averaged velocities for Storm conditions (wind NW, 150 km/h)

Flow of water and sediment over the SE rim of the atoll. For wind of 150 km/h in the lagoon, from the NW, the height of the waves along the SE and S rim of the atoll, was estimated to be 5 metres (Ref. [4]). The height of the rim above the mean water level is between 1 and 3 metres. It is expected that a cyclone would cause overtopping of the rim, with significant discharge of both water and sediment over the rim into the ocean. This is confirmed by observations from similar atolls. Lachenaud [4] describes observations on several atolls during cyclones, mostly with the "greatest part of the atoll submerged". A combination of data from the aforementioned paper and an empirical procedure from Sylvester [5] was used to determine the water discharge and another procedure from [3] was used to determine the sediment concentration and further on the sediment outflow.

Basic data and assumptions for the calculations are given in Fig. 6. The mean height of the rim above the mean water level is taken to be 1.5 m. As the wind shear stress would cause storm surge of about 0.5 m at the SE side of the lagoon, we have taken into account the average height of the rim over the storm water level (SWL in Fig. 5) to be only 1 meter. According to [4], it was calculated that the wave run-up over the mean water level is Ro = 3.5 m. Due to the fact, that the atoll is about 300 to 400 m wide, we estimated the effective wave run-up to be diminished to 2.7 m. (Together with a 0.5 m rise of the water level due to storm surge the total rise of the water level over the mean is 3.2 m which is in agreement with observations from [4]). According to Sylvester [5], with the aforementioned values of parameters we obtain a discharge of water of 0.5 m3/s/m. Over the length of 20 km and over 6 hours of the storm duration, a volume of $2.16 * 10^8$ m3 of water would flow over the rim.

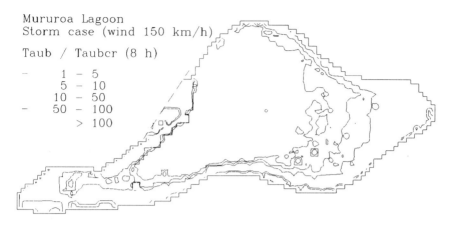

Fig. 5. Ratio of bottom shear stress to critical shear stress: Storm conditions.

For the conditions described, the sediment concentration near the bottom of the lagoon would be about 50 kg/m3. As high mixing over the depth is to be expected, we have assumed that the depth-averaged concentration will be 0.6 of this, i.e. 30 kg/m3. This results in a mass of $6.5 * 10^6$ tons of sediments washed out of the lagoon over the 20 km of the rim during one storm event. However, this outflow of sediment could only be realistic if the whole area of the shallow sea near the SE to S rim, the beach and the nearby part of the atoll were all completely erodible. As there are parts of the described region which are not, or are only partly erodible (corral reefs, carbonate, structures...) we estimated that only 50 % of the calculated sediment mass would be washed out. Hence the total mass of sediments, washed over the rim during one storm event would be **$3.3 * 10^6$ tons.** This is roughly 4.5 times the mass, which would be washed out of the lagoon *through the pass* during the same storm event.

A simple calculation shows that this would cause a beach erosion about 0.5 m deep, which is calculated over the whole 20 km length of the rim and over 300 m width (this would be partly in the shallow sea near the beach, along the beach itself and partly on the atoll). In fact the erosion would be concentrated along some weaker, more erodible sections. Cases of total erosion of some parts of atolls by cyclones confirm that such erosion is realistic (The Caye Caulkner Atoll in Belize was recently cut in two by a cyclone and a channel was formed in between).

In total, the amount of sediment washed out of the lagoon during one storm is roughly **4×10^6 tons.** This is equivalent to a removal (erosion) of a layer about 2 cm thick *over the entire lagoon bottom*. In reality this erosion would be distributed very non-uniformly.

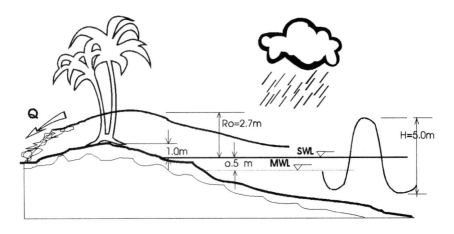

Fig. 6. Parameters for computation of overflow over the atoll rim: Storm conditions. MWL: mean water level, SWL: storm water level, Ro: wave run-up, Q:water discharge over rim.

4 Outflow of radionuclides

The average $^{239+240}$Pu concentration in the top 10 cm of the bottom sediment in the Mururoa lagoon has been estimated, on the basis of French measurements and results obtained from the Study [9], to be 500 Bq/kg. For the regions near the pass and in the vicinity of the SE rim, the concentration is estimated to be much lower, about 20 Bq/kg. As during both cases (normal and storm conditions) the sediment is partly, but not entirely, mixed over the lagoon, we estimated the concentration near the pass and in the vicinity of the SE rim to be about 100 Bq/kg.

The final result is: The amount of radioactive material washed out through the pass by permanent action of the trade winds and tides would be about **8 GBq/year**, and the total outflow of plutonium during one storm event (through the pass and over the rim) is **0.7 TBq/storm**.

The estimated outflow of sediment and plutonium from the Mururoa lagoon to the open ocean is summarised in Table I. The dominant release of plutonium will occur during storm conditions. Due to some not precisely known data (thickness of sediment cover on the bottom, topography of the atoll rim) and to some rough estimations in the calculations (empirical formulae for calculation of the overtopping of the rim due to storm surge and waves) the estimates of the outflow of sediments and plutonium are considered to be accurate to within a factor of three.

Table 1. Sediment and plutonium outflow from Mururoa lagoon.

	Wind-tide case		Storm case	
	Sediment (t/year)	$^{239+240}$Pu (Bq/y)	Sediment (t/storm)	$^{239+240}$Pu (Bq/storm)
Through pass	*8 x 10^4*	*8 x 10^9*	*7.2 x 10^5*	*3.6 x 10^{11}*
Over SE rim			*3.2 x 10^6*	*3.3 x 10^{11}*
TOTAL	**8 x 10^4**	**8 x 10^9**	**3.9 x 10^6**	**6.9 x 10^{11}**

5 Conclusions

To determine all the aspects of impact of the underground tests on the environment, besides the local modelling of the conditions in the lagoon, regional and global modelling has also been carried out in the study. For regional modelling, with the goal to determine radiological conditions in the nearby islands of the Tuamotu group, compartment models were used. Further on global models were applied to determine the transport and dispersion of radionuclides over the Pacific Ocean.

The final conclusions of the Study [9] showed clearly, that even for the worst scenarios (earthquakes, undersea landslides) the radioactive pollution due to underground tests is negligible. The radiation dose to biota would everywhere be far beyond the natural levels of radiation.

References

[1] Rajar, R. and Četina, M. Hydrodynamic and Water Quality Modelling: An Experience. *Ecological Modelling*, 101, (1997), 195-207.

[2] Rajar, R., Četina, M., Sirca, A. Hydrodynamic and Water Quality Modelling: Case Studies. *Ecological Modelling*, 101 (1997), 209-228.

[3] Van Rijn, *Principles of Sediment Transport in Rivers, Estuaries and Coastal Sea*, Aqua publications, Amsterdam, (1993) 405 pp.

[4] Lachenaud, R. Atoll et cyclone tropical. Sur les conditions dans un lagon par gros temps. *Met Mar*, No. 133, (1986).

[5] Sylvester, R., *Coastal Engineering,* I. Elsevier Sc. Publ., Amsterdam, (1974), 457 pp.

[6] Masse, L., And Mussa, C. Etude granulometrique des sediments de surface du lagon de Mururoa. Service mixte de surveillance radiologique et biologique de l'homme et de l'environment. *Report, Ministere de la Defense, Republique Francaise* (1988).

[7] Tartinville, B., Deleersnijder, E., And Rancher, J. The Water Residence Time in the Mururoa Atoll Lagoon: A 3D Model Sensitivity Analysis. Submitted to *Coral Reefs* (1997).

[8] Gabrie, C, And Salvat, B., General features of French Polynesian Islands and their Coral Reefs, *Proc. of the Fifth Intern. Coral Reef Congress*, Tahiti, (1985).

[9]*The Radiological Situation at the Atolls of Mururoa and Fangataufa*, Main report, (Radiological assessment reports series). Vienna: International Atomic Energy Agency, 1998.

Interpolating H&R wave overtopping coefficients

T.S. Hedges
Department of Civil Engineering, University of Liverpool, UK

Abstract

Various expressions have been developed for predicting the mean rate at which seawalls are overtopped by random waves. All of these expressions contain empirical coefficients derived from experimental data. The values of the coefficients are dependent on, amongst other parameters, the front slope of the structure. However, data are usually collected in physical model studies on seawalls with a limited range of front slopes (see, for example, Owen [1, 2]). Thus, it is generally necessary to interpolate the values of these coefficients for the purposes of design.

The H&R overtopping model [3] is the latest proposal for predicting mean overtopping discharges. It is based more soundly than earlier models on a physical understanding of the processes involved. It has also been shown to provide the most promise for future development [4]. The present paper reports one such development, a method for interpolating the empirical coefficients contained within the model. The method turns out to be very simple, a consequence of the model's sound theoretical basis.

Introduction

Owen [1, 2] has described an extensive series of model tests to determine the overtopping discharges for a range of seawall designs subjected to different random wave climates. The modelled seawalls were all of the same general type: a flat-topped embankment fronted in some cases by a flat berm. The modelled seawalls possessed uniform seaward slopes of 1:1, 1:2 and 1:4. The tests were aimed at establishing the impact on overtopping discharge of the wave climate, the seawall slope, the crest and berm elevations, and the berm width.

Hedges and Reis [3] reanalysed Owen's data and proposed a model to describe the relationship between the mean overtopping discharge and the seawall freeboard. The model contains empirical coefficients determined from the data. This paper presents a means of interpolating the coefficients contained within the theoretical model in order to deal with seaward slopes of other than 1:1, 1:2 and 1:4.

An overtopping theory for regular waves

Before we consider overtopping from random waves, let us address the simpler case of regular waves of height H approaching normal to a seawall. We will assume that the instantaneous discharge of water over unit length of the seawall, q, is given by the weir formula [5]:

$$q = C_d \frac{2}{3} \sqrt{2g} \left(\eta - R_c \right)^{3/2} \qquad \text{for } \eta > R_c \qquad (1)$$

in which η is the water surface elevation above still-water-level at the seawall (a periodic function of time), R_c is the freeboard of the structure and C_d is a discharge coefficient. Obviously, overtopping occurs only when the water surface is above the structure's crest.

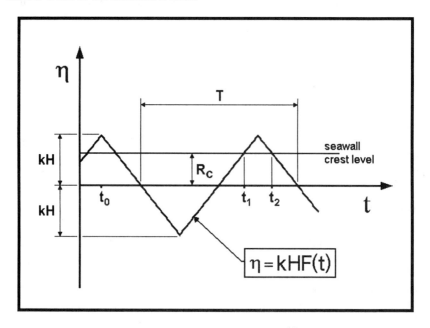

Figure 1: Form of function $F(t)$

We will also assume that:

$$\eta = kHF(t) \tag{2}$$

$F(t)$ denotes a function of time, t. For simple, sinusoidal, progressive waves, $k = 0.5$ and $F(t) = \cos(2\pi t/T)$ where T is the wave period. However, following Kikkawa et al [6], we will adopt the simpler form for $F(t)$ shown in Figure 1; k remains a coefficient determined by the particular wave and wall details. The mean discharge, Q, is determined as follows:

$$Q = C_d \frac{2}{3}\sqrt{2g}\frac{1}{T}\int_{t_1}^{t_2}\left\{kHF(t) - R_c\right\}^{3/2}dt \tag{3}$$

in which $t_1 < t < t_2$ corresponds to the interval during each wave period for which $kHF(t) > R_c$. Using the form for $F(t)$ given in Figure 1 then yields:

$$\frac{Q}{\sqrt{g(kH)^3}} = C_d\frac{2\sqrt{2}}{15}\left\{1 - \frac{R_c}{kH}\right\}^{5/2} \quad \text{for } 0 < R_c < kH$$

$$= 0 \qquad\qquad\qquad\qquad \text{for } R_c \geq kH \tag{4}$$

Note that overtopping occurs only when $R_c < kH$. In other words, kH represents the run-up on the face of the seawall. Since wave run-up is a function of the incident wave height and steepness, and of the seawall slope, the overtopping discharge can be expected also to depend upon these parameters.

The Hedges and Reis (H&R) overtopping model

Analogous to eqn (4), the Hedges and Reis [3] overtopping model for random waves may be written:

$$\frac{Q}{\sqrt{gR_{max}^3}} = A\left(1 - \frac{R_c}{R_{max}}\right)^B \quad \text{for } 0 \leq \frac{R_c}{R_{max}} < 1$$

$$= 0 \qquad\qquad\qquad\qquad \text{for } \frac{R_c}{R_{max}} \geq 1 \tag{5}$$

Since we are no longer dealing with regular waves, kH in eqn (4) has now been replaced by R_{max}, the maximum run-up on the face of the seawall under the specified random wave conditions. Coefficients A and B have replaced

$C_d \dfrac{2\sqrt{2}}{15}$ and the exponent 5/2, respectively. A and B must be determined from the experimental data.

Eqn (5) gives a finite discharge when the freeboard is zero. It gives zero discharge when the crest of the seawall is above the maximum run-up, R_{max}. Unless R_{max} exceeds the freeboard, R_c, there is no overtopping (apart from wind-blown spray). Note that R_{max} may be replaced by CH_s in which H_s is the significant height of the incident waves; C is a coefficient which is governed by the seawall profile and the incident wave characteristics. The value of C may be estimated from run-up measurements for random waves acting on slopes for which there is no overtopping. These run-up measurements complement Owen's overtopping results, allowing the H&R model to be applied outside the range of his experimental data.

A number of equations describing random wave run-up are available. For example, the CIRIA/CUR manual [7] gives two equations for evaluating the significant wave run-up, R_s, on smooth slopes without overtopping. It notes that these equations are probably conservative. Rewritten in our notation and allowing for a printing error, the expressions are:

$$\frac{R_s}{H_s} = 1.35\xi_p \qquad\qquad \text{for } 0 < \xi_p < 2$$
$$\frac{R_s}{H_s} = 3.00 - 0.15\xi_p \qquad\qquad \text{for } 2 < \xi_p < 12 \tag{6}$$

Here, ξ_p is the surf similarity parameter calculated using the period of peak spectral density, T_p ($\xi_p = \tan\alpha / \sqrt{H_s / L_{op}}$ in which $L_{op} = gT_p^2 / 2\pi$).

Adopting the common assumption that run-up may be described by a Rayleigh distribution [7] then the p% confidence value of maximum run-up (defining a level below which p% of the cases should lie) is related to the significant wave run-up by:

$$(R_{max})_{p\%} = \left[\frac{1}{2}\left(\ell n\, N - \ell n\left(-\ell n\frac{p}{100}\right)\right)\right]^{1/2} R_s \tag{7}$$

in which N is the number of run-up values.

Owen recorded his overtopping discharges during tests involving sets of five different runs, each of 100 waves, characterised by the same value of H_s. Making the conservative assumption that the number of run-up values equals the number of incident waves, then the most probable maximum run-up during each run (the value not exceeded in 37% of the cases for a Rayleigh distribution of run-ups) is:

$$(R_{max})_{37\%} = \sqrt{(\ell n \, 100)/2} \; R_s = 1.52 \, R_s \qquad (8)$$

In none of Owen's cases were there overtoppings for freeboards greater than $(R_{max})_{37\%}$ if R_s was evaluated using eqns (6). In fact, all nine reported cases of zero overtopping were for freeboards of less than this value. Hence, setting $C = (R_{max})_{37\%}/H_s$ is conservative in this instance and the following expressions for C then arise from eqns (6) and (8):

$$\begin{aligned} C &= 1.52 \, (1.35 \xi_p) & &\text{for } 0 < \xi_p < 2 \\ C &= 1.52 \, (3.00 - 0.15 \xi_p) & &\text{for } 2 < \xi_p < 12 \end{aligned} \qquad (9)$$

The fact that these expressions for C are conservative may be a result either of the conservative nature of eqns (6) or of deficiencies in the assumptions relating to the distribution of run-ups. However, setting $C = (R_{max})_{37\%}/H_s$ may not always be appropriate. Note that the value of C to be adopted in the regression model depends both upon the level of confidence associated with the prediction of R_{max} and upon the duration of the incident wave conditions. If C is changed then there will be corresponding changes in the values of A and B.

Table 1. Values of coefficients A and B in the H&R model

		H&R MODEL (LAD regression)	
		$C = (R_{max})_{37\%}/H_s$	$C = (R_{max})_{99\%}/H_s$
Slope 1:1	A	0.00703	0.00515
	B	3.42	6.06
Slope 1:2	A	0.00753	0.00542
	B	4.17	7.16
Slope 1:4	A	0.0104	0.00922
	B	6.27	10.96

Table 1 contains the values of A and B determined by Hedges and Reis [3] for eqn (5) using the regression method of least absolute deviations (LAD) when $C = (R_{max})_{37\%}/H_s$. Also shown are the values of A and B when $C = (R_{max})_{99\%}/H_s$. Figure 2 illustrates how coefficients A, B and C influence the overtopping rate.

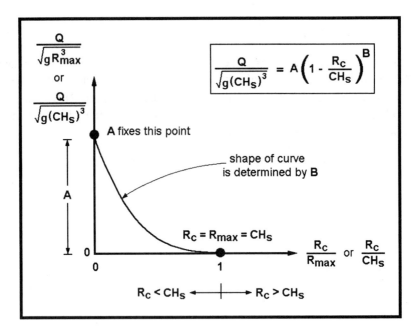

Figure 2: Coefficients A, B and C in the H&R wave overtopping model

Interpolating A and B values

The H&R wave overtopping model has a number of advantages over other overtopping models available in the literature [3].

1. It satisfies the requirements that $Q = 0$ when $R_c = R_{max}$ and Q is finite when $R_c = 0$, a feature which is especially important when the model is used near these freeboard conditions.

2. It explicitly recognises (through its foundations in a simple theoretical model for regular waves) that regression coefficient A depends upon the shape of the structure since the shape, particularly at its crest, affects the discharge coefficient. Coefficient A represents the dimensionless discharge when the dimensionless freeboard is zero.

3. Coefficient B depends upon the detailed behaviour of the water surface on the seaward face of the structure; it increases as front slopes (in the range 1:1 to 1:4) become flatter, linked to increasing non-linearity in the wave conditions.

4. Coefficient C relates the maximum run-up to the significant height of the incident waves and may be chosen to allow for the influences of the seawall slope, the surface roughness and porosity, and the incident wave steepness. Coefficient C can also account for storm duration in influencing R_{max}. Finally, it may be chosen so that there is a specified confidence level associated with R_{max}.

5. Hawkes [4] has stated that the model shows the greatest promise for future development when dealing with combinations of wind-driven waves and swell.

Figures 3 and 4 show the behaviour of the coefficients given in Table 1. Here, the front slope is denoted 1:m (where m = 1, 2 and 4 in Owen's experiments). The ease with which coefficients A and B may be interpolated is clear. The linear trend lines shown in the figures are based upon the coefficient values obtained using the regression method of least absolute deviations, as recommended by Hedges & Reis [3]. However, the values of A and B obtained by the least squares (LS) method are also shown for completeness. Table 2 gives the equations of the trend lines, valid for front slopes in the range 1:1 to 1:4.

Table 2. Equations for coefficients A and B in the H&R model

H&R MODEL (LAD regression)	
$C = (R_{max})_{37\%} / H_s$	$C = (R_{max})_{99\%} / H_s$
A = 0.0012m + 0.0056	A = 0.0014m + 0.0033
B = 0.96m + 2.37	B = 1.67m + 4.16

Concluding remarks

The paper presents a simple means of interpolating empirical coefficients A and B contained within the H&R wave overtopping model. The coefficients behave in a rational fashion, an advantage of the model's sound theoretical basis. Providing explicit relationships for A and B, as well as for C (eqn 9), simplifies the estimation of overtopping rates for seawalls with front slopes in the range 1:1 to 1:4. This range encompasses those front slopes most commonly employed in seawall construction.

Acknowledgement

Some of the ideas in this paper were formulated during discussions with my friend and former PhD student, Maria Teresa Reis. I am grateful for her help.

$$C = (R_{max})_{37\%} / H_s$$

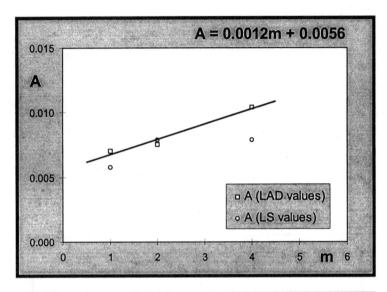

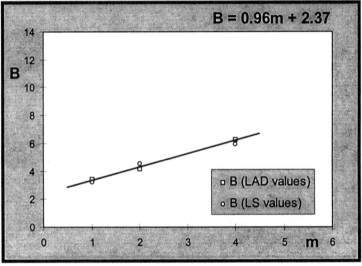

Figure 3: Variation of coefficients A and B with front slope ($C = (R_{max})_{37\%} / H_s$)

$$C = (R_{max})_{99\%} / H_s$$

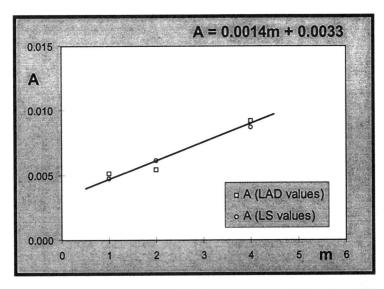

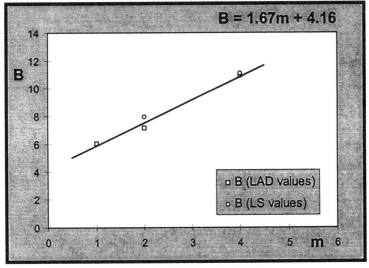

Figure 4: Variation of coefficients A and B with front slope
$(C = (R_{max})_{99\%} / H_s)$

References

[1] Owen, M.W. *Design of Seawalls Allowing for Wave Overtopping*, Report No. EX 924, Hydraulics Research Station, Wallingford, UK, 1980.
[2] Owen, M.W. The hydraulic design of seawall profiles, *Proc. Conf. On Shoreline Protection*, Institution of Civil Engineers, London, pp.129-136, 1982.
[3] Hedges, T.S. & Reis, M.T. Random wave overtopping of simple seawalls: a new regression model, *Proc. Instn Civ. Engrs, Wat., Marit. & Energy*, 130, pp.1-10, 1998.
[4] Hawkes, P.J. Mean overtopping rate in swell and bi-modal seas, *Proc. Instn Civ. Engrs, Wat., Marit. & Energy*, 136, pp.235-238, 1999.
[5] Streeter, V.L. & Wylie, E.B. *Fluid Mechanics*, McGraw-Hill Kogakusha Ltd, Tokyo, 1979.
[6] Kikkawa, H., Shi-igai, H. & Kono, T. Fundamental study of wave overtopping on levees, *Coastal Engineering in Japan*, 11, pp.107-115, 1968.
[7] CIRIA/CUR. *Manual on the Use of Rock in Coastal and Shoreline Engineering*, Special Publication 83, Construction Industry Research and Information Association, London, 1991.

Survey of small fisheries in Japan

M. Takezawa & Y. Maeno
*Department of Civil Engineering, College of Science & Technology
Nihon University, 1-8 Kandasurugadai Chiyodaku, Tokyo, Japan*

Abstract

The coast in Japan is dotted over with many fishing ports.   The former coastal fishery had been flourishing remarkably in Japan. However, many fishing villages have been declining gradually because of successors falling off and other industries developing. As a result, an aging and an under-population are making progress in the present circumstances of fishing villages in Japan.   The fishing villages have enough motivation to be developed as a base of marine resort and recreation, since they have sufficient human and physical resources.   The actual conditions of population movements, industrial structures, welfare facilities, public services, medical facilities, first-aid and rescue works, commerce and business, transportation, water works and sewer, etc. in fishing villages along the coast of Chiba and Kanagawa prefectures closing by the Tokyo Metropolitan area is investigated in this study.

1 Introduction

Recently, fishing ports and municipalities (fishing) in Japan are becoming depopulated due to the employment shift of the young population from primary industries to tertiary industries and the growth of the aged population. This phenomenon can also be seen in Chiba and Kanagawa Prefectures next to Tokyo. Chiba has fishing ports scattered on the Kujukurihama Coast and on the coast inside and outside the Boso Peninsula as well as fishing municipalities behind the ports. Fishing districts in Kanagawa used to flourish mainly along the coasts of the Miura Peninsula and supplied fresh fish to Edo (old Tokyo) but these are now becoming urbanized and extended. The opening of the Tokyo Bay Aqua Line is expected to promote exchange and distribution activities between Chiba and Kanagawa and to influence the two prefectures in many different ways. In the present study, the authors selected main municipalities with fishing ports from both prefectures and conducted materials analysis and a field survey in order to investigate the movement of residents around the fishing ports and changes in the living environment.

2 Survey Method

To find out facts about fishing districts, a materials analysis and a questionnaire survey were conducted as follows:

Objects of survey: Choshi Fishing Port (Specific Class 3), Katsuura Fishing Port (Class 3), Wada Fishing Port (Class 2), Iioka Fishing Port and Tomoura Fishing Port (Class 1) along the coasts of Chiba, and Misaki Fishing Port (Specific 3 Class), Odawara Fishing Port (Class 3), Hiratsuka Fishing Port (2 Class), Iwa Fishing Port of Manazuru and Kita-shimoura Fishing Port of Yokosuka (1 Class) along the coasts of Kanagawa. (Fig.1)

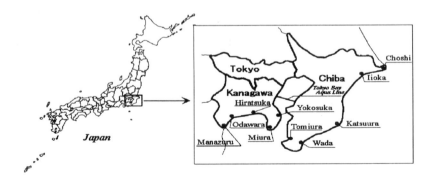

Fig.1: Objects of survey

Items of survey: Population transition (increase or decrease, population by ages, and population by industries), basic configuration of fishing establishments, welfare services (installation of welfare facilities and number of home helpers), medical institutions (medical facilities and number of doctors), educational institutions, waste treatment plants, established water supply systems, citizen strengths, financial statuses, tourist and sightseeing facilities, business spheres (average scale of retailing stores and number of financial institutions), and transportation (shift of workers and means of transportation).

3 Survey Result

3.1 Population Transitions and the Number of Workers by Industry [1),2)]

In both prefectures, the population is increasing every year. However in Chiba, the population is decreasing in most areas within fishing districts. In the aging society, the percentage of aged people has exceeded 25% in Wada-machi and is currently around 20% and increasing in other areas. Of the fishing districts in Kanagawa, the population is increasing in Odawara City and Miura City but is decreasing in other areas where the percentage of aged people is increasing but not as remarkably as in Chiba. When both prefectures are compared in terms of industrial structure, the percentage of workers engaged in the primary industries is higher in Chiba than in Kanagawa. On the other hand, this is not the case for tertiary industries where the percentage of workers is higher in Kanagawa than in Chiba. This is also true for the fishing districts. In Katsuura City in Chiba, however, the percentage of workers engaged in the tertiary industries is the same as in Kanagawa. In Miura City in Kanagawa, the percentage of workers engaged in the primary industries is high but that in the tertiary industries is also high. In general, however, the fishing districts in Kanagawa have the average prefectural industrial structure. In Chiba, the industrial structure differs greatly between municipalities. The percentage of aged people and industrial structures of the fishing districts of Chiba and in Kanagawa are shown in Table 1 and Table 2.

Table1: The percentage of aged people

Year	1980	1985	1990	1995
Iioka	12%	14%	16%	19%
Tomiura	16%	18%	21%	23%
Wada	16%	19%	23%	27%
Katsuura	15%	16%	18%	22%
Choshi	11%	12%	15%	18%
Manazuru	11%	12%	15%	19%
Yokosuka	8%	9%	12%	14%
Hiratsuka	6%	7%	9%	11%
Odawara	8%	10%	12%	14%
Miura	8%	10%	12%	15%

Table 2: The percentage of industrial structures

Industries	Primary	Secondary	Tertiary
Iioka	20.5%	29.5%	49.9%
Tomiura	31.3%	18.4%	50.2%
Wada	27.5%	22.3%	50.2%
Katsuura	12.5%	24.7%	62.7%
Choshi	11.0%	34.1%	54.9%
Manazuru	3.4%	29.0%	67.6%
Yokosuka	1.1%	28.3%	70.3%
Hiratsuka	2.3%	37.3%	60.0%
Odawara	3.4%	34.1%	62.1%
Miura	12.7%	23.4%	63.6%

3.2   Basic Configuration of Fishing Establishments [3], [4]

According to analysis of fishing establishments in both prefectures, fishermen in the fishing districts of Chiba work in family run operations rather than as employees of non-family run operations.   In Kanagawa, however, most fishermen are employees of non-family run operations. Table3 shows the number of fishing establishments, fishing boats, family run operations, and employees of non-family run operations in the fishing districts of Chiba and Kanagawa Prefecture.   Table 4 shows a change of the fishing catches and costs in the fishing districts of Chiba and Kanagawa Prefecture in 1992 ~1996.

Table3: Fishing establishment

	Establish	Boats	Family	Employees
Iioka	19	27	31	34
Tomiura	120	105	169	17
Wada	79	34	81	50
Katsuura	44	30	53	13
Choshi	260	255	331	598
Manazuru	38	41	99	7
Yokosuka	16	10	19	18
Hiratsuka	7	8	11	29
Odawara	53	55	88	59
Miura	50	88	29	1014

Table 4: A change of the fishing catches and costs

	1992	1995	1996	1997	1998
Iioka	19489	15405	12309	8507	19277
	1345454	1575339	1317065	1459475	1341044
Tomiura	713	723	656	584	577
	572160	554096	538761	444697	483320
Wada	2216	2328	1234	1251	1881
	425000	408000	369000	394000	406000
Katsuura	12784	16381	14123	15359	23488
	6184684	4868324	4725939	4160008	10446024
Choshi	306348	305506	226953	184154	208770
	22293975	20780781	19069145	21689673	20999575
Manazuru	1683	1504	1651	1373	1767
	543052	531703	551260	462475	427141
Yokosuka	648	621	440	471	292
	198000	278000	306000	293000	325000
Hiratska	806	684	438	389	543
	17100	296000	202000	217000	211000
Odawara	1500	1443	1247	1233	1506
	690000	539844	510546	539994	549863
Miura	21317	23294	24947	26256	23837
	21627384	25827736	26072516	25557753	24754857

Upper: fishing catches (ton):Lower: total costs of fishing catches (1000yen)

3.3 Number of Homes for the Aged and Medical Institutions [3]

If fishing districts in both prefectures are compared in terms of homes for the aged and home helpers, the fishing districts in Chiba were found to be far behind those in Kanagawa. The numbers of medical institutions and doctors in the fishing districts are greater in Chiba than in Kanagawa. Table 5 shows the number of homes, home helpers, medical institutions (clinics) and doctors in the fishing districts of Chiba and Kanagawa prefecture.

Table 5: Number of homes, home helpers, medical institutions and doctors

	Homes	Helpers	Clinics	Doctors
Iioka	0	3(1.4)	4	5(0.4)
Tomiura	0	2(0.5)	2	1(0.2)
Wada	0	2(1.2)	2(1)	3(0.5)
Katsuura	2(100)	3(0.5)	16(2)	26(1.1)
Choshi	4(290)	9(0.6)	52(6)	121(1.5)
Manazuru	0	2(1.0)	3	4(4.2)
Yokosuka	12(990)	407(6.2)	289(14)	623(14.3)
Hiratsuka	7(500)	322(10.5)	171(11)	315(12.4)
Odawara	7(548)	149(5.5)	160(14)	288(14.4)
Miura	2(100)	21(2.3)	22(2)	57(10.5)

( ) : the number limit of homes, the number of home helpers per a thousand aged people, the number of hospitals and the number of doctors per 10 thousands of population.

3.4 Number of Educational Institutions

The number of educational institutions in fishing districts is somewhat smaller in Chiba than that in Kanagawa. There are no senior high schools in Iioka-machi or Tomiura-machi.

3.5   Environmental Facilities

With the problem of depopulation, this produces an environment in which it difficult for young people to live. The number of cleaning agents waste treatment plants is greater in the fishing districts of Kanagawa but differs greatly between districts in Chiba. Water supply system in fishing districts in Chiba are not so well established, compared to those in fishing districts in Kanagawa. The diffusion index of sewage is zero in some districts in both prefectures. These areas may use independent or combined treatment septic tanks or privies, but also drain domestic wastewater directly into the sea.

3.6   General Index of Citizen Strength and Financial Coefficient [3]

The general index of citizen strength indicates the critical living and management energies of people living in an area, such as populations, the number of households, and amount of taxable income. The fishing districts have a smaller general index in Chiba than in Kanagawa. The financial coefficient in fishing districts is much smaller in Chiba than in Kanagawa. The average income in fishing districts in Chiba is equal to or greater than the national average but 70% to 80% of the national average in fishing districts in Kanagawa. Table 6 shows the general index of citizen strength (per a hundred thousand), the level of citizen strength for each person (the national average 100), the gap of income (the national average 100), the average value of residential area (yen per m²), and the financial coefficient.

Table 6: General index of citizen strength and financial coefficient

	Index	Level	Gap	Land	Financial
Iioka	7.6	84.4	71.9	28100	0.32
Tomiura	3.9	83.0	69.6	47100	0.26
Wada	4.2	87.5	78.9	32900	0.25
Katsuura	16.9	88.9	72.6	57600	0.56
Choshi	61.1	93.6	81.5	71400	0.57
Manazuru	6.7	87.0	98.8	145300	0.65
Yokosuka	329.7	95.7	113.0	207400	0.92
Hiratska	201.2	100.1	114.5	234200	1.16
Odawara	160.3	101.5	112.7	196300	1.14
Miura	37.5	87.2	100.0	201000	0.83

3.7   Number of Tourists

The number of tourists in fishing districts is much smaller in Chiba than in Kanagawa.   This is probably because public transportation networks are more developed and the percentage of workers in the tertiary industries is higher in Kanagawa than in Chiba as shown in Table 2.   However, the amount of money spent by each tourist is high because the percentage of lodging tourist is great in the fishing districts of Chiba while the majority of tourists are day trippers in the fishing districts of Kanagawa.   Table 7 is the number of tourist and the amount of money by each tourist, and Table 8 is various facilities of hotels, restaurants, fish shop, sea products.

Table7: Number of tourist and amount of money spent by each tourist

	Total tourist	Day tripper	Lodging	Amount
Iioka	565000	490000	13.3%	3411yen
Tomiura	657000	467000	28.9%	3886yen
Wada	157000	143000	8.9%	1444yen
Katsuura	2585000	2051000	20.7%	4562yen
Choshi	2541000	2183000	14.1%	3734yen
Manazuru	1674000	1585000	5.3%	980yen
Yokosuka	8632000	8393000	2.8%	670yen
Hiratsuka	6300000	6272000	0.4%	270yen
Odawara	4064000	3875000	4.7%	480yen
Miura	5147000	4549000	11.6%	2020yen

Table 8: Various facilities

( ):fishing shop

	Hotel	Restaurant	Fish store	Boat shop	Sea product
Iioka	19	16	13	19(4)	10
Tomiura	4	17	13	6(5)	5
Wada	5	3	4	0(0)	6
Katsuura	34	34	35	30(11)	43
Choshi	35	164	111	0(15)	196
Manazuru	73	64	24	25(3)	24
Yokosuka	59	322	147	91(40)	37
Hiratsuka	19	222	82	8(15)	11
Odawara	50	283	134	23(21)	54
Miura	28	49	69	67(17)	81

3.8   **Business Sphere** [3)]

In terms of the average scale of retail stores, both the sales per store and the annual sales per worker in fishing districts are slightly higher in Kanagawa than in Chiba.   The number of financial institutions is almost equal in the fishing districts of both prefectures but the average population covered by each store is greater in Kanagawa than in Chiba.   Table 9 shows the sales per store and the annual sales per worker , and Table 10 shows the number of financial institutions, the average population and aging population covered by each store in the fishing districts.

Table 9: The sales per store and the annual sales per worker

	Sales per store	Sales per worker
Iioka	45,000,000 yen/year	15,000,000 yen/year
Tomiura	30,000,000 yen/year	8,000,000 yen/year
Wada	35,000,000 yen/year	11,000,000 yen/year
Katsuura	50,000,000 yen/year	15,000,000 yen/year
Choshi	59,000,000 yen/year	15,000,000 yen/year
Manazuru	58,000,000 yen/year	15,000,000 yen/year
Yokosuka	102,000,000 yen/year	18,000,000 yen/year
Hiratsuka	110,000,000 yen/year	19,000,000 yen/year
Odawara	118,000,000 yen/year	21,000,000 yen/year
Miura	65,000,000 yen/year	16,000,000 yen/year

Table 10: Financial institutions

	Number	People per store	Aging per store
Iioka	6	185	351
Tomiura	5	1177	275
Wada	7	856	233
Katsuura	23	1057	232
Choshi	55	1494	276
Manazuru	5	1911	370
Yokosuka	116	3711	552
Hiratsuka	64	3976	466
Odawara	76	2636	380
Miura	15	3579	574

3.9   Transportation

The number of workers in the fishing district at all times is 60 ~80% in Chiba Prefecture and 40 ~60% in Kanagawa Prefecture.   The number of private cars per each people in the fishing districts of Chiba is more than that in the fishing districts of Kanagawa, because the transportation of Chiba is poor in comparison with that of Kanagawa.

4 Discussion and Conclusion

In general, the fishing districts in Chiba Prefecture are behind those in Kanagawa Prefecture.   The area differences, especially significant in public transportation networks, educational institutions, and medical institutions, are one of the factor promoting an aging population and depopulation.   In particular, compared with the medical and welfare facilities for the aged in cities, those in fishing districts are under no control despite an aging population.   Since an increasing aging population is anticipated not only in fishing districts but also throughout Chiba and Kanagawa, welfare services and social foundations (houses, roads, rivers, ports, and railways) should be improved greatly and those fishing districts becoming urbanized and extended should work together.   In addition, the Aqua Line is expected to solved the area differences between fishing districts in both prefectures.

References

1) Management and Coordinate Agency: 1995 Population Census of Japan, 1998.
2) Management and Coordinate Agency: Statistical Observations of Shi, Ku, Machi, Mura, 2000.
3) Toyo Keizai: Digest of Local Economy, 1999.
4) Ministry of Agriculture, Forestry, and Fisheries: The Ninth Fishery Census, 1995

Modelling of trajectory and fate of spills

C. Ambjörn
The Swedish Meteorological and Hydrological Institute, Sweden

Abstract

Purpose
A forecast of an oil spill needs a fully operational and extremely user-friendly system. This has been developed at SMHI (The Swedish Meteorological and Hydrological Institute). It is an Internet-based system for calculation of drift and spreading of oil and chemicals. A user has access to the system via a password and is able to make forecasts 2 days ahead or a hindcast of 5 days. The area on which the system is applied is the Baltic Sea, the Kattegatt, the Skagerrak and partly the North Sea. The system is named STW, Seatrack Web.

Results
The forecasts of the currents are simulated in a 3-dimensional numerical model, which is in full operational drift. Currents from 16 layers are used. Concerning oil; evaporation, emulsification, sinking, stranding and dispersion are taken into account. For chemicals; sinking, mixing and spreading are calculated. Floating objects can also be forecasted and then knowledge about their depth penetration is important. It is also possible to trace where oil and other substances are coming from. The result can be shown as trajectories, animation or spot at every hour.

Conclusions
The system can be used by all countries around the Baltic Sea. The calculations are fast and being an Internet solution makes it very easy for SMHI to make changes in the code, which means that every country gets access to improvements immediately. The system contains models on a high theoretical level, it is easy to get access to at any time and it is extremely user-friendly.

Purpose

The authorities responsible for cleaning up oil that has come out into the sea need a good tool in forecasting the trajectory and the fate of the oil. This is necessary in order to minimise the damages on flora and fauna, the costs and the efforts. The long time effects are also possible to calculate with the same calculation system.

A good forecasting system is necessary to give information about where the threats for beaches and birds are, as well as finding oil that might have got out of sight. Knowing the characteristics of oil at different times in order to choose the most efficient method in the clean-up process.

The calculation has to be fast and within a few minutes there is a 24 - or 48 hour forecast ready to examine.

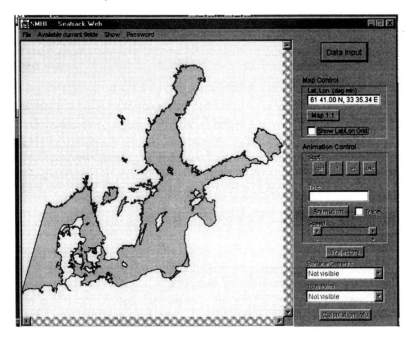

Figure 1 The map of the area

User friendly

The forecasting possibility has to be very user-friendly, which means that it can be used at any time of the day and also be used by persons with limited experience of the system.

Therefore the settings of variables and layout must be self-instructive. As few variables as possible, that are unique, shall need to be given. The rest are default values, which often are good enough for a first quick and rough calculation. In a phase when there is more than one minute's time of planning the calculated forecast or a quite experienced person is making the decision of input data, then it is possible to alter the input radically.

The system also must be fully operational, it is continuously updated with the latest forecasts of winds and currents. The forecast covers 48 hours ahead and 5 days of historical data.

Manual

A detailed pedagogical manual is elaborated, where the user can follow the different steps of a forecast. All information about the technical system needed is also described here.

How the demands are fulfilled

SMHI has since about 20 years had an operational oil drift forecasting system. This is continuously improved, which means that the latest technique is applied. Since the 1st of January 2000 we have a web solution. Anybody having a password has access to this system. The Swedish responsible authorities as well as Denmark, Estonia, Latvia and Lithuania have today access to the forecasting possibilities. The area on which the system is applied is the Baltic Sea, the Sounds, the Kattegatt and the Skagerrak. The system is named STW, Seatrack Web.

Some of the real advantages of the Internet solution are the following:
- Each user always has the latest version, as the code is updated and in the house of SMHI.
- The forecasts of winds and currents are available as soon as they are computed.
- A special News Page is updated continuously
- A new user is easily incorporated
- The forecast is produced in the headquarters by the responsible persons in charge

The different parts of the system

The hydrodynamical model of the North Sea/Baltic Sea

A three-dimensional model called HIROMB (HIgh Resolution Operational

Model for the Baltic Sea) has been developed together with the countries around the Baltic Sea in a co-operation project. The model is in daily operational use and one of its main purposes is to provide current fields to be used in oil drift forecasts. The grid size is 12 nautical miles in the North Sea. Coupled to this and included in the model is a grid of resolution 3 nautical miles, which covers the area from longitude 6 degrees east and eastwards to the Gulf of Finland. In the very near future the model will be run using a spatial resolution of 1 nautical mile, which will increase the quality and the calculation time.

The model is baroclinic, i.e. it simulates salinity and temperature conditions in the entire area. There are 24 layers and the surface layer has a thickness of 4 meters. The layer thicknesses increase successively towards the bottom. The layers are fixed in depth, which means that they do no not follow the density levels. The model is forced by tides, wind and air pressure (provided by the meteorological forecasting model HIRLAM run in daily routine at SMHI), daily forecasted values of fresh water input from 73 rivers are given. A hydrological model is in daily operational use at our institute and those forecasts are given to the model. This is a heavy improvement of the reliability. The time step of the calculations is 10 minutes.

Today, daily operational simulations are made at SMHI and forecasts cover the next 48 hours. It is the results from this model, which among others, are used by the Swedish Coast Guard when making forecasts of oil drift. The forecast routines have a very high security.

As this hydrodynamical model is a co-operation project between the countries surrounding the Baltic Sea all those countries are being members of owning the model and have full access to all the results. That also means that each country has the possibility to make improvements of the code, which is beneficial for every member.

A great amount of money and efforts are within this project and the main purpose is to have a high quality tool to forecast the drift and fate of oil spills.

This also gives other positive effects. The forecasts can be used for chemicals with many different kinds of behaviour, for search and rescue operations. They can also be used for the reverse mode, i.e. using historical data and make a drift calculation backwards in order to construct the trajectory the substance followed before it came to the point where it was found.

The oil drift model

The oil drift model run by the user contains a spreading and weathering routine (Dick, Soetje, 1990) which uses calculated currents from the hydrodynamical model, as well as water temperature and properties of the oil. Oil, or other substances are represented by a large amount of Lagrangian particles. The precalculated current field transports each particle and the turbulence is included through Monte Carlo processes. Wind drift close to the surface is added separately. As the surface layer in the model is 4 m thick, the water moves with the vertically integrated velocity, which is about 1 per cent of the wind velocity. Oil on the surface normally drifts with 3 per cent of the wind velocity. Hence an extra velocity of 2 per cent of the wind velocity is added to particles in the surface layer. This velocity is calculated from a logarithmical current profile. Mixing, transport and distribution can be calculated for oil, soluble substances and floating objects.

Today there are 8 different kinds of oil with different kinds of characteristics. For oil; evaporation, emulsification, sinking, stranding and dispersion are taken into account. Evaporation is a function of wind velocity and water temperature. Sinking is dependent of the size of the droplets. Stranding is coming into account when an oil droplet reaches the shore or feels the sea bed. Dispersion is calculated from the wind and current velocities and a random effect is added.

For chemicals, mixing and spreading are calculated. The different kinds of specific behaviour that at the moment are taken care of are sinking velocity, floating chemicals and passive behaviour.

The transport from the hydrodynamical model is added and depending of where the particle is, regard is taken to the weight of the velocities in the neighbouring cells.

Floating objects are also forecasted and knowledge about their vertical penetration is necessary. A scheme is constructed, where different wind factors are recommended for different kinds of objects. This scheme is included in the manual.

It is also possible to trace where oil and other substances are coming from. This is an extremely valuable possibility as this can lead to prosecution of offenders. Oil found out in the sea is followed backwards as long as we have calculated current information. This makes it possible to see the point where the oil trajectory crosses an anticipated route of a suspected ship. By checking the time of this point one can get a good picture of which ship that has let out the oil. This information is valuable in court when trying to let the shipping company pay the cleanup and the damages. This calculation does not take into account turbulent movements, as these can not be calculated as a diminishing spreading. Therefore it is the three-dimensional information about the currents, which is used here.

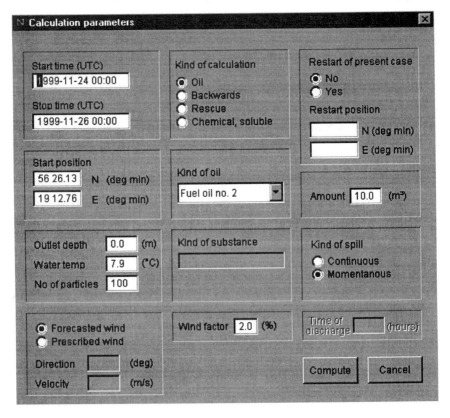

Figure 2 The sheet, where to fill in wanted input data

The calculation

The forecasts of the currents from the 3-dimensional numerical model, Hiromb are used. The model gives forecasts 48 hours ahead. The current information will be available for every 3:rd hour. Currents from 16 layers out of 24 are used, which means that the deepest ones are not used here.

A choice is here made between different data input. The water temperature is given as a default value and this can be altered. Also is an amount of particles suggested which is recommended to make the simulation fast and reliable, this is easily changed. One has a possibility not to use the forecasted wind, but instead a constant wind by one's own choice. This is useful when wanting a special scenario. The choice is then made if one wants to simulate the drift of oil (where 8 different kinds can be chosen), a chemical, rescue or a backward calculation. The amount is given, a choice is made if it is a continuous or an instant spill. If continuous, then also the duration time is given. Having made a forecast today gives the possibility to continue on the same forecast one or two days later. Then one chooses this earlier made result and continues with new forecasts exactly where the other one ended.

Presentation of results

The simulated results can be shown in many different ways, such as trajectories, animation or spot at every hour. The trajectory links together the centre positions of the spots every hour.

Spots are shown at any hour wanted and then a good documentation is given of the situation at any time of the forecast. It is also possible to plot the currents and wind velocity in the same map in which the oil drift is shown. This is very valuable in order to get an understanding of the factors giving the drift and spreading results. There is also an animation facility, which gives a live picture of the course during the whole forecast.

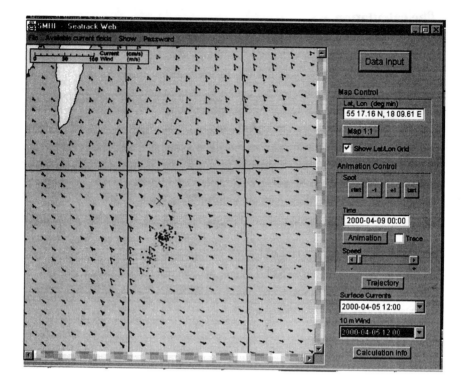

Figure 3 An example of the spot, winds and currents.

Future plans

There are plans to introduce data assimilation in the 3-dimensional hydrodynamical model. There will also be made more validation and of course continuous improvements are made.

Reference

Dick Stephan, Soetje Kai Christian 1990. An operational oil dispersion model for the German Bight. Deutsche Hydrographische Zeitschrift 16, 1990. Bundesamt fur Seeschiffart und Hydrographie, Hamburg. Germany.

Coastal management supported by modelling: optimising the level of treatment of urban discharges into coastal waters

R. Neves[1], A. Silva[2], J. Delfino[2], P. Leitão[1], J. Leitão[2], P. Pina[1], F. Braunschweig[1], R. Miranda[1] & H. Coelho[3]
[1]*MARETEC- Instituto Superior Técnico, Lisbon, Portugal.*
[2]*Hidromod, Modelação em Engenharia, Lda, Oeiras, Portugal*
[3]*University of Algarve, Faro, Portugal*

Abstract

During the 20th century, the very fast growth of urban population created environmental problems in rivers, streams and in some coastal areas (estuaries and coastal lagoons). A Directive for Urban Residual Waters has been issued by DG Environment in 1992 and is being implemented.

This directive imposes a minimum level of treatment, for each size of the agglomeration. Agglomerations with more than 150 000 inhabitants must have secondary treatment. Exceptions to this directive can only be considered in special circumstances and it must be demonstrated that this level of treatment does not produce any environmental benefit compared with a primary treatment.

In the framework of different projects (*e.g.* OPCOM) a modelling system including hydrodynamics, sediment transport and water quality is being used as a tool to help understanding the functioning of coastal wastewater disposal systems. As a result of the implementation of such a system it is possible to evaluate the efficiency of the wastewater treatment method as a function of the receiving water properties.

In this paper it is shown how this integrated modelling system, complemented by monitoring programs and a web database, can be used to demonstrate the existence of the special circumstances mentioned in the European Directive for Residual Waters.

1   Introduction

The recent growth of urban areas led to major problems in what concerns water supply and wastewater management systems. Agriculture's industrialisation also resulted into an increased use of chemical fertilisers. Those fertilisers, together with urban discharges, stimulated the eutrophication of streams and poorly refreshed coastal areas (coastal lagoons and estuaries). The levels of dissolved oxygen in many rivers and in some lakes decreased below the critical values necessary to maintain local ecosystems.

As a remedy to this situation it was imposed that the urban discharges of wastewater must be submitted to primary and secondary levels of treatment in order to remove solids, dissolved organic matter and nutrients. The aim of this policy is to increase the light penetration and to avoid eutrophication in receiving waters. Its success is clear in what concerns inland waters, especially in small stream and lakes.

For a matter of equity between inland and coastal areas, or just for a simple generalisation tendency, the same treatment policy for discharges in inland waters tends to be imposed also in coastal areas where eutrophication is not a problem. In coastal areas, European Union Residual Water Directive obliges agglomerations with more than 150.000 inhabitants to have secondary treatment, unless exceptional circumstances are verified. In these cases it must be demonstrated that this level of treatment will not produce any environmental benefit compared with primary treatment.

In this paper it is shown how modelling can be used to demonstrate the existence of those special circumstances. It is also shown how modelling can be a tool to optimise the degree of treatment of the wastewater as a function of the receiving water properties. For a typical situation in the western Portuguese coast results are shown for nutrients, phytoplankton and dissolved oxygen. The irrelevance of the secondary treatment is a consequence of the high initial dilution rate promoted by a submarine outfall and of the low concentration of nutrients and suspended matter in the receiving waters.

2   The modelling system

MOHID2000 modelling system is based on a set of coupled models including hydrodynamic models, wave propagation models, eulerian and lagrangian advection-diffusion models, sediment transport models and ecological models. MOHID2000 is a 3D-baroclinic model using an object oriented programming philosophy, making use of all the FORTRAN 95 potential [1,2,3]. The system has four main modules: hydrodynamics, eulerian transport, lagrangian transport and water properties.

The system is based on the finite volume concept [4]. In this approach the discrete form of the transport equations are applied macroscopically to the cell control volume in the form of flux divergence. As a consequence this method automatically guarantees the conservation of the transported properties [5].

The number of dimensions of the simulation (1D, 2D or 3D) and the grid to be used (cartesian, sigma or lagrangian) is defined in the hydrodynamic module [2]. This module computes flow properties (velocities, surface elevation, turbulent viscosity, water fluxes) and controls grid deformation, according to the type of grid defined by the user.

Each module has its own initialising procedures. Input records in the datafile are controlled by keywords. In this way each user is completely free to add or eliminate records in the input files used by his module, without compromising any other module. This aspect is particularly important for the water quality module. This module simulates many state variables, with highly empirical formulations. Different tests can consider a variable number of state variables or different parameterisations, which may need different sets of parameters.

2.1 The water quality model

The water quality module provides services to the transport module, calculating source and sink terms. The transport module also uses results calculated by the hydrodynamic module and the air-sea interface module, which calculates heat and mass exchange (water, oxygen, nitrogen) at the free surface.

Source and sink terms in the water quality module are calculated using a "Zero-Dimensional" philosophy. The information is provided to the module in the form of an array, where each line represents a calculating point and each column contains variables relevant for the calculation of the sources/sinks in each point. One point per state variable is calculated, plus one per environmental variable relevant to compute rates: temperature, salinity, light or others.

This methodology leads to a water quality module independent of the number of dimensions in Eulerian calculations and independent of the formalism used in the transport module (Eulerian or Lagrangian).

The water quality module is based on a simplified two level trophic structure, although planktonic food web is unstructured [6]. The module simulates nutrient cycles, organic matter mineralisation, phytoplankton and zooplankton [7]. Dissolved oxygen and biochemical oxygen demand, sources and sinks are calculated explicitly from other variables. Organic mater decomposition due to denitrification is also considered.

3   Wastewater disposal systems in coastal areas

This paragraph describes the steps and results obtained in cases studys developed for different locations in the Portuguese coast. They all had in common a wastewater disposal system composed by a primary treatment plant and a submarine outfall. The flow conditions are influenced by local constrains, such as tide and wind regime, outfall depth and diffuser design and vertical density structure. These features determined the transport and dispersal conditions for each location.

In order to evaluate the benefits of a secondary treatment, when compared with a primary treatment simulations considering the two resulting effluents were performed. The discharge of $1m^3$/s of a typical urban effluent was considered.

Table 1 shows example concentrations, in the presence of the two scenarios, considering an initial dilution of 1:300.

Table 1 - Concentrations after an initial dilution of 1:300 (winter conditions)

Property	Primary Treatment	Secondary Treatment
Phytoplankton (mgC/l)	0.04	0.04
Zooplankton (mgC/l)	0.004	0.004
Particulate Nitrogen (mgN/l)	0.01	0.01
Refractory Nitrogen (mgN/l)	0	0
Labile Nitrogen (mgN/l)	0.001	0.001
Nitrate (mgN/l)	0.1207	0.1207
Nitrite (mgN/l)	0.0299	0.0299
Ammonia (mgN/l)	0.145	0.119
Biol. Oxy. Demand (mgO$_2$/l)	0.555	0.2085
Dissolved Oxygen (mgO$_2$/l)	7.992	7.922

3.1 Western Iberian Continental shelf budget

As an eastern ocean boundary, the coastal transition zone of the Iberian Peninsula is characterised by a marked seasonality related to the large-scale wind climatology. Both satellite imagery and in situ observations of the coastal ocean reveal a surface poleward current as a persistent feature of the winter circulation when winds relax or are downwelling favourable [8,9].

A well defined upwelling season occurs between March and September, [10,11] induced by northerly winds associated with the northward displacement of the Azores high-pressure cell and the weakening of the Iceland low. Upwelling causes the surface dynamic height to decrease towards the coast, and the resulting equatorward geostrophic current is enough to counter the poleward slope current at and near the surface establishing a southward flow. At deeper levels the flow continues poleward. The presence of a meridional pressure gradient off western Iberia all year, balanced during the summer by the southward wind stress, was hypothesised by [7].

During the upwelling season the exchanges between the continental shelf and the open ocean are enhanced. The mechanism is well known. The wind driven transport within the Ekman layer (*e.g.* the upper 100 m) is offshore oriented, contributing to export material produced in the continental shelf. Estimates made during OMEX I-II (EU funded project to study exchanges between shelf and deep ocean) indicate that 30% of the net primary production over the shelf is exported during summer. Bellow the surface layer, the transport is onshore, to compensate the offshore flow in the surface.

Associated with this onshore transport is an inflow of nutrients from the open ocean that are upwelled to the photic zone over the shelf leading to increased primary production. Measures done in the framework of OMEX I-II indicate 400 gC m^{-2} yr^{-1} over the shelf and 200 gC m^{-2} yr^{-1} in the open ocean. On the other hand, the poleward flow is very constrained by bottom topography. There is some evidence that in some places the current can overshoot, forming offshore jets. This seems to be the case in the Setúbal-Lisbon shelf where pronounced capes (Espichel, Roca and Raso) are able to drive the current offshore. This also provides a mechanism to exchange water between the shelf and the open ocean.

As a consequence of the physical processes occurring in the shelf break, of the wind forcing on the shelf and of its short width, the residence time of the water on the shelf is small and the nutrients budget is mainly controlled by the deep ocean import. On the other hand, a large part of the primary production is directly exported to the deep ocean by surface currents perpendicular to the coast. These aspects make the land discharge of nutrients of secondary importance for primary production, eliminating the danger of eutrophication of those areas. Taking into account the fact that urban discharges from coastal cities are responsible only for a small part of the nutrients discharged into the coastal areas, the continental shelf must be expected to be insensitive to the level of treatment of urban effluents from coastal cities.

3.2   Primary production evaluation

The eventual benefits for the receiving waters of a secondary treatment of urban sewer, instead of a primary treatment, were evaluated using MOHID 2000 modelling system.

Long-sea submarine outfalls are designed to promote strong initial mixing of sewer water discharged within the receiving waters. The stronger the initial dilution, the smaller is the environmental impact of the discharge, because the concentration of eventual pollutants is decreased, but also because biological activity is enhanced in the plume. Discharges in coastal waters must be done in zones as deep as possible, to enhance initial dilution and where currents are strong, both for initial dilution reasons and to reduce residence time.

Marine bacteria carry on mineralisation of organic matter and phytoplankton consumes ammonia and nitrate. Initial dilution increases the concentration of those organisms in the plume and consequently increases the rate of consumption of the substances they feed on. High dilution also contributes to an increase in light penetration and, through plume area increase, to promote the exchange of oxygen between the water and the atmosphere.

Estimates of initial dispersion were performed using CORMIX model ("CORnell MIXing System"), developed at the University of Cornell, USA.

Evolution of Biochemical Oxygen Demand, dissolved oxygen, nitrates, ammonia, phytoplankton and faecal coliforms concentrations were assessed for primary treatment and secondary treatment scenarios. Simulations were carried out using MOHID2000 lagrangian transport module.

The water discharged by long-sea submarine outfalls (after initial mixing in the near field) is released at different levels in the water column and labelled by the model. These labelled water masses (often called particles) are then transported by the mean flow velocity. A random velocity accounts for dispersion by turbulent eddies larger than the water masses while smaller eddies are responsible for their increasing volume.

Winter and summer conditions were simulated for diferent case studies to account for seasonality of biological activity. It is shown that environment's reaction is not substantially different in summer and winter.

Even in winter conditions the nutrients are limiting factors to primary production in the Portuguese coast. Results confirm the conclusions of some authors who claim that in southern latitudes even in winter, light is not a major factor limiting primary production.

Results have shown typical maximum differences of BOD, nitrate, amonia and phytoplankton of the order of 1%. The results of the mathematical simulations also lead to the conclusion that the choice of secondary treatment of the effluent, instead of primary treatment, would not lead to any visible benefits for the receiving waters. This situation derives from the high dilution rate of the discharge and of the exceptional renewal capacity of the receiving waters (low residence time).

3.3 Microbiological concerns

Pathogens still are a matter of concern in sea-discharges of urban sewage. Faecal coliforms are sometimes pointed as poor indicators of bathing-water quality and/or their mortality rates in the marine environment questioned. As a consequence of those uncertainties, disinfection is often claimed as essential prior to ocean disposal close to bathing water areas.

Ultra-violet (UV) and ozone disinfections are becoming more and more popular when compared to chlorination. They kill microorganisms without creating other environmental problems.

During the last decade UV technology strongly improved. Former disinfection systems needed clear water to be efficient. Prior to disinfection a secondary treatment plant was then required. Nowadays the effluent of a primary treatment plant, or even row sewage, can be disinfected using UV. It is a matter of design of the disinfection plant.

Nowadays, the technology of sewage treatment plants can be optimised according to receiving waters requirements. Combination of a long-sea submarine outfall and a treatment plant designed to remove harmful substances for the specific receiving environment usually results in a less costly and environmentally most effective system.

4 Alternatives to ocean deposition

The degree of treatment of urban wastewater it is not just an economical matter. It is also a matter of global environmental management. A treatment plant is essentially a separation plant. It separates the carrying water suspended matter, in case of primary treatment, and also part of dissolved matter in case of secondary treatment. The resulting effluents of a treatment plant are proper water (liquid phase) and sludge (solid phase).

Nevertheless the sludge must be disposed somewhere. The water is discharged in the ocean or in a river or, if it is pure enough, can be reused. The sludge can be deposited in landfills, used in agriculture, incinerated or disposed in the ocean in selected areas.

Treatment plants are built to separate liquid and solid phases. Liquid phase is disposed in the ocean (in a river in non-coastal areas) or inland through irrigation or recycling. Solid phase can be incinerated, disposed in a landfill or used as a fertiliser. Only the use of solid phase as a fertiliser is a real alternative to ocean disposal. Other alternatives are just a different way of ocean deposition or introduce just a time lag to direct ocean disposal.

The substances disposed in landfills are slowly removed by groundwater and may become a source of contamination of that water, while the use of the sludge in agriculture can also be a source of contamination of the soil and through it a source of contamination of food and of groundwater.

Agricultural use is really a recycling procedure of nutrients. The problem is that sludge from treatment plants also contains toxic substances (mainly heavy metals) and pathogens. For that reason consumers are less and less accepting the use of solid phase into agriculture. Three reasons can be pointed out to show that land disposal of sludge is more dangerous that wastewater ocean disposal:

(1) the humans are direct consumers of agricultural products, which will contain those substances;

(2) the dilution in the ocean is many orders of magnitude higher than land dilution;

(3) the residence time in the soil is many orders of magnitude higher than in a specific oceanic area because of soil adsorption and of the slow movement of underground water.

Another option is the incineration. This process disperses the products of combustion over a very large area, but it is expensive and generates other pollutants during combustion. In any case, all substances not adsorbed by the soil or used by the biota are transported by water (in the ground or on the surface) into the rivers and streams and by them into the sea.

The point is the treatment plants only protect receiving waters at a local scale. If quantities or concentrations of substances discharged are not big enough to endanger receiving waters these become part of the treatment system. To make sense environmentally and technically a wastewater policy general rule (and not the exceptional one) must take into account the local environmental conditions and should be aimed at protecting the environment as a whole and not a part of it alone (water, soil or atmosphere).

5   Conclusions

In this paper the management of wastewater and the role of modelling as a decision support tool was addressed. It is contested the decision of imposing a uniform degree of treatment independently of the final disposal of treatment plant effluents. It is shown that final disposal of primary or secondary treatment effluent in the sea, using a long-sea submarine outfall is indistinguishable for the marine environment. High dilution rate, small residence time and low concentration of nutrients justify the small difference between both degrees of treatment.

It is also discussed the need for an integrated management of the environment and final disposal of sludge is addressed. The rationale of wastewater management is based on the logic that substances transported by the wastewater are harmful for the aquatic environment and then must be removed (in the form of sludge) and are less harmful in sludge disposal systems. This rationale is globally acceptable for fresh water systems (rivers or lakes) but not for the marine system.

As a conclusion one can say that considering only environmental protection concerns, legislation have to be different according to the receiving waters. Open Atlantic coasts are not submitted to eutrophication and consequently it is environmentally wrong to perform secondary treatment of wastewater. Extra sludge will create extra environmental problem in the deposition area and will have extra costs that could be used to remedy others environmental problems.

Acknowledgements

This work was partially carried out within the OPCOM research project sponsored by EU trough the MAST research program (contract n° MAS3-CT97-0089).

References

[1]   Miranda, R., Braunschweig, F., Leitão, P., Neves, R., Martins, F., Santos, A., *MOHID 2000 - A coastal integrated object oriented model.* Accepted for presentation in Hydrosoft 2000, Lisbon, 2000.

[2]   Martins, F., Leitão, P., Silva, A., Neves, R., *3D modeling in the Sado estuary using a new generic vertical discretization approach.* Accept for publication on Oceanologica Acta.

[3]   Neves, R., Leitão, P., Braunschweig, F. Martins, F., Coelho H., Santos, A., Miranda, R., *The advantage of a generic coordinate approach for ocean modelling.* Accepted for presentation in Hydrosoft 2000, Lisbon, 2000.

[4]   Chippada S., Dawson C., Wheeler M.: *A godonov-type finite volume method for the system of shallow water equations,* Computer methods in applied mechanics and engineering. 151(01):105-130. 1998

[5]   Ferziger, J., Perić, M.: *Computational methods for fluid dynamics,* Springer, 1995.

[6]   Isaacs, J.D.: *Potential Trophic Biomasses and Trace Substance Concentrations in Unstructured Marine Food Webs*. Marine Biology, 22, 97-104. 1973

[7]   Miranda, R.: *Nitrogen biogeochemical cycle modeling in the North Atlantic Ocean*. MSc Thesis, Instituto Superior Técnico, Lisboa. 1999

[8]   Frouin, R., Fiúza, A.F.G., Ambar, I. and Boyd, T.J.: *Observations of a poleward surface current off the coasts of Portugal and Spain during winter, J. Geophys. Res., 95(C1), 679-691, 1990.*

[9]   Haynes, R.,and Barton, E. D.: *A poleward flow along the Atlantic coast of the Iberian Peninsula*, J. Geophys. Res., 95(C7), 11,425-11,442. 1990

[10] Wooster, W.S., Bakun,A. and McLain, D. R.: *The seasonal upwelling cycle along the eastern boundary of the North Atlantic*, J. Mar. Res., 34, 131-141. 1976

[11] Fiúza, A. F. G.: *The Portuguese coastal upwelling system, in Actual Problems of Oceanography in Portugal*. pp. 45-71, Junta Nacional de Investigação Científica e Tecnológica, Lisbon. 1982.

A refined three-dimensional layer-integrated model of flows in estuarine and coastal waters

Yan Wu & R.A. Falconer
School of Engineering, Cardiff University, UK

Abstract

A refined three-dimensional layer-integrated model for predicting free surface flows in estuarine and coastal waters has been developed. Instead of using the depth-integrated equations for predicting the water elevation field, a set of equations derived by layer-integrating the continuity and momentum equations over the total water depth have been used to calculate the water elevation fields. The refined model was firstly used to calculate the wind-driven flow in a rectangular basin to verify the model performance. A comparison between the model predictions and an analytical solution for this simplified test case has been made, with identical results being obtained. The model was then applied to simulate tide-induced velocity profiles in an ideal harbour, with the experiments being undertaken in the Delft Tidal Flume at Delft Hydraulics. Reasonable agreement between the model predictions and experimental data was obtained.

1 Introduction

There-dimensional shallow water flow models have been used extensively to study the hydrodynamic processes in estuarine and coastal waters over the past two decades. According to the choice of the vertical co-ordinate system, three-dimensional circulation models are mainly divided into two categories. The simplest uses a regular fixed grid based on a Cartesian co-ordinate system, for example: Leendertse et al. [1], Casulli and Cheng [2], Jin and Kranenburg [3], Casulli and Stelling [4]. The other approach uses a boundary fitted grid structure based on sigma co-ordinates in the vertical, for example: Freeman et al. [5], Sheng [6], Stelling and van Kester [7]. The main advantages of using sigma co-ordinates are that the grid can fit both the moving free surface and the bed, and refinements of the vertical resolution can be achieved near the free surface and near the bed. However, the sigma co-ordinate transformation has certain disadvantages on the other hand. For example, additional terms involving cross-derivatives are introduced by the non-conformal sigma co-ordinate transformation, spurious density-induced motions can arise as result of truncation errors in the approximation of horizontal density gradients in the sigma co-ordinates [8], and the inability of the model to resolve the flooding and drying processes on floodplains [9].

To achieve computationally more efficiency for relatively shallow coastal and estuarine waters, Lin and Falconer [10] refined the three-dimensional layer-integrated model of Falconer *et al.* [11] and included a robust algorithm for modelling the flooding and drying processes. However, recent studies have shown that inconsistencies in using two-dimensional depth-integrated and three-dimensional layer-integrated equations in three-dimensional layer-integrated models can cause stability problems for severe tidal conditions.

In this paper, effort has been focused on eliminating the inconsistencies in solving the two-dimensional depth-integrated equations for the pressure field (or water elevations) in the three-dimensional layer-integrated model to improve model stability. The refined model is firstly used to study the wind-driven flow in a rectangular basin, then applied to simulate tide induced velocity profiles in an ideal harbour, with the experiments being undertaken in the Delft Tidal Flume at Delft Hydraulics.

2 Mathematical model

The governing hydrodynamic equations describing flows in coastal and estuarine waters are generally based on the 3-D Reynolds equations for incompressible and unsteady turbulent flows [12]. The hydrodynamic layer-integrated equations can be derived through integrating the 3-D continuity and momentum equations over layers. A sketch of the layers and the relative variable locations in the $x - z$ plane used in the 3-D layer-integrated model is illustrated in Fig. 1.

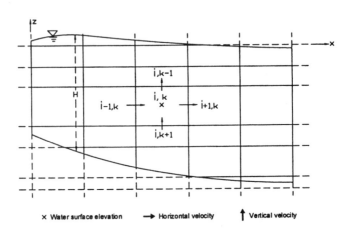

Figure 1: Sketch of layers and variable locations in the vertical plane

(i) Continuity equation for layer k :

$$w_{k-\frac{1}{2}} = -\sum_{i=k}^{K} \left\{ \frac{\partial(\Delta z \bar{u})}{\partial x} + \frac{\partial(\Delta z \bar{v})}{\partial y} \right\}_i \qquad (1)$$

where w is vertical velocity; Δz is layer thickness; K is total number of layers; $\bar{u}$ and $\bar{v}$ are layer-averaged velocities, defined as:

$$\bar{u} = \frac{1}{\Delta z} \int_{k+\frac{1}{2}}^{k-\frac{1}{2}} u(x,y,z,t)dz , \qquad \bar{v} = \frac{1}{\Delta z} \int_{k+\frac{1}{2}}^{k-\frac{1}{2}} v(x,y,z,t)dz \qquad (2)$$

For the first layer, which describes the free surface, eqn (1) becomes:

$$\frac{\partial \zeta}{\partial t} + \sum_{k=1}^{K} \left\{ \frac{\partial(\Delta z \bar{u})}{\partial x} + \frac{\partial(\Delta z \bar{v})}{\partial y} \right\}_k = 0 \qquad (3)$$

where ζ is water elevation above (or below) datum.

(ii) Momentum equations for layer k :

$$\frac{\partial q_x}{\partial t}\bigg|_k + \left[\frac{\partial \bar{u} q_x}{\partial x} + \frac{\partial \bar{v} q_x}{\partial y} \right]_k = fq_y\big|_k - g\Delta z \frac{\partial \zeta}{\partial x}\bigg|_k + (w\bar{u})_{k+\frac{1}{2}} - (w\bar{u})_{k-\frac{1}{2}}$$

$$+ \left\{ \frac{\partial}{\partial x} \varepsilon_h \Delta z \left[\frac{\partial \bar{u}}{\partial x} + \frac{\partial \bar{u}}{\partial x} \right] + \frac{\partial}{\partial y} \varepsilon_h \Delta z \left[\frac{\partial \bar{u}}{\partial y} + \frac{\partial \bar{v}}{\partial x} \right] \right\}_k + \frac{1}{\rho} \left(\tau_{xz}\big|_{k-1/2} - \tau_{xz}\big|_{k+1/2} \right) \qquad (4)$$

$$\frac{\partial q_y}{\partial t}\bigg|_k + \left[\frac{\partial \bar{u} q_y}{\partial x} + \frac{\partial \bar{v} q_y}{\partial y} \right]_k = -fq_x\big|_k - \Delta z g \frac{\partial \zeta}{\partial y}\bigg|_k + (w\bar{v})_{k+\frac{1}{2}} - (w\bar{v})_{k-\frac{1}{2}}$$

$$+ \left\{ \frac{\partial}{\partial x} \varepsilon_h \Delta z \left[\frac{\partial \bar{u}}{\partial y} + \frac{\partial \bar{v}}{\partial x} \right] + \frac{\partial}{\partial y} \varepsilon_h \Delta z \left[\frac{\partial \bar{v}}{\partial y} + \frac{\partial \bar{v}}{\partial y} \right] \right\}_k + \frac{1}{\rho} \left(\tau_{yz}\big|_{k-1/2} - \tau_{yz}\big|_{k+1/2} \right) \qquad (5)$$

where $q_x = \bar{u}\Delta z$ and $q_y = \bar{v}\Delta z$ are layer-integrated velocities per unit width in x, y directions respectively; f is the Coriolis parameter; g is acceleration due to gravity; ε_h is horizontal eddy viscosity; ρ is fluid density; $\tau_{xz} = \rho \varepsilon_v \left[\frac{\partial \bar{u}}{\partial z} + \frac{\partial w}{\partial x} \right]$ and $\tau_{xz} = \rho \varepsilon_v \left[\frac{\partial \bar{v}}{\partial z} + \frac{\partial w}{\partial y} \right]$ are shear stresses, ε_v is vertical eddy viscosity.

At the free surface (where $k = 1$), the terms $(w\bar{u})_{k-1/2}$ and $(w\bar{v})_{k-1/2}$ can be eliminated using the kinematic free surface condition and Leibnitz rule, and the shear stresses $\tau_{xz}\big|_{k-1/2}$ and $\tau_{yz}\big|_{k-1/2}$ are equated to the wind stresses. At the bed, the terms $(w\bar{u})_{k+1/2}$ and $(w\bar{v})_{k+1/2}$ become zero due to the no-slip boundary condition, and the shear stresses $\tau_{xz}|_{k+1/2}$ and $\tau_{yz}|_{k+1/2}$ are equated to the bed shear stresses [10].

In solving the three-dimensional layer-integrated equations above, the water elevation fields calculated from the 2-D depth-integrated equations were normally used. This inconsistency of using the two-dimensional depth-integrated equations in a three-dimensional layer-integrated model was found to cause stability problems for severe tidal conditions. To eliminate the need to use the 2-D depth-integrated equations, eqns (4) and (5) are summed up for all layers to give the following layer-integrated momentum equation over the total water depth:

$$\frac{\partial Q_x}{\partial t} + \sum_{k=1}^{k=K}\left(\frac{\partial \overline{u}q_x}{\partial x} + \frac{\partial \overline{v}q_x}{\partial y}\right)_k = fQ_y - Hg\frac{\partial \zeta}{\partial x} + \frac{\tau_{xz}^w - \tau_{xz}^b}{\rho}$$

$$+ \sum_{k=1}^{k=K}\left\{\frac{\partial}{\partial x}\varepsilon_h\Delta z\left[\frac{\partial \overline{u}}{\partial x} + \frac{\partial \overline{u}}{\partial x}\right] + \frac{\partial}{\partial y}\varepsilon_h\Delta z\left[\frac{\partial \overline{u}}{\partial y} + \frac{\partial \overline{v}}{\partial x}\right]\right\}_k \qquad (6)$$

$$\frac{\partial Q_y}{\partial t} + \sum_{k=1}^{k=K}\left(\frac{\partial \overline{u}q_y}{\partial x} + \frac{\partial \overline{v}q_y}{\partial y}\right)_k = -fQ_x - Hg\frac{\partial \zeta}{\partial y} + \frac{\tau_{yz}^w - \tau_{yz}^b}{\rho}$$

$$+ \sum_{k=1}^{k=K}\left\{\frac{\partial}{\partial x}\varepsilon_h\Delta z\left[\frac{\partial \overline{u}}{\partial y} + \frac{\partial \overline{v}}{\partial x}\right] + \frac{\partial}{\partial y}\varepsilon_h\Delta z\left[\frac{\partial \overline{v}}{\partial y} + \frac{\partial \overline{v}}{\partial y}\right]\right\}_k \qquad (7)$$

where $Q_x = \sum_{k=1}^{k=K}q_x|_k$ and $Q_y = \sum_{k=1}^{k=K}q_y|_k$ are total discharges per unit width in x, y directions, respectively; H is total water depth; τ_{xz}^w, τ_{yz}^w are components of wind stress at the free surface; τ_{xz}^b, τ_{yz}^b are components of bed shear stress.

3 Numerical method

The governing hydrodynamic equations were solved using a combined explicit and implicit finite difference scheme. A space-staggered grid was used in the horizontal plane. Firstly, eqns (3), (6) and (7) were solved using an alternating direction implicit scheme, based on Falconer's model [13], to give the water elevation field. The layer-integrated momentum equations (4) and (5) were then solved to obtain the layer-averaged velocities, using the water elevation field predicted in the first step. Finally, the vertical velocity was solved from the continuity equation (1). The Crank-Nicolson scheme was used to solve the layer-integrated hydrodynamic eqns (4) and (5), with the vertical diffusion terms being treated implicitly and the remaining terms being treated explicitly [10].

The components of the wind stress at the free surface were calculated as [11]:

$$\tau_{xz}^w = C_d \rho_a W_s W_x, \qquad \tau_{yz}^w = C_d \rho_a W_s W_y \qquad (8)$$

where C_d is a resistance coefficient; ρ_a is the air density; W_x and W_y are the wind velocity components in the x and y directions respectively; and W_s is the wind speed at an elevation of $10m$ above the free surface.

Assuming a logarithmic velocity profile with the bottom layer, the components of the bed shear stress were calculated as [14]:

$$\frac{\tau_{xz}^b}{\rho} = \overline{u}_b \sqrt{\overline{u}_b^2 + \overline{v}_b^2} \left[2.5 \ln\left(\frac{30d}{2.72 k_s} \right) \right]^{-2}$$

$$\frac{\tau_{yz}^b}{\rho} = \overline{v}_b \sqrt{\overline{u}_b^2 + \overline{v}_b^2} \left[2.5 \ln\left(\frac{30d}{2.72 k_s} \right) \right]^{-2}$$

$$(9)$$

where $\overline{u}_b$ and $\overline{v}_b$ are bottom layer-averaged velocities in x and y directions respectively; d is the thickness of the bottom layer; k_s is the roughness length.

In this study, the horizontal eddy viscosity was assumed to be constant over the water column, and calculated using a simple mixing length approach [15]; the vertical eddy viscosity was computed using a two-layer mixing length model (see Rodi [16] for details).

4 Model Applications

Firstly, the refined model was applied to calculate the wind-driven circulatory flow in a rectangular basin with a flat bottom, where an analytical solution exists, to verify the model performance. Then the model was applied to simulate tide-induced currents in an ideal harbour, with the experiments being undertaken in the Delft Tidal Flume at Delft Hydraulics.

4.1 Wind-driven currents

To verify the refined three-dimensional layer-integrated model performance, the model was used to calculate the steady wind-driven flow in a channel with a flat bottom. For this ideal case, after assuming that all the dependent variables vary only along the vertical, and neglecting the advection, Coriolis and horizontal diffusion terms, then the following analytical solution [17] can be derived:

$$u = C_d W_s W_x \frac{H}{\varepsilon_v} \left(0.75 \frac{z^2}{H^2} + \frac{z}{H} + 0.25 \right) \qquad (10)$$

A rectangular channel of $4000m$ in length and $1000m$ in width was used in this test case with the water depth being $20m$. In the computation a mesh of 40×10 with grid size of $100m$ was used in the horizontal plane, and in the vertical plane both 10 and 20 layers were used for comparison. The wind speed was assumed to be 20m/s, the vertical eddy viscosity ε_v was set to 0.0158m^2/s, the wind resistance coefficient was set to 1.78$\times 10^{-6}$ and a time step of 60s was used in the computation. Figure 2 shows a comparison between the analytical solution and model predictions. It can been seen that the model results agree with the analytical solution very well, and the model results with 20 layers in the vertical are almost identical to the analytical solution.

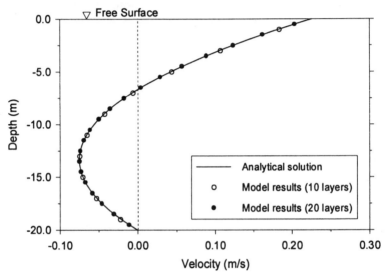

Figure 2: Comparison of predicted velocity profile of wind-driven flow with analytical solution.

4.2 Tide-induced flow in an ideal harbour

The numerical model was then applied to simulate the tide-induced flow in an idealised harbour, which has been studied experimentally in the Delft Tidal Flume at Delft Hydraulics and within the framework of a European Large Installation Plan project. The model square harbour had a plan-form dimensions of 1m$\times$1m and was located along the 1m wide tidal flume. In the experiments a tidal period of 650s and a tidal amplitude of 0.05m were used. The mean water depth was 0.25m. Velocities were measured at 80 locations with an EMS (i.e. 16 locations horizontally and 5 locations vertically) and 20 locations with a WSM (i.e. 4 locations horizontally and 5 locations vertically), and water elevations were measured at 6 locations with a WAVO [18].

Figure 3 shows the modelling domain and the experimental set-up. The open boundaries were situated where measured water elevation or velocity data were available. A mesh of 140$\times$40 grid squares, with a uniform grid size of 0.05m was used. Five layers were used in the vertical, with the thickness of the top layer being 0.09m at mean water level and with the other layers being 0.04m thick. At the open boundaries measured water elevation and velocity time series were applied. The time step was set at 0.02s and the closed boundaries were treated as partial-slip boundary conditions.

Figures 4 and 5 show the experimental and numerical velocity patterns at various stages of the tide, for the top and bottom layers respectively. It can be seen that the model predictions are in good agreement with the measured data, apart from a slight under-estimation of the velocity close to the wall.

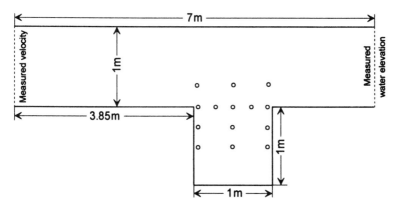

Figure 3: Dimensions of modelling domain and experimental set-up.

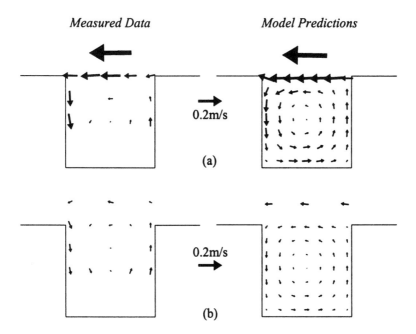

Figure 4: Measured and predicted flow patterns for top layer: (a) t = 162.5s; (b) t = 305.5s; (c) t = 409.5s; (d) t = 637s.

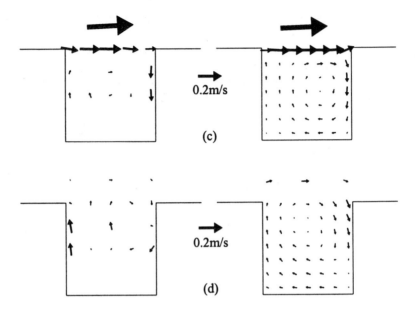

Figure 4(Cont.): Measured and predicted flow patterns for top layer:
(a) t = 162.5s; (b) t = 305.5s; (c) t = 409.5s; (d) t = 637s.

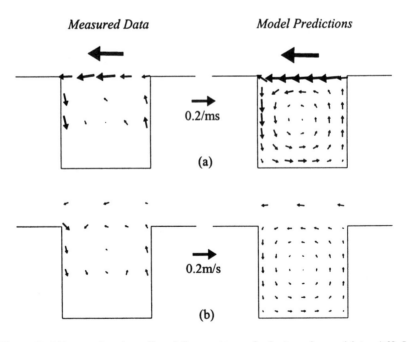

Figure 5: Measured and predicted flow patterns for bottom layer: (a) t = 162.5s;
(b) t = 305.5s; (c) t = 409.5s; (d) t = 637s.

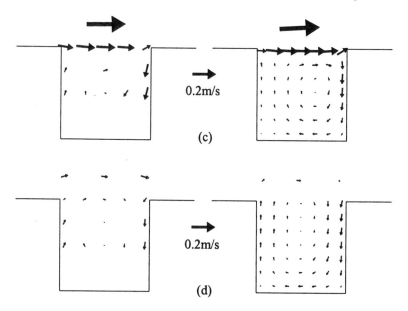

Figure 5(Cont.): Measured and predicted flow patterns for bottom layer:
(a) t = 162.5s; (b) t = 305.5s; (c) t = 409.5s; (d) t = 637s.

5 Conclusions

A refined three-dimensional layer-integrated model for predicting free surface flows in estuarine and coastal waters has been presented. The present model has eliminated the inconsistencies arising in using the two-dimensional depth-integrated equations for the pressure field, which have been found to cause stability problems for severe tidal conditions.

The model has been tested for two case studies, an analytical solution and experimental data acquired in the Delft Tidal Flume at Delft Hydraulics. For the analytical solution the model agreed perfectly with the analytical velocity profile and for the acquired data the agreement between both sets of results was encouraging.

References

[1] Leendertse, J.J., Alexander, R.C. & Liu, S.K. *A three-dimensional model for estuaries and coastal seas: Volume I. Principles of Computation.* Report No. R-1417-OWRR, Rand Corporation, Santa Monica, 1973.

[2] Casulli, V. & Cheng, R.T. Semi implicit finite-difference methods for three dimensional shallow water flow. *International Journal for Numerical Methods in Fluids,* **15(6)**, pp. 629-648, 1992.

[3] Jin, X. & Kranenburg, C. Quasi-3D numerical modelling of shallow-water circulation. *Journal of Hydraulic Enginerring, ASCE,* **119(4)**, pp. 458-472, 1993.

[4] Casulli, V. & Stelling, G.S. Numerical simulation of 3D quasi-hydrostatic free-surface flows. *Journal of Hydraulic Engineering, ASCE,* **124(7)**, pp. 678-686, 1998.

[5] Freeman, N.G., Hale, A.M. & Danard, M.B. A modified sigma equations' approach to the numerical modeling of Great Lakes hydrodynamics. *Journal of Geophysical Research,* **77(6)**, pp. 1050-1060, 1972.

[6] Sheng, Y.P. On modelling three-dimensional estuarine and marine hydrodynamics. *Three-Dimensional Models of Marine and Estuarine Dynamics* (Nihoul, J.C.J. & Jamart, B.M., eds). Elsevier Oceanography Series, Elsevier, Amsterdam, pp. 35-44, 1987.

[7] Stelling, G.S. & van Kester, J.A.T.M. On the approximation of horizontal gradients in sigma co-ordinates for bathymetry with steep bottom slopes. *International Journal for Numerical Methods in Fluids,* **18(10)**, pp. 915-935, 1994.

[8] Paul, J.F. Observations related to the use of the sigma coordinate transformation for estuarine and coastal modeling studies. *Estuarine and Coastal Modeling Studies,* pp. 337-350, 1993.

[9] Cheng, R.T. & Smith, P.E. A survey of three-dimensional numerical estuarine models. *Estuarine and coastal modelling* (Spaulding, M.L. ed.), pp. 1-15, 1989.

[10] Lin, B. & Falconer, R.A. Three-dimensional layer-integrated modelling of estuarine flows with flooding and drying. *Estuarine, Coastal and Shelf Science,* **44(6)**, pp. 737-751, 1997.

[11] Falconer, R.A., George, D.G. & Hall, P. Three-dimensional numerical modelling of wind-driven circulation in a shallow homogeneous lake. *Journal of Hydrology,* **124(1-2)**, pp. 59-79, 1991.

[12] Falconer, R.A. An introduction to nearly horizontal flows. *Coastal, Estuarial and Harbor Engineers' Reference Book* (Abbott M.B. & Price, W.A.,eds). E & FN Spon Ltd., London, pp. 27-36, 1993.

[13] Falconer, R. A. A two-dimensional mathematical model study of the nitrate levels in an inland natural basin. *Proceedings of the International Conference on Water Quality Modeling in the Inland Natural Environment,* BHRA Fluid Engineering, Bournemouth, England, Paper J1, pp. 325-344, 1986.

[14] French, R.H. *Open Channel Hydraulics.* McGraw-Hill Book Company, Singapore, 1986.

[15] Fischer, H.B., List, E.J., Koh, R.C.Y., Imberger, J. & Brooks, N.H. *Mixing in Inland and Coastal Waters.* Academic Press, New York, 1979.

[16] Rodi, W. *Turbulence models and their application in hydraulics*, Second Edition, IAHR Publication, Delft, 1984.

[17] Matsoukis, P.F. & Papadopolis-Dezorzis, A. Three-dimensional characteristics model of wind-generated turbulent flow. *Journal of Engineering Mechanics, ASCE*, **118(8)**, pp. 1526-1545, 1992.

[18] Delft Hydraulics. *LIP1 harbour studies: Part I presentation results from homogeneous tests*. Report No. VR323/Z352, 1992.

Section 2
Environmental Impact Assessment

Gravel beach swash zone dynamics

A.T. Williams[1] & M.R. Phillips[2]

[1] *Air Terra Water Ltd., Environmental Consultants, UK*
[2] *Faculty of Applied Design and Engineering, Swansea Institute of Higher Education, UK*

Abstract

A swash flow transducer was constructed to record wave energies and swash velocities measured via a system of marker poles. Long (4m) metal poles were installed on a gravel beach located at Nash Point, South Wales, UK, a rugged, high energy, coastal environment. These were secured to three 30kg concrete anchoring legs per pole buried in the beach. A Rustrak Event recorder with manual trip pen device recorded the time taken by a series of swash edges to pass each marker pole. Twenty such consecutive swashes were monitored over various tidal stages. The event recorder also obtained wave period, swash period and run up. Profiles and surface sediments were also recorded both before and after each experiment. Non-parametric testing of the sediment samples pre and post experiments were carried out in order to identify any significant differences in gravel size. No specific point of peak velocity was found. A scatter of velocity value points, which increased up beach, correlated with the swash tip. Swash values were an order of magnitude higher than those produced by other workers. Average swash velocity was > 4m/sec. A correlation (0.7) was found between t/T_b (swash period/breaker period) - the 'phase difference' and H_b (breaker height), reflecting the association between wave height and the amount of water brought onto the beach surface. For a given wave period, an increase in height induced an increase in phase difference. No such relationship was found between swash velocity and H_o (deep-water wave height) or H_b (breaker height).

Introduction

It is an acknowledged axiom that there is an immense variability in the forms of response taken by beach sediment to sea waves. Although the beach at Nash Point (Figure 1) could be mathematically described and classified, lateral variation under the same conditions impedes accurate modelling. In addition, despite success in identifying differing sediment types (Caldwell and Williams [1]), along and down beach variation often defied predictive facies responses. Most sea wave studies have been concentrated on deep water waves and many years ago, van Dorn [2] (p21) succinctly summed this up as, *'there is no self-consistent mathematical description of the behaviour of a wave system of finite height, originating in deep water...and terminating in uprush and/or reflection from the shoreline.'* Instead there exists a number of piece-wise solutions. Little has changed in the past 30 years and cross-shore sediment and wave activity is still a frontier zone for coastal scientists. Swash and backwash physical processes appear to be qualitatively/quantitatively different from the periodic surface waves from which they are derived. As shoaling waves approach breakpoint an increasing proportion of the total flow energy becomes associated with harmonic components of the basic wave motion which complicates theoretical descriptions of resultant water motion.

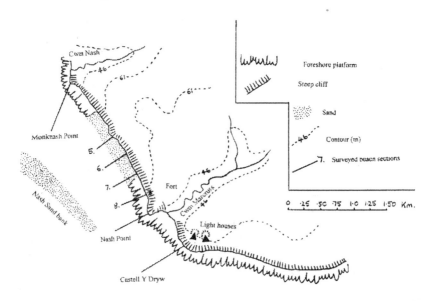

Figure 1: Location of Nash beach.

Swash velocity depends upon wave height and breaker ι backwash velocity is a function of beach slope and rate of groundwa from the beach, both being wave motion of a translatory type. Palmer first noted the importance of wave phase conditions on shingle beaches ..ıl he ascribed erosion/accretion processes; '*the difference between the twc uctions was determined by the rapidity in succession of the waves upon the shore.*' Kemp [4] formalised this aspect by coining the term 'phase difference', i.e. the relationship between breaker period (T_b) and swash period (t), being (T_b/t). The relationship between the phase of a wave at its breakpoint and the swash limit is indicative of the stability/instability of the beach profile (Kemp [4]). Swash period (t) is described as the time taken between a wave breaking and the completion of its run up on a beach face. When the phase difference (T_b/t) is < 0.5, each wave is able to complete its run up and return as backwash prior to the next wave breaking. These rhythmic motions operate independently of one another. However, when the swash period approaches the wave period, the backwash cannot clear the foreshore before the next wave arrives. This enhances beach erosion and Kemp [4] identified three phase levels as being important:

Low, termed 'surge' which enhanced foreshore deposition.
High, termed 'surf' associated with erosion.
Intermediate, termed 'transition' responsible for the development of discrete areas of erosion/deposition.

Kirk [5] suggested that a better terminology would be:

Low for T_b/t < 0.6
Medium for $0.6 > T_b$/t < 1.0
High for T_b/t > 1.0

Phase difference is controlled by swash length, which in turn is controlled by breaker height (Figure 2). As natural wave trains are irregular and various phenomena is associated with the surf/swash interaction, e.g. wave 'set up', 'set down', rips, waves pass in and out of 'phase' over a characteristic time cycle according to conditions currently being experienced. Waves and backwash could be 'in phase' when outgoing water was cleared before the next breaker occurred and the converse would be true when collision between onshore and offshore moving masses took place. The swash zone can act as a filter, which permits passage of certain waves. Many studies have emphasised the centrality of phase relationships with respect to morphological responses of, for example, bed scour, sediment entrainment, sorting etc. Beach gradient is of vital importance in governing particle entrainment. The force P_1 necessary to dislodge a submerged pebble of mass m up a beach having a gradient $\delta\Omega$ and limiting angle of repose $e\tau$ being given by:

$$P_1 = mg\, (\square_s\, /\, \square_w -1)\, \sin\, (e\tau + \delta\Omega)$$

where $\square_s$ and $\square_w$ are the specific weights of the pebble and water respectively.

The force P_2 necessary to dislodge a pebble down beach is:

$$P_2 = mg\,(\square_s / \square_w - 1)\sin(\mathit{er} - \mathcal{\alpha})$$

For gentle slopes the ratio of these pressures is near unity but for steep slopes the difference is appreciable implying considerable flow asymmetry.

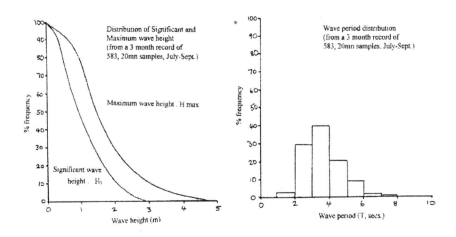

Figure 2: Wave records off Nash Point.

The velocity of the leading edge of the swash reflects the application of swash energy across a beach. Correlations between swash velocity and deep water wave height were first established by Dolan and Ferm [6], who noted that the correlation weakened as energy conditions increased. They attributed this to the influence of phase relations as wave height controls swash length which in turn controls swash period. Kemp [4] gave swash length $(l) = 90.5$, k, g, $H_b)^{0.5}$ and swash velocity $(C_x) = [k, g, H_b\,(1-X_1^{-1})]^{0.5}$ where, k is a coefficient, g is the acceleration due to gravity, and X_1 is the distance landward from the breaker zone to the point being considered. A value for k of 1.28 has been shown to exist (Kirk [5]). He also showed that there existed a rapid increase in 'equilibrium' swash length with breaker height, especially pronounced for longer period waves.

The beach

Several profiles taken at spring tides and sampling of gravel was carried out on Nash beach, but the swash velocity experiments were all carried out at profile section 6 (Figure 1). Nash Point headland acts as a natural groyne and as the longshore drift is in a SE direction, coarse gravels accumulate at the headland. Usually, no complete contact exists at Nash Point as reflected waves push the abraded gravels back to the NW. The beach is 40m wide and 7.5m OD high at its crest (averaged) at the bay's centre, thinning to 30m in width and 8.5m in height near to Nash Point. Vertical cliffs surround the bay (35m to 50m in height), whilst offshore, a large sandbank runs ESE-WNW from a position 1km west of the Point. It lies around the -3m OD at the south-eastern end falling to -5.5m OD further out to sea and this creates an energy shadow across the centre of the bay. Significant and maximum wave heights together with wave periods are shown in Figure 2).

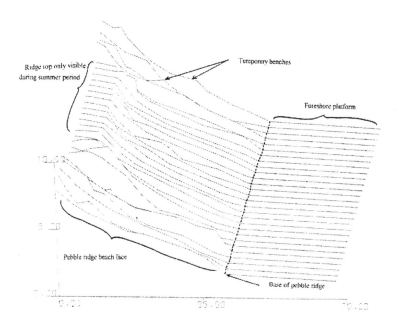

Figure 3: Annual spring tide profiles at section 6, Nash beach.

Methodology

Because of the large tidal range (> 8m) and high wave energy, six long (4m) heavy-duty metal poles were each inserted into three 30kg concrete/metal anchoring legs buried in the pebble beach face. These were bolted to a steel collar through which the 30mm diameter pole was passed. The anchoring legs were found on the gravel beach via a metal detector. The equipment was bolted in place on the beach and the exact distance between poles recorded pre monitoring. A Rustrak event recorder with a manual trip pen device recorded the time taken by a series of swash edges to pass between each marker rod. A sample of 20 consecutive swashes was monitored at various tidal stages. The event recorder was also used to obtain breaking wave period, swash period and run up period. Pebble axes measurements of thirty pebbles at each marker pole were taken pre and post each experiment.

Results and Discussion

Eighteen quantitative measurements of across beach swash zone hydraulics were initiated at Nash beach during 1997-98, of which three were abandoned due to equipment malfunction. The ridge base was used as a reference point and mean velocity values obtained between each pair of marker rods were averaged and average values plotted in Figure 4 using the distance between ridge base and the mid-point between marker rods as a means of locating it on the x axis. For example, for the mean velocity values recorded between rods 1 and 2 (2.89m/sec and 4.55m/sec, Figure 5) 2.89 was multiplied by the sample number of 11, and 4.55 by the sample number 20, the results were summed and divided by 31 to obtain a value of 3.96m/sec. This was plotted on Figure 4 at 3.5 m from the ridge base. Consequently, all data accrued from the 15 experiments was used to construct Figure 4. No points of specific velocity were found, instead a range of values was recorded throughout the profile length. The point scatter upbeach reflected a fall in sample size and a greater variability in swash values associated with the swash tip. Recorded values were of an order of magnitude higher than those encountered in the classic pioneer studies of Schiffman [7] and Kirk [8]. The former quoted modal values of 1.3 and 1.9 m/sec; the latter up to a maximum of 2.0m/sec. The average value at Nash was just above 4.0m/sec (Figure 4). Schiffman [7] worked on two Californian sand beaches, whilst Kirk [8] located his experiments in mixed sand/gravel New Zealand beaches. The steep pebble beach gradient and general wave regime probably accounts for the difference in values.

Significant translation in tide levels could have obscured any sought after variation in swash velocity across the profile. Therefore all data contained in all the experiments in which Figure 5 is but one example, was re-plotted in Figure 6 where individual velocity scores were plotted according to whether they fell into five distinct phases of high tide. (The less than 1hr and more than 0.5hr; less than 0.5 hr results have been omitted in Figure 6 for brevity). Again, no significant peaks/troughs could be seen. Localised ones occurred but these were associated with experiments that gave higher/lower average swash values.

Therefore no generalisations could be garnered from the results as to whether there was an increase/decrease in velocities as each swash moved upbeach, or if there was a velocity peak/trough between the ridge and swash limit. Neither was it possible to identify different distributions of velocity throughout the flooding tide. Figures 4 and 6 agree with Kirk's [8] findings i.e. swash velocities remain relatively unchanged throughout the bulk of their passage over the beach face and terminate abruptly at the swash limit. Figure 7 shows an example of swash velocity results from one experimental run.

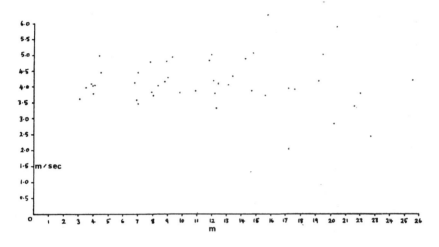

Figure 4: Mean swash velocity results across the beach profile.

Base of ridge

SAMPLE																									metres
1	2.89 ⑪																								
2	4.55 ⑳	3.59 ⑱	4.03 ③																						
3		4.76 ㉒	4.78 ⑲	3.84 ⑧																					
4		3.46 ⑳	4.18 ②																						
5			3.78 ⑨																						
6			5.56 ①																						
7																									

Figure 5: Field book values for mean swash velocity in relation to tide.

In Figure 5, sample numbers are ringed and inset. Vertical divisions on the right hand sides of these diagrams indicate marker rod positions in relation to

In Figure 5, sample numbers are ringed and inset. Vertical divisions on the right hand sides of these diagrams indicate marker rod positions in relation to the ridge base. Various marker rods collapsed at times and their loss is indicated by omission from the figures.

Swash velocity is the average recorded between marker rods during the experiments. H_b was visually recorded from a staff; T_b obtained from the trip pen recording results after sampling 50 consecutive waves; H_o and T_o were obtained from wave rider buoys in the Severn Estuary (Table 1). A marked disparity was seen between T_b and T_o. The reason is unclear although 'red shift' due to interference in the shoaling zone could have caused this. Swash period (t) calculated from the pen recorder enabled calculation of the 'phase difference (t/T_b)' as well as run up period (r) representing the mean period between arrival of a sample of consecutive swash tips. This is a useful value as in conjunction with T_b it can ascertain the level of swash interference on any date. A regular incoming wave field gives minimal interference between consecutive translatory swashes and T_b and r would be roughly equivalent. With an irregular wave field, swash/swash - as opposed to swash/backwash, interference could happen. The resulting 'red shift' would increase r with respect to T_b.

Table 1. Various wave parameters experienced in the experiments.

Velocity (cm/sec)	H_b (cm.)	H_o (cm)	T_b (secs)	T_o (secs)	t (secs)	t/T_b (secs)	r (secs)
4.6	106	-	8.2	-	-	-	-
3.6	56	-	8	-	-	-	-
5.0	91	74	10	4.5	-	-	-
4.7	152	73	8.6	5.9	-	-	-
4.8	30	35	4	4.3	-	-	-
5.0	91	74	10.1	4.4	-	-	-
4.2	53	15	7	2.3	-	-	-
4.0	76	15	5	4.1	-	-	-
4.5	60	64	5.3	3.3	1.9	0.36	6.9
4.0	46	84	10.9	3.0	2.2	0.20	-
3.9	122	71	8.2	5.4	10.8	1.32	8.4
3.7	46	32	5.7	2.3	3.2	0.56	10.1
-	122	-	7.6	-	3.9	0.51	9.3
-	61	-	7.6	-	3.1	0.4	-
3.9	6130	-	7.2	-	2.4	0.33	7.3

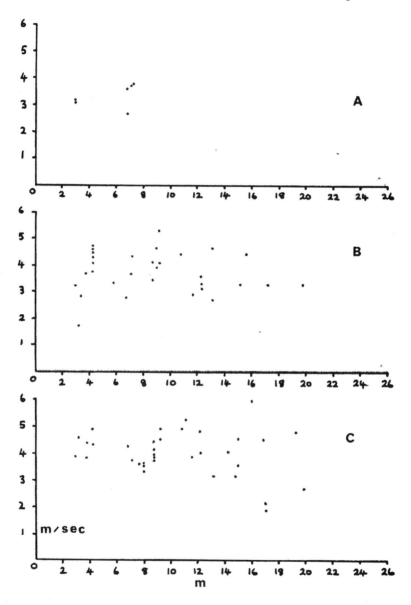

Figure 6: Distribution of mean swash velocity across the beach during 3 phases before high tide. (Horizontal axis represents the base of the ridge. A > 2 hours; B < 2 hr > 1.5 hours; C <1.5 hours > 1 hour.)

Correlation tests of the parameters given in Table 1 gave variable results. Correlations between H_o and T_b (0.74) and T_o and H_b (0.82) differed surprisingly from H_o and T_o (0.46) and H_b and T_b (0.36). However, the sample number was low. Dolan and Ferm [6] found a relationship between H_b and H_o with swash velocity. The Nash experiments gave correlation values of 0.22 and 0.36 respectively for these parameters. The relationship advocated by Kemp [4] regarding H_b and phase difference had a correlation value of 0.7 at n=6, and indicated the relationship between wave height and the amount of water brought onto the beach, but overall, parameter correlations were poor.

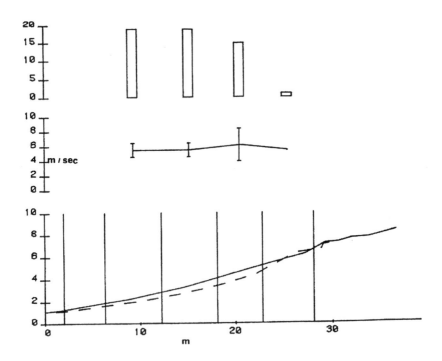

Figure 7: Selected example of swash velocity results

(The lower third shows the profiles (m) before (solid line) and after (broken line) during the experiment. Vertical lines indicate the position of the beach rods. The middle third shows the velocity data (m/sec.). The bar centre gives the mean swash velocity, the ends one standard deviation either side of the mean. When the centre lines are joined it shows at a glance an increase/decrease in mean velocity. The top third shows vertical blocks indicating the sample number. Sequences of 20 waves were monitored not all of which might have reached the marker rods.)

Measurement of 30 pebbles at each pole was taken at the beginning and end of each experiment. Despite considerable profile alterations, sediment size comparisons apart from spasmodic instances did not produce any statistically significant results. Time sequences are therefore involved pre the building up of facies sediments on a beach, as advocated by Williams and Caldwell [9].

Conclusions

No appreciable swash velocity variations either spatially across the Nash pebble beach profile, or temporally was found in 15 distinct experiments. Results indicated that velocities were reasonably constant throughout the passage of incoming swash, averaging some 4m/sec. with rapid deceleration limited to a very short distance behind the limit of swash run up. Correlations found between the various wave parameters were not statistically significant, although this could be a function of the low number of experiments involved. No change in pebble size was found before and after the various experiments.

References

[1] Caldwell, N.C & Williams, A.T. The role of beach profile configuration in the discrimination between differing depositional environments affecting coarse clastic Beaches. *J Coastal Research*, 1 (2), 129-139, 1985.

[2] van Dorn, W.G. Theoretical and experimental study of wave enhancement and run up on uniformly sloping impermeable structures. *Final report to Congress. Univ. of California, Scripps Institute, ONR, Washington DC.* Nonr. 2216(16), 101pp, 1966.

[3] Palmer, H.R. Observations on the motion of shingle beaches, *Phil. Trans. Royal Soc., London*, 124, 567-576, 1834.

[4] Kemp, P.H. A field study of wave action on natural beaches. *Proc. 10th Conf. Int. Assoc. Hydraul. Res.* 132-138, 1963.

[5] Kirk, R. Aspects of surf and run up processes on mixed sand and shingle beaches, *Geog. Annaler*, A57, 117-133, 1975.

[6] Dolan, R and Ferm, J. Swash processes and beach characteristics, *Prof. Geog.* 18, 210-213, 1966.

[7] Schiffman, A. Energy movements in the swash zone, *Limnol. Oceanog.*, 10, 255-260, 1965.

[8] Kirk, R. *Swash zone processes.* Unpub. Ph.D. thesis. Canterbury University, New Zealand, 1970.

[9] Williams, A.T and Caldwell, N.E. Particle size and shape in pebble beach sedimentation. *Marine Geology*, 82, 199-215, 1988.

Web-based pollution monitoring system for the Lake Ohrid

K. Mitreski, D. Davcev & Z. Koneski
Faculty of Electrical Engineering, Skopje, Macedonia

Abstract

In this paper, we will present a Web-based pollution monitoring system for Lake Ohrid. Our intention is to describe some of the processes that contribute to the eutrophication of Lake Ohrid. Our work is related to the ecological state of Lake Ohrid. Tectonic by origin, Lake Ohrid is situated in the southwestern part of the Republic of Macedonia. The size of the lake is 358 square kilometres. Its average depth is 151m and the maximum depth is 289m. "Man-made" eutrophication, in the absence of control measures, proceeds much faster than the natural phenomenon and is the major reason for pollution of this lake. A Dynamic Integrated Monitoring System (DIMS) was previously presented in another paper. The monitoring system is based on on-line measurements. One of the main objectives of DIMS is to provide a real-time monitoring and efficient protection of the environment. In this paper, we present a Web-application for handling the data for the pollution monitoring system. Data acquisition and analysis is performed by a DBMS server and then presented in HTML format in regular pre-defined time intervals. The Web server and corresponding software are used for this presentation. In this way the data and its analysis is performed in one place and they are available to any user having only a web browser. The Web-based pollution monitoring system is implemented using UNIX/RDBMS environment and GIS software.

1 Introduction

Our work is related to the ecological state of Lake Ohrid, which represents rare natural ecosystem inhabited by many endemic and relict species. For its outstanding natural surroundings, Lake Ohrid has been placed on the UNESCO world natural heritage list. The lake faces accelerated deterioration of its waters and a change of the trophic state of this aquatic ecosystem, mainly caused by nutrient load. "Man-made" eutrophication, in the absence of control measures, proceeds much faster than the natural phenomenon and is the major reason for pollution of this lake, e.g. World Bank [1], OECD report [2].
We proposed in the paper by Mitreski et al. [3], [4] and Davcev et al. [5] a mathematical model that gives general picture of the level of eutrophication in the Lake Ohrid and shows general trends of the lake behaviour. The specific nature of the lake has been taken into consideration and embodied in the model. Such dynamic model, which tends to predict future eutrophication and the trophic state lake-wide, is the first attempt undertaken in describing the ecological state of Lake Ohrid.
In this paper, we will present a Web-application for handling the data for pollution monitoring system. Data acquisition and analysis is performed by DBMS server and then presented in HTML format in regular pre-defined time intervals.
The paper is organized as follows. Section 2 gives a pollution monitoring system architecture for the Lake Ohrid. In section 3, the design and implementation of Web-based pollution monitoring system for the Lake Ohrid is presented, while section 4 will conclude the paper.

2 Pollution monitoring system architecture

In the paper by Mitreski et.al [6], we presented our scheme for planning a DIMS for the Lake Ohrid. The structure of the pollution monitoring system (PMS) includs three levels: (Fig.1)
- *Data acquisition level* responsible for systematization, validation, comparison of data with alarm limits and creation of dynamic part of the relational database system: the static part of the database with information about standards, limits, models etc.
- *Data handling level* responsible for supervisory control and specialized data analyses and model evaluations; it's also responsible for integration of on-line with off-line information to make statistical reports and data presentation.
- *Management level*, which is responsible for short-term decision and long-term strategies for the pollution reduction approach.

Our PMS consists of 6 measurement points along the coast of Lake Ohrid and 4 measurement points along the rivers. PMS was used for measurement of the following parameters: temperature, pH, total phosphorus, dissolved oxygen, total alkalinity etc.

For example, temperature measurements were made by Pt100 sensor; pH measurements were made by *HI 1910B* sensor for calm water and with EURO2015 sensor for rivers. Measurement sensors are connected with intelligent microprocesor system PH500122 in local stations. The data are measured on-line by the instruments and transmitted to the central station where they are processed by the computer system.

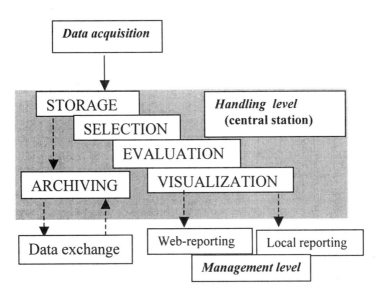

Fig.1 Process steps of monitoring system

The measured values are organized as relational database, which is suitable for statistical analysis using different OLAP (on-line analitical processing) tools. One example of graphical presentation of pH measurements if given on Tab.1

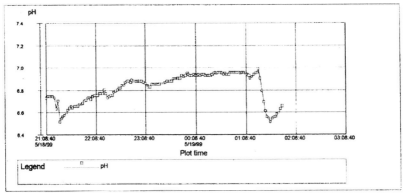

Tab 1. Graphical presentation for the pH measurements

Aquatic ecosystems are extremely complex and the basic objective of most PMS is to determine the status of the ecosystem. Mathematical model [3] are build and validated with empirical data from our PMS. Web-based pollution monitoring system will gave more possibility for predictive models about Lake Ohrid in the future.

Web-based pollution monitoring system architecture for the Lake Ohrid is based on client-server technology (Fig.2). DBMS, WEB and GIS servers are connected with Intranet/Internet network.

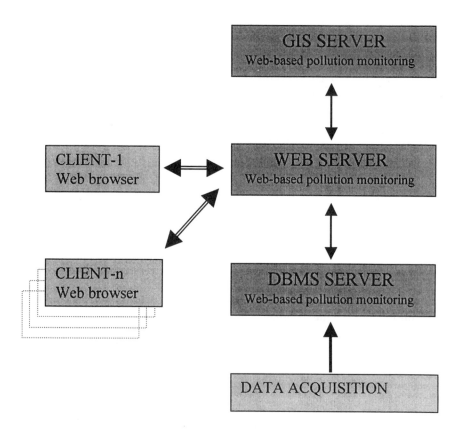

Fig.2 Pollution monitoring system architecture

3 The design and implementation of Web-based pollution monitoring system for the Lake Ohrid

The interface of the Web application is presented on Fig.3. Web-based pollution monitoring system is implemented using Red-Hat Linux and SQL RDBMS. Clicking on the picture or on the hyperlink could choose the location. (Fig.4).

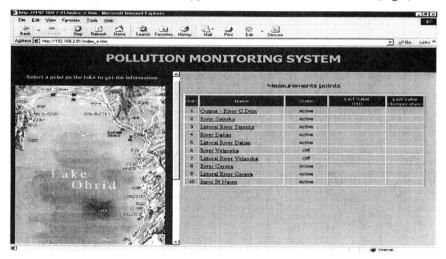

Fig.3 Main screen of the Web interface of the PMS

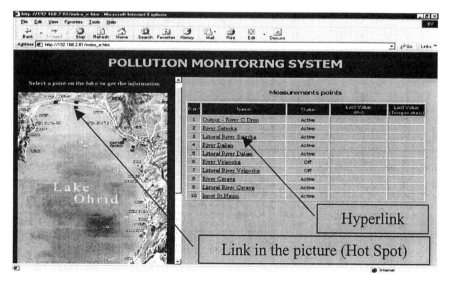

Fig.4 You can chose location by clicking on the picture or on the hyperlink

After choosing the location and query from date to date, application will generate dinamicly from the database analitical and graphical presentation for all measured parameters. (Fig.5 and Fig.6).

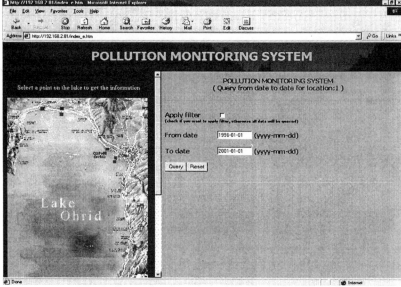

Fig.5. Query for the selected location (from date to date)

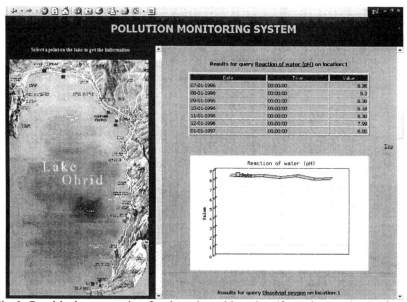

Fig.6. Graphical presentation for the selected location (from date to date) - for all measured parameters

PHP Graph modul dinamicly create graphical presentation from DBMS for all available parameters. PHP3 WEB generator dinamicly read measured data and using HTML orders, web browser will display all presentation to the client.

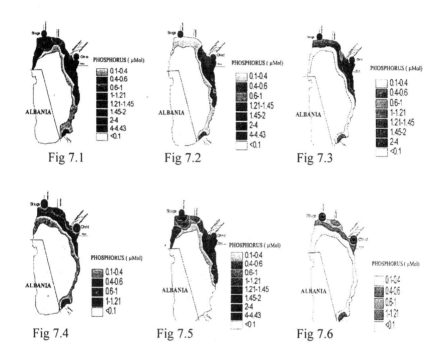

Fig 7.1 Fig 7.2 Fig 7.3

Fig 7.4 Fig 7.5 Fig 7.6

Fig.7. Graphical presentation for phosphorus concentration with GIS software

We have used selected data from PMS and ESRI GIS software for virtual modelling and visualization. Fig.7.1 shows present state of pollution of Lake Ohrid. Four main tributaries-rivers Sateska, Koselska, Velgoska and Cerava are shown a model of phosphorus pollution with value classification is presented. On Fig.7.2-7.6, the virtual models of the Lake Ohrid are presented. Fig.7.2 shows virtual model in the case when the river Sateska doesn't enter the Lake Ohrid, i.e. it is redirected or the river is totaly cleaned from the phosphorus. Fig.7.3, 7.4, and 7.5 shows virtual models in the case when the rivers Velgoska, Koselska and Cerava, do not enter the Lake Ohrid respectively. Fig.7.6 shows virtual models in the case when the rivers Velgoska and Koselska (together) are totally cleaned from the phosphorus.

The virtual model shows that if the influence of rivers Velgoska and Koselska are minimized, or if they are totally cleaned from phosphorus, the average pollution of the lake will be evidently decreased, especially in the litoral zone.

4 Conclusion

In the paper, we present a Web-application for handling the data for pollution monitoring system. Data acquisition and analysis is performed by DBMS server and then presented in HTML format in regular pre-defined time intervals. Web server and corresponding software are used for this presentation. In this way the data and its analysis is performed on one place and they are available to any user having just a web browser.

Also using selected data from PMS and GIS software, virtual modelling and visualization of the pollution are presented.

References

[1] World Bank, *Feasibility study on the lake Ohrid conservation project*, Erhst Basler & Partners, Zollikon, 1995.

[2] OECD, *Eutrophication of waters. Monitoring, assessment and control*, OECD Report, Paris, 1982.

[3] Mitreski, K., Koneski, Z., Davcev, D., Naumoski, T., Mitic, V., Sumka, S. & Guseska, D., An ecological model of Lake Ohrid, *Proceedings of the IASTED Conference, AMS'98*, Honolulu, pp.32-35, 1998.

[4] Mitreski, K., Koneski, Z., Davcev, D., Naumoski, T., *Mathematical model for the available phosphorus, phytoplankton and zooplankton in the Lake Ohrid* , *Proceedings of VII Internacional Conference-"Envirosoft"* , Las Vegas, pp.209-216, 1998.

[5] Davcev, D., Mitreski, K., Boskovska, V. & Koneski, Z., Management and analysis of the phosphorus data in the lake Ohrid by multimedia presentation, *Proceedings of the IASTED International Conference, Advanced Technology in the Environmental Field*, Editor: M.H.Hamza, Gold Coast, pp. 178-181, 1996.

[6] Mitreski, K., Davcev, D., Jordanovski, M., Naumoski, T., Pollution monitoring system for the Lake Ohrid - a planning scheme, *IV Proceedings of the International Conference "Water pollution 97", Bled, Slovenia, pp.341-348, 1997.*

Preliminary studies on environmental impact of cage aquaculture in Canary Islands

J.M. Vergara Martín[1], N. González Henríquez[2], R. Haroun Trabaue[1], L. Molina Dominguez[2] & M.I. García Rodríguez[2]

[1] *Departamento de Biología. Universidad de Las Palmas de Gran Canaria, España*

[2] *Instituto Canario de Ciencias Marinas, España*

Abstract

More than 50% of the present marine finfish aquaculture production in the Canary Islands (800 Mt for 1999) is carried out in off-shore cage farms, being the predominant technology involved in all new commercial projects in the Archipelago. This economical sector has experienced a significant growth in recent years at a Regional level, with excellent prospects for expansion due to different geographic and climatic advantages. However, this is giving rise to an increasing social concern about environmental implications, being frequently used as an argument to restrain this activity. The present work aimed to develop a protocol for environmental impact assessment of these farms, by identifying those physical, chemical, and biological parameters more sensitive to the operation of this type of facilities. A comprehensive sampling program was carried out since 1994 at a cage farm producing 500 metric tons per year of gilthead seabream (*Sparus aurata*) and seabass *(Dicentrarchus labrax)*, located at the East Coast of Gran Canaria Island. The studies included water quality, retention and discharge of nitrogen and phosphorus by cultured fish, sediments, hydrology, fouling, and biological assessment of benthic ecosystems.

As a whole, none of the studied parameters showed negative effects on the surrounding ecosystems, the results suggesting that both local currents regime and carrying capacity of this particular Bay seem enough to disperse solid organic waste deposits and to transform ammonia excreted by cultured fish. The presence of fouling organisms resulted in appositive biofilter effect.

1 Introduction

The culture of the marine fish gilthead seabream (*S. aurata*) and European seabass (*D. labrax*), predominant in the Mediterranean area has experienced a dramatic growth in recent years, from production figures amounting a few thousands in 1984, until around 100 000 metric tons estimated for 2000, totally commercialised in European markets.

These two species are also produced in the Canary islands, the main advantages being a large coastline (1291 km), and a warm range of surface water temperature, shortening the ongrowing period of these fish species when compared with Mediterranean farms. Thus, the required ongrowing times for seabream and seabass in the Mediterranean are 16 and 20 months, respectively, while in the Canaries these periods are reduced to 13 and 16 months. This fact also allows to economically produce bigger (one kg) fish with no competence in the Mediterranean. However, other factors derived from the long distance to markets represent negative counterweights in the total profitability of local farms, despite their positive whole balance. Another advantage lays in the absence of heavy water polluting industries, giving to coastal waters the high quality required for the culture of the above marine species.

Funded by the EU program STRIDE, our research group installed a pilot farm in May 1994, consisting of six polystyrene, floating cages of 900 m^3 each (Figure 1), which were moored in the bay of Melenara, at the Southwest coast of Gran Canaria island. Simultaneously, an Consortium Agreement was signed with the private company ADSA, in order to allow the firm to operate the facilities and to fund a variety of research work to be done, including environmental impact assessment. Fish were firstly stocked in the cages on June 1994, and the environmental studies were iniciated a month earlier. In addition, our research group was recently granted with funds from a national and European program (CICYT-FEDER) that will allow us to improve and continue these studies at least until 2001.

2 Environmental impact studies

The environmental impact study was designed to identify those parameters more sensitive to the operation of this type of farms, thus allowing to produce an environmental management program which could be useful both to farmers and to public institutions implied in the regulation of aquaculture at the Canary islands. The work included three categories of studies: 2.1) nutritional – mass balance of nutrients; 2.2) water quality; and 2.3) ecological – sediments, fouling, and benthic ecosystems. Simultaneously 2.4), the hydrology of the zone was also studied using an automatic currentmeter.

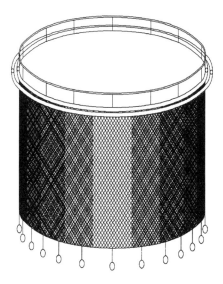

Figure 1: Polystyrene floating cage

2.1 Nutritional studies

The nutrient budget or mass balance model (Gowen et al., 1987) was studied by a series of biochemical analyses carried out on a significant fish population and the fishfeeds used during the growout period of both fish species, then the nitrogen and phosphorus retained by cultured fish was determined, as well as the nitrogen and phosphorus fractions released to the environment. Results showed that yearly total nitrogen and phosphorus inputs to the farm via feeds amounted for 39.5 and 5.5 Mt, respectively, resulting in a release of 32.1 and 3.8 Mt, respectively, in both dissolved and particulate forms (Figure 2).

2.2 Water quality

Water quality was studied by means of sampling and subsequent analyses of the water column at different influence zones, including one control and three different points, and two different depths for each point. On site data on water temperature and dissolved oxygen were obtained at each point with a portable electrode probe, while water samples were filtered (0.45 μm) and frozen at –20°C for further analysis.

Water temperature ranged between 17.5°C -January-March- and 24.1°C -September-November-, and dissolved oxygen values ranged between 6.3 and 11.7 mg/l. Phosphate values ranged between 0.0 and 1.1 $\mu mol.l^{-1}$, with an average value below the average value, 0.45 $\mu mol.l^{-1}$, reported for this bay (García, 1999). Silicate values ranged between 0.4 and 2.6 $\mu mol.l^{-1}$, with an average value of 1.1 $\mu mol.l^{-1}$. Nitrate+Nitrite values ranged between 0.0 and 1.2 $\mu mol.l^{-1}$, always below the average value, 3.52 µmol.l⁻¹, reported for this bay (García, 1999). Ammonia values ranged between 8.77 and 21.7 µmol.l⁻¹, with an average value, 14.9 µmol.l⁻¹, above the average value, 6.96 µmol.l⁻¹, reported for this bay (García, 1999, Table 1).

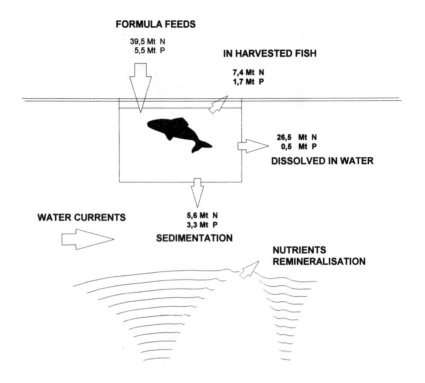

Figure 2: Yearly nitrogen (N) and phosphorus (P) mass balance in the studied farm, with calculated dissolved and solid fractions.

Table 1. Range of values of water quality in the studied cage farm.

	T ª (°C)	O₂ (mg.l⁻¹)	PO₄³⁻ (µM)	SiO₃²⁻ (µM)	NO₃⁻ + NO₂⁻ (µM)	NH₄⁺ (µM)
Cage Installations Values Ranged	17.5-24.1	6.3-11.7	0.0-1.1	0.4-2.6	0.0-1.2	8.77-21.7
Bay of Melenara Average Values			0.45		3.52	6.96

Nitrate+nitrite, and ammonia values were the parameters showing highest differences between points inside the cages and those outside the facilities, particularly during the autumn season. This can be explained as a consequence of highest water temperature values occurring during this season, leading to an increased food consumption by fish and corresponding ammonia excretion. As a consequence of nitrifying bacterial processes, nitrate+nitrite values were also highest inside the cages in this season. However, these values were always well below those values considered as normal for this area, suggesting that the carrying capacity of this bay was enough to transform the ammonia excreted by all fish stocked at the cages, and to keep normal nitrate+nitrite values.

2.3 Ecological studies

Sediment samples were collected with cores in three different influence zones: underneath the cages (two sampling points), 60 meters from the cages (four sampling points), and 200 meters from the cages (four sampling points). Analyses included particle size, organic matter, and total nitrogen and phosphorus contents. No significant differences were found for all parameters between points located within each sampling zone. When different zones were compared, no significant differences were found for particle size, organic matter and phosphorus content. However, the average nitrogen content of sediments underneath the cages was significantly higher than those of the other sampled zones. A marked seasonal trend was observed for values in both nutrients and organic matter, similar in all zones and characterised by periodical recovering of initial values in the three zones. These results suggest that the water currents regime of the area was enough to whash away organic sediments.

Studies on the fouling occurring on cage installations (floats, frames, ropes, moorings, nets, etc.), in terms of biomass, species diversity, and biochemistry analyses, together with results from chemical water analyses, suggest that these organisms act as filters of dissolved organic and inorganic products resulting from fish loads, diminishing the amount of nutrients finally released to the environment.

Two 150 meters length transepts (NE and SW), croosing over the vertical of the cages, were marked on the sediments, and used to perform periodical visual counts by divers, in order to assess the benthic ecosystems. By the end of the third year of the cage operation, a significant decrease in populations of the seagrass *Cymodocea nodosa* was apparent in the zone underneath the cages, as well as an increase in the population of the polychaete *Diopatra neopolitana*. An increase in benthic fishes was also observed in this area, with average values of 150 individuals per 20 meters length.

In general terms, some negative effects could be appreciated in the benthic ecosystems directly underneath the cages, although such effects were less apparent as showed by the sediments analyses and by the increase in species diversity observed in the fouling studies.

2.4 Hydrology

With respect to water current velocities, maximum registered values (60 cm/sec, with average values of 25-30 cm/sec) coincided with storm episodes, and were below maximum values recommended by the cages manufacturer (100 cm/sec). Minimum average velocities (10 cm/sec) are generally considered as enough to allow for a good water and oxygen renewal in cages, as well as for whashing away organic sediments.

Predominant directions of currents arriving the cages were NE and SW, this pattern being determined by the general Canarian Current (North-South direction), above which the locally predominant tidal currents are superimposed. Its direction NE or SW will vary depending on high or low tides.

3. Results of the studies

As a whole, results obtained up to present suggest that the currents regime is the main factor affecting the interaction between this type of farms and the surrounding ecosystems, and that in this particular case, both released organic matter in solid form to the sediments and dissolved nutrients have produced no significant negative effect on the surrounding environment. These results have been submitted every year to the Local Government of the Canary Islands, as a requisite of the permission granted in 1994. The firm, which operates the farm also, receives these results with similar periodicity.

4. Socio-economical effects

At the present, the production of these two finfish species in the Canary islands amounts 1000 Mt per year. If a reasonable number of those new commercial farm projects (all off-shore cages based) which nowadays are waiting for permission start to operate and consolidate in the next three to five years, it can be estimated that total local production could then reach 5000 Mt per year. This would imply an overall 4000 million peseta a year business, directly employing between 100 and 150 people and near 1000 indirectly. These production figures would also allow, by scale economies, local feed manufacturing and hatchery industries to open, which would contribute to dramatically reduce production costs, and hence, to improve competitiveness of local farms. In the long term, a conservative estimation of the potential for this sector could be near 15 000 Mt per year, three times as big as the previous estimates.

On the other hand, reports opposing to this and other similar farms have been regularly published in the local press, including a variety of arguments on their negative effects, among which can be summarised:

- Attraction of dangerous species (sharks, etc.) by cages.
- Negative visual impact.
- Negative environmental impact, including water and sediment pollution, damages to benthic seaweed species, fish scapes, the use of dangerous anti fouling products and medicated feeds, or the presence of oil films in the surface of surrounding waters.

The origin of this opposition is also varied, including local town councillors, community associations, and green groups. Regularly, and when these opposition campaigns were more continuous, local R & D institutions have released reports to the press with the above results of environmental impact studies, where different arguments against these farms were invariably contradicted.

All the above indicates:

a)   That this new industry generates a series of conflicts with different local users of coastal resources, as with local fishermen.

b)   That these conflicts become apparent in the form of opposition to settling of this industry within the local community, generally expressed in the form of environmental arguments.

c)   That an evident difficulty to transfer the results of environmental impact studies to the community where these farms intend to operate exists.

5. Discussion and conclusions

One of the possible reasons for the above described situation could be the fast growth of this industry in the Canary islands, interfering in the short and medium term with their integration within local economies. This fact can be also the cause of an image problem, which become apparent in a variety of conflicts with the local communities.

If we assume this hypothesis, maybe this industry should consider an approach of corporative image improvement targeted towards local communities where they intend to settle, including the public divulgation of environmental management programs carried out at present in every, as part of this strategy.

In addition, local authorities implied in the regulation of this sector must make a substantial effort in promoting this industry as strategic for the economy of the Islands, as well as supporting the wide use of environmentally sound practices by farmers.

References

García-Rodríguez, M.I. *Utilización de Ulva como biofiltro en Acuicultura. Estudio comparativo de la retención de compuestos nitrogenados*, Ph. D. Thesis. Universidad de Las Palmas de Gran Canaria, 1999.

Gowen, R.J. & Bradbury, N.B. The ecological impact of salmonid farming in coastal waters. A review. *Oceanogr. Mar. Biol. Ann. Rev.*, **25**, pp. 563-575, 1987.

Section 3
Harbours, Ports and Marinas

Air pollution prediction by artificial neural networks

M.I.V. Furtado & N.F.F. Ebecken
COPPE/Federal University of Rio de Janeiro, Brazil

Abstract

In this work, an artificial neural network strategy is presented to predict suspended particulate matter in the atmospheric air. The considered area is the Rio de Janeiro metropolitan region, a brazilian urban center. The topography, the nonuniform occupation, the Guanabara Bay and the meteorological peculiarities contribute to the need of a complex model. Backpropagation, generalized regression neural networks and genetic algorithms are used to evolve neural network structures to search for significant input variables and maximize the predictive accuracy of the results. The input data corresponds to 14 meteorological variables on different locations. The predicted values are satisfactory and the methodology presented promises perspectives.

1 Introduction

Artificial Neural Networks are a mathematical paradigm inspired by the functioning and organization of the human brain, being capable to play computational tasks in diverse scientific and technological areas. They possess features as learning for experience, generalization from a previous example and abstraction of essential features from inputs with noise [1].

In this study, an artificial neural network is applied to make predictions, concerning with the presence of particle in suspension in the troposphere, consequence of the association of the polluting sources with the meteorological conditions. The region in study is the city of Rio de Janeiro, Rio de Janeiro state capital, according to economic polar region of Brazil, with raised degree of urbanization and industrialization. This region possesses 17 cities and has the emission of pollutants as one of the main determinative factors of the life quality of the population.

2 Artificial Neural Networks

To establish the neuron network architecture, the parameters to be defined are: the number of intermediate layers; the number of processing units (neurons); the transference function; and the learning process [1, 2, 3, 4].

In this paper the Backpropagation and Generalized Regression Neural Network (GRNN) are adopted.

The backpropagation algorithm is a well-known supervised approach feedforward type with any function of activation derivable. This algorithm allows to reduce the values of synapses in the output to optimize the neural network [1, 2, 4, 5, 6].

The GRNN is a feedforward type of neural network and is based in the evaluation of the probability function. It makes intensive use of memory, requires that all the training examples are stored for future predictions, and needs a relatively great data set [7, 8].

The GRNN topology consists of just one hidden layer.

A GRNN can be understood as an interpolation of the function, called factor of smoothing (the only parameter that modifies).

3 Choosing the Neural Network Architecture

The systems based on Artificial Neural Networks strongly depend on the topology of the networks (size, structure, and connections), as well as of its parameters (learning tax, moment). As a result, the determination of the architecture of the network affects its performance [5].

The design of the network depends on the type of function to the approximated, and the trial-and-error approach is most used. It is initiated with 1 layer but, if the error is great, a 2 layers network is tried and so on [2].

The number of neurons in the hidden layer should be, in principle, between the number of output and input vectors. It can be tried to reduce the number of neurons in the intermediate layer.

Genetic Algorithm can be applied to determine good neural architectures. For such, each individual corresponds to the number of layers, number of neurons for layers, standards of connectivity and parameters of learning [9, 10, 11, 12].

The crossover operator acts as the learning tax η with high value, and the mutation as this same tax, however with low value.

The interesting thing of this method is the discovery of connectivity standards that are generally not investigated in processes that use trial-and-error approach.

With the training data, one evaluates the performances in the test set, evaluating the performance of the network for accuracy criteria, quadratic average error under the data, size of the network, number of training cycles and test then. It is possible to estimate the current state of the population and evaluate the performance of each network, in agreement with its fitness.

4 Atmospheric Pollution and Meteorological Factors

The pollution represents one of the most serious problem resultant of the technological development and the process of urbanization. In the urban areas, the biggest problem is consequence of the pollution of atmospheric air, caused by the features of the atmospheric circulation and the permanence of some pollutants in the atmosphere for long periods of time [13].

The natural features of the atmosphere can be modified by the action of atmospheric pollutants (gases or particles) that, when gifts in enough amounts in the atmosphere, can be dangerous for human, animal, vegetation and materials, or degrade the life.

The main pollutants of atmospheric air are emitted in gaseous form (organic and inorganics) or particulates (solid or liquid finely divided).

The main sources of atmospheric pollution had appeared with the urban and industrial growth, involving sectors of the economic activity, as industries and carriers. The particulate materials easily are perceived, as much in suspension as deposited. It is presented in three ways: smoke, particles and total particles in suspension [13, 14, 15].

The pollutants emitted by sources located in the surface of the land and by atmospheric air are carried and dispersed in the environment.

The quality of air is influenced by complex factors, being strongly associated with weather and climate. The concentration of the atmospheric pollutants is directly associated with the dilution and the capacity of dispersion of the troposphere, beyond the roughness of the terrestrial surface.

The precipitation is one of the most important mechanisms of cleaning many pollutants, mainly if associated to the wind.

The increase of the irregularity of the surface affects the vertical profile of the wind and the temperature. The urban areas possess different roughness and thermal features of the agricultural areas, causing a particular circulation of the wind between buildings. The insolation intervenes directly and indirectly with the pollution, inducing the formation of atmospheric oxidants. When heating the terrestrial surface, it generates turbulence and mixes the pollutants closed to the terrestrial surface [13].

The solar radiation causes a temperature increase and is dependent of other factors as cloudiness, duration of the daylight, or either, the season of the year.

There is also the occurrence of the thermal inversion that delays the dilution of the pollution next to the surface. This phenomenon can be active while to continue under the effect of high pressures and low vertical speeds of the air mass.

The least atmospheric pressures contribute for the quality of air, allowing the formation of ascending heaves that decrease the concentrations next the surface.

The atmospheric humidity is resultant of the evaporation and the evapotranspiration of the plants; it depends on factors as geographic location, season of the year and climate. The water vapor is the raw material for the cloud formation and carrier of heat in atmosphere.

The establishment of national and international standards air quality defines legally a maximum limit for the concentration of an atmospheric component. It is established by organizations and countries, as the World-Wide Organization of the Health and, in Brazil, by the National Advice of Environment (CONAMA) [16].

The standards contemplate the following pollutants: total particles in suspension, particles, smoke, sulphur dioxide, nitrogen dioxide, carbon monoxide and ozone.

5 Related Aspects of the Region

The Metropolitan Region of Rio de Janeiro, presents factors that aggravate the problem of the pollution. This problem is caused by rough topography, not uniform occupation of the soil, presence of the sea and the Guanabara Bay, that produce a complex and heterogeneous air stream in relation to the distribution of dispersion of the pollutants [14,16].

The region has as meteorological peculiarity, the occurrence of strong thermal inversions, high frequency of calmness, presence of favorable areas for the ventilation and the features of the tropical climate.

It also has the presence of three great bulks that create areas of isolation for the circulation of winds and air masses. This confers to the city a significant climatic complexity and harms the dispersion of atmospheric pollutants in some parts.

This paper considers the Monitoring Campaign of the Air Quality, supported by FEEMA (State Foundation of Environment Engineering) in the years of 1991 the 1995, and collected in a height of 2 to 4 meters, beyond data of meteorological monitoring [16].

The meteorological evaluation and the local topography are necessary to the better understanding of the relations of the pollutant concentrations in the atmosphere.

The meteorological data contains the following measured parameters: direction and wind speed, precipitation, atmospheric pressure, total of clouds, direction of clouds, horizontal visibility, relative humidity, temperature of air (dry bulb, humid bulb) and dew point.

The pieces of data employed do not present fixed intervals, occurring intervals that varies from one day until 6 months between two consecutive measures.

To bypass this problem of incomplete data, a linear trend between two contiguous measurements was adopted.

6 Methodology

The first methodology chooses the neural network architecture using the considered practical approach (M1)[17].

The second method (M2) [18] is based on a software, that uses genetic algorithms to evolve architectures of neural networks, while, simultaneously, it looks for expressive variable of input to maximize the predictive accuracy. The best 10 architectures are selected.

In both methods, the same criteria of normalization and handling of the data has been adopted; the same standards were presented to the input and the same mode to define the data set for test is used.

To make the forecast using artificial neural network, the inputs correspond to the meteorological parameters and the outputs are the particles in suspension in nine locations.

The set of pairs input/output that form the data test, were obtained by the selection of 1 vector to each 5 of the presented to artificial neural network.

This way, the data of each set, is representative for all the seasons of the year.

In both methods, the calculation of the accuracy for each network is made by the difference between the output and the output of the trained network. The objective is to minimize the root mean square error:

$$RMSE = \sqrt{\frac{\sum \left(y_{real} - y_{predict}\right)^2}{n}}$$ (1)

7 Results and Discussion

M1

In general, the backpropagation networks have presented greater divergence between the measurements and the predicted values. GRNN Networks have shown better results than backpropagation networks and are faster in the training/test cycle.

M2

The backpropagation networks were not selected. It can be concluded that networks of GRNN type have better capacity for mapping the studied problem.

All the neural architectures of GRNN type selected for genetic algorithms have presented well satisfactory results, mainly in function of the time used in the training/test. In terms of the expended computational time the M1 methodology is 6 times more expensive than M2. To compare the employed methods and the architectures, the Root Mean Square Error is shown in table 1.

Table 1: Percentage of the Root Mean Square Error for each architecture of the backpropagation type (M1).

	% RMSE
Bonsucesso	8.9
Centro	5.5
Copacabana	4.5
Inhaúma	7.1
Irajá	6.3
Itaguaí	7.1
Jacarepaguá	6.3
Maracanã	4.5
São João de Meriti	7.1

The comparison including both methodologies is indicated in table 2.

Table 2: Percentage of the Root of the Quadratic Average Error for
each architecture of GRNN type using M1 and M2

	% RMSE		
	M1	M2 - BEST	M2 – 10TH
Bonsucesso	6.3	8.9	8.4
Centro	5.5	5.5	6.3
Copacabana	3.2	8.9	3.2
Inhaúma	5.5	6.2	7.5
Irajá	4.5	7.1	7.1
Itaguaí	5.5	6.3	7.7
Jacarepaguá	4.5	4.5	6.3
Maracanã	3.2	3.2	4.5
São João de Meriti	5.5	5.5	7.1

In some locations, as Copacabana for example, the best architecture presents a
bigger RMSE. This also occurs because the choice is based not only on the
accuracy, but according to the topology of the network, represented for a lesser
number of neurons, used parameters and the consumed computational time.

The neural architectures evolved by the genetic algorithm have presented a
lesser accuracy, probably caused by the lesser time of training, that diminishes
the capacity of generalization of the network.

Prediction

Figures 1 and 2, indicate some values predicted through M1 and M2
methodologies considering the Bonsucesso Station and Itaguaí, respectively.

Figure 1: Values predicted for the Bonsucesso Station

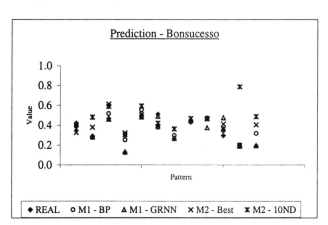

Figure 2: Predicted values Itaguaí Station

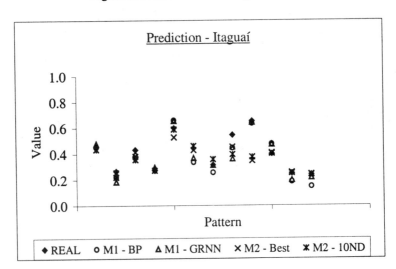

Conclusion

The studied methodologies proved to have a good potential to be used in this type of problem, in spite of the diversity of situations found in the modeling of the data. The encouraging results come from the ability of the neural networks in inferring nonlinear mappings, of these complex systems.

The obtained results must be considered preliminary, since it had the necessity of making many smoothing that, possibly, are not full acceptable.

However, the uncertainties are minimized by the certainty that the neural networks are sufficiently effective in the learning of standards from nonlinear, incomplete data, with noises and contradictory examples. This ability of data manipulation makes the networks extremely efficient in the processing of information.

The search for excellent architectures is a difficult task. To construct networks, a great variety of architectures must be tested, making this method of high cost. To guarantee the optimization of the solution, since the performance criterion is based on a complex combination of factors, it was adopted evolutionary method, through the use of genetic algorithms. This provided a more efficient method for the optimization of neural architectures and produced excellent solutions.

Following the restrictions imposed for the available data, the developed prototype reached good results, making possible for the deriving results to be used as element keys for elaboration of an ambient control strategy.

References

[1] Haykin, S., *Neural Networks: a Comprehensive Foundation.* 2nd ed. New Jersey, Prentice Hall, 1999.

[2] Calôba, L. P., "Neural Computer Introduction". In: IX *Automation Brazilian Congress,* pp. 25-38, Vitória, ES, Brazil, Sept. 1992.

[3] Tafner, M. A., "Artificial Neural Network: Learning Plasticity". *Journal of Brain and Mind.* March/May, 1998. <http://www.epub.org/cm/n05/tecnologia>.

[4] Wasserman, P. D., *Neural Computing: Theory and Practice.* 1nd ed. New York, Van Nostrand Reinhold, 1989.

[5] Carvalho, A. C. P. L. F., *Tutorial of Artificial Neural Network, Specially about Multi Layer Perceptron Networks Trained with Backpropagation.* USP, Brazil, 1998. <http://www.icmsc.sc.usp.br/~prico/neural1.html#topicos>.

[6] Martino, M. B., *Neural Networks.* Rio de Janeiro, RJ, Brazil, 1996. <http://www.cepel.br/~martino/rn.htm>.

[7] Chemometrics Research Group, *GRNN.* Naval Research Laboratory. <http://chem1.nrl.navy.mil/~shaffer/grnn.html>.

[8] Neural Computing, *A Technology Handbook for NeuralWorks.* Professional II / Plus and NeuralWorks Explorer. Neural Ware. USA, 1998.

[9] Carvalho, A. P. L. F., Neural Architecture Evolutionary Project. USP, Brazil, 1997. <http://www.icmsc.sc.usp.br/~andre>

[10] Harp, S. A., Samad T., "Genetic Synthesis of Neural Network Architecture". In: L. Davis (Ed.), *Handbook of Genetic Algorithms,* New York, USA, Van Nostrand Reinhold, pp. 202-221, 1994.

[11] Schaffer, J. D., Whiley, D., Eshelman, L. J., "Combinations of Genetic Algorithms and Neural Networks: A Survey of the State of the Art". In: *International Workshop on Combinations of Genetic Algorithms and Neural Networks,* pp. 1-37, IEEE Computer Society Press, 1992.

[12] Tanomari, J. "Motivation, Fundamentals and Applications in Genetic Algorithms". In: *II Brazilian Congress in Neural Networks / III School of Neural Networks,* Curitiba, Brazil, Oct. 1995.

[13] Stern, A. C., Boubel, R. W., Turner, D. B., Fox, D. L., *Fundamentals of Air Pollution.* 2nd ed. Orlando, Academic Press, Inc., 1984.

[14] Cabral, S. D., *Methodological Proposal Monitoring Atmospheric Pollution Caused by Transport System – The Case of Carbon Monoxide.* Thesis M.Sc., COPPE/UFRJ, Rio de Janeiro, RJ, Brazil, 1997.

[15] *CETESB – Technological Company and Environmental Control.* <http://www.cetesb.br/index.htm>.

[16] Fraga, I., Pasenau, H., *Air Quality in the Metropolitan Region of Rio de Janeiro,* State Foundation in Environmental Engineering – FEEMA, Rio de Janeiro, Brazil, 1995.

[17] NeuroShell 2, Version 1.5, Ward Systems Group, Inc. "Neural Network Products for the Next Century"[TM], 1993

[18] NeuroGenetic OptimizerTM (NGO), Version 2.5, BioComp Systems, Inc, Professional Edition, 1997.

Modelling of pollution transport in the marine environment of Sakhalin shelf

I.E. Kochergin, A.A. Bogdanovsky & S.I. Rybalko
Far Eastern Regional Hydrometeorological Research Institute (FERHRI), Vladivostok, Russia

Abstract

The described methods have been developed to assess the spatial-temporal scales of the impact produced by pollutants discharged into the sea over the shelf of Sakhalin Island (Far East Russia region), where oil and gas fields are being developed. Modelling technique includes the software for preliminary processing of initial information and the databases of hydrometeorological observations and pollution source characteristics, engineering methods and models, and the software for analysis and visual interpretation of modelling results.

The described computation technique of pollution transport applies the three-dimensional advective-diffusive model with the Lagrange description based on the Monte-Carlo method. Peculiar of the model is realization of turbulent pulsation mechanism describing the diffusive indeterminate distribution of pollutants.

The described modelling technique has been successfully tested in the shelf zone of Sakhalin.

1 Introduction

Oil production on the Northeast Sakhalin shelf potentially impacts the adjacent marine environment. The impact can be caused by sea-water pollution, ground removal and damping, drilling mud, cutting, and produced water discharges, those are designed impacts, and by accidents. To assess the spatial-temporal scales of such impacts on sea-water and sea-floor the modelling technique has been developed.

This technique includes the software for preliminary processing of the initial information, databases of pollution source characteristics and hydrometeorological information, a set of engineering methods and models, and the software for analysis and visual interpretation of modelling results. The technique of pollution transport and settling applies the three dimensional advective-diffusive model using the Monte-Carlo method [1,2]. The model developed in the Lagrange coordinates accounts interaction at media boundaries. Peculiar of this model is realization of mechanisms of turbulent pulsations, boundary conditions, and settling processes [3]. Besides, it accounts the effects of "jet sinking" at the initial stage and "flocculation".

2 Model description

Pollution transport and settling are calculated over the three-dimensional diffusive-advective model based on the Monte-Carlo method and using the generator of random numbers to initiate non-determined processes (in accordance with the general ideology of Ozmidov [2]). Its analogue is OOC Mud Discharge model [4]. It is better to use the models based on the Monte-Carlo method for numerical realization of the practical calculations [2,3].

The below equations describe transport and settling of markers:

$$\frac{dx_i}{dt} = u(x_i, y_i, z_i, t) + u'(x_i, y_i, z_i, t),$$

$$\frac{dy_i}{dt} = v(x_i, y_i, z_i, t) + v'(x_i, y_i, z_i, t), \tag{1}$$

$$\frac{dz_i}{dt} = w(x_i, y_i, z_i, t) + w'(x_i, y_i, z_i, t) + w_0(x_i, y_i).$$

Where: x_i, y_i, z_i - coordinates of i marker; u, v, w - zonal, meridional and vertical current constituents; u', v', w' - velocity constituents of turbulent pulsations; w_0 - pollutant settling rate.

The initial coordinates, start time and physical-chemical properties are determined for each individual marker. Concentration of pollutants in each element of the discrete spatial grid is determined in accordance with the number of markers and their weight. Concentration error is the ratio of the marker's weight to the volume of discrete element in the grid.

The conditions are set at the boundaries, they are partial settling, evaporation, or reflection being dependent on the specific characteristics of boundaries and pollutants. Reflection and settling values at the boundary are conditioned by adhesion coefficients.

Velocity constituents of currents, u, v, w, for a modelled shelf area, are set up by different methods depending on the task. These methods are described in section 3 given below.

In eqns (1) the settling rate of total suspended solids (TSS) w_0 for fine and medium fractions is calculated over the three-force-balance equation for the ball-shaped particles of the given equivalent radius [5]:

$$W_0 = \frac{g(\rho_i - \rho(z))l_i^2 \sqrt{1 + 0.862 \lg k}}{\mu \rho(z)(18 + 0.61 \sqrt{g(\rho_i - \rho(z))l_i^3 / (\mu^2 \rho(z))})}. \tag{2}$$

Where: g - acceleration of gravity (9.82 m/s^2); ρ_i - specific weight of a particle; $\rho(z)$ - water density at depth z; l_i - characteristic diameter of a particle; k - geometric shape coefficient; μ - viscosity coefficient determined from the empiric equation, for fresh water it is:

$$\mu = \frac{0.01775}{1 + 0.0337 T_w + 0.000221 T_w^2}. \tag{3}$$

Where: T_w - temperature ($C°$).

In eqn (2) the mechanism of l_i and k determination for the particles of irregular shape and certain degree of roughness is described in [5]. For engineering calculations geometric-shape coefficient k is selected from 1.17-1.67 values characteristic of the sands of various types.

Turbulence parameterization is described in [2]. Turbulence is represented in terms of vortex structures of various spatial-temporal scales dependent on oceanic processes. The calculation scheme accounts up to three scales of the velocity of turbulent pulsations set up by the discreteness of statistic tests. Velocity constituents of turbulent pulsations u', v', w' in eqns (1) are found using the calculated σ_u, σ_v, σ_w dispersions and the concept of normal distribution of the spectrum of oceanic turbulent pulsations.

Total equations of root-mean-square deviations are derived from the assumption saying that the basic dispersion value is directly conditioned by the current proper together with wind and near-bottom corrections. Both are produced by in-a-row expansion of parameterization of the wind wave effect τ on turbulence $\alpha(z) \approx k_1 \cdot z \cdot \tau \cdot \exp(-k_2 \cdot z)$ and parameterization of the near-bottom friction effect (it is proportional to $(z-h)^2$, where h - sea depth).

Thus, the equations of turbulent pulsation rate look as follows:

$$\sigma_u(z) = \sqrt{2 + 0.196 v_x^2 + 0.076 v_y^2} * (K_1 + K_2 \exp(-\alpha z) + \frac{K_3}{\beta(h-z)^2 + 1}),$$

$$\sigma_v(z) = \sqrt{2 + 0.196 v_y^2 + 0.076 v_x^2} * (K_1 + K_2 \exp(-\alpha z) + \frac{K_3}{\beta(h-z)^2 + 1}), \tag{4}$$

$$\sigma_w(z) = \gamma \sqrt{\sigma_u^2 + \sigma_v^2} * f(Ri).$$

Where: v_x, v_y - current velocity constituents; K_1, K_2, K_3 - relative contribution of various processes; α - scale of the surface disturbance effect (wind waves); β - effect of near-bottom conditions (sea-floor features); γ - transition parameter characterizing the average horizontal/vertical diffusion ratio (0.083 is used); h - sea depth; $f(Ri)$ - function determined by the Richardson's criterion, reducing turbulent mixing under stable stratification and intensifying it under unstable stratification.

Empiric coefficients were selected from eqns (4) using the criteria determined in accordance with the scales of turbulent pulsations. Parameter α was chosen as the value inversely proportional to the characteristic scale of the wind wave effect - $0.25L$, where L is the characteristic length of wind waves. For deep waves the length depends on $L=T^2g/2\pi$ relation, where T - wind wave period. β parameter describes the thickness of the near-bottom boundary layer, it is calculated for the mean velocities of near-bottom currents, $\beta=0.16$. K_1 coefficient ranging as 0.4-1.2 is experimentally determined. K_2 ranges as 0.1-0.6, it is conditioned by wind characteristics (assumed being equal to 0.03 of wind speed). K_3 coefficient assumed equal to 0.05-0.3 and depends on the sea-floor relief.

The following semi-empiric parameters are used to calculate turbulent dispersion of vertical velocities. γ coefficient of characteristic transit from horizontal to vertical scales is determined over the vertical/horizontal velocity ratio. The determined 0.083 value does not contradict the values stated in scientific literature [2]. The function of the Richardson number in eqns (4) was determined by means of various polynomial and exponential approximations. Exponential formula for stable stratification $F(Ri)=exp(-1.1Ri)$ gained the best reputation.

Characteristic dispersion profiles of turbulent pulsation rate for various seasons are described in [3].

At the initial stage of pollution transport modelling the processes in a jet, such as jet destruction due to layer instability, turbulent pulsations, and other factors; "jet sinking" due to the density being higher than the density of the surrounding sea-water; destruction of colloid structure of drilling wastes, and TSS settling due to gravity, are taken into account.

The effect of "jet sinking" works in case of intensive discharge of heavy drilling wastes. It means that the bulk of pollutants settle with the rate that is much greater than the settling rate in case of less intensive discharge [5]. As the concentration of drilling mud in sea-water drops by 100 times and more, the jet structure destructs. The jet stops sinking and suspended matter continues settling due to gravity. The duration of "jet sinking" effect depends on current velocities, discharge rate and other factors. In average it lasts 0.5-1.5 min. Best of all this effect is manifested in bulk dumps.

In case of drilling wastes it is important to account the effect of "flocculation", that is sticking of mud and cutting solids into combined cohesive particles [6]. Flocculation is a function of colloid solutions, that is formation of suspension cover on solid particles that holds them combined after mixing with sea-water. A cohesive particle (floc) settles quicker. Adhesion process intensifies when drilling wastes contain flocculators. Assessments of drilling mud and cutting discharges assume that 25% of particles in light muds and up to 70% of particles in heavy barium drilling muds with cuttings participate in the process of adhesion.

The calculation technique of pollution transport and settling is realized by means of specialized software. Calculation results are produced in the form of tables, diagrams, graphs, trajectory maps, two- and three-dimensional images with isolines and isosurfaces, and colored images showing pseudoreal plumes in sea-water and sedimentation zones on sea-floor. The recent "V-7.0" computer version of the model is capable to process 20 sources operated simultaneously, with up to 500 000 markers.

3 Initial data

The initial data are subdivided as pollution source characteristics and hydrometeorological information for the environmental model. The pollution source characteristics include geometric size, location, detailed discharge regime, composition, and physical-chemical properties of discharged pollutants. Physical-chemical properties describe their concentration in the discharged wastes, specific weight, non-conservative properties, toxicity, and, for TSS, distribution over fractions. The initial hydrometeorological information includes representative data collected during the periods provided by significant statistics, such as: instrumental current and wave data series, wind parameters, sea-water temperature and density, background pollution characteristics. The initial information is analyzed and processed for the purpose of environmental model construction, that includes the profiles of currents and density, parameters of turbulent pulsation rate, background pollution level, etc.

According to eqns (1) currents are the main factor the transport of pollutants in marine environment depends on. Current scenarios are developed on the basis of historical instrumental observations. Calculations and experiments proved that the impact produced by typical discharges is localized within the zone up to several kilometers in size. That allows to use the given vector profiles of currents, density, and temperature for the areas of open shelf. In calculations realized for impact assessments and ecological monitoring data processing purposes the following current data are set up:

1) The observed data series describing the currents at several horizons or the series-analogue is used to calculate TSS settling on the sea-floor under durable discharge regime, or the averaged characteristics of sea-water pollution.

2) The profile of lower current velocities (with 95% cumulative probability of current speed) is constructed to determine the control zone of mixing for the purpose of waste discharge normalization under unfavorable hydrological conditions. Current velocity profiles are constructed using the frequency of occurrence tables of the combined data series for the given area.

3) Model fields of tidal and nontidal currents are calculated over the wind characteristics of pollution transport. Tidal currents are calculated using the harmonics constructed over the related instrumental data series. Nontidal currents are calculated over the diagnostic model of the Ekman type [7].

4) Typical current profiles are used in calculation of the averaged and the most probable characteristics of sea-water pollution. Tolerant criteria of vector characteristics (velocity and direction) combining the corresponding velocity and direction intervals help to determine which current profile corresponds to a concrete typical situation.

Typical current profiles are constructed as it is described below. At the first stage the series are represented as a set of synchronous current vectors at three horizons. Percentage of each vector combination in this set is determined. At the second stage using the similarity criteria 30 the most significant situations are added with the situations answering the given criteria. At the third stage the toleration zones are extended and the 30 situations obtained by means of similarity criteria are reduced to 6-10 the most probable typical situations that include over 90% of all situations. Fig. 1 illustrates the characteristic typical current profiles for the Piltun-Astokh oil field on Sakhalin shelf.

4 Model verification and discussion

The model has been verified for Sakhalin shelf conditions in course of monitoring of the wells drilled during the summer-autumn navigation season. Verification included: 1996-1998 experiments with discharges into marine environment to assess sea-water pollution parameters in the area of the Arkutun-Dagi oil field; sampling of sediments to assess the sea-floor impact parameters in 1996-1998 in the Arkutun-Dagi oil field and in 1998 in the Piltun-Astokh oil field (see Fig. 1).

Fig. 2 shows the results of drilling mud discharge modelling realized during the 1996 experiment at the Arkutun-Dagi oil field. Preliminary calculations helped to forecast the direction of pollution migration and the sampling sites. Current and meteoparameters for this experiments were measured from the rig. Fig. 2 demonstrates the results of concentration calculation showing the increase of concentrations at the sampling sites.

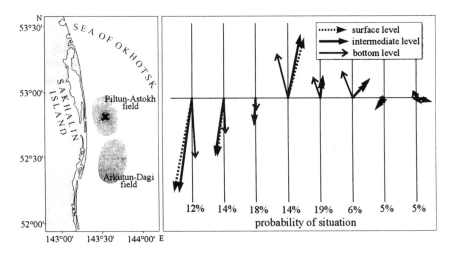

Figure 1:   The example of current velocity vectors at the surface, intermediate, and near-bottom horizons for the typical situations at the Piltun-Astokh oil field.

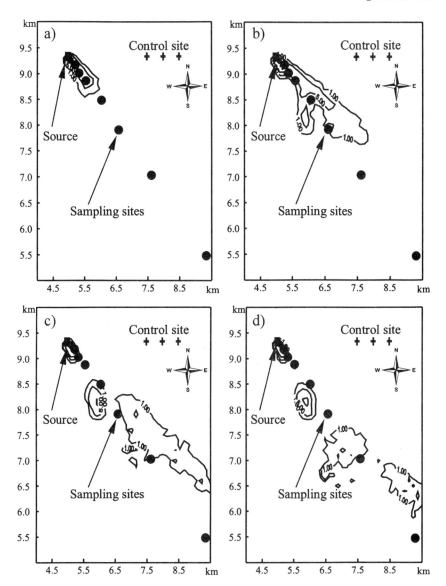

Figure 2: Calculated integral isolines of drilling mud concentrations in sea-water obtained in course of 1996 experiments correspondingly: 1 hour (a), 3 hours (b), 6 hours (c) and 8 hours (d) after discharge beginning (concentration isoline gradation is given in 1, 5, 20 and 50 g/m^3).

The complex experiment with drilling mud discharges was carried out in 1998 at the Arkutun-Dagi oil field. The samples were collected by three small vessels, and stationary current and turbidity meters. The discharged drilling mud included: TSS - 11.5%, water containing dissolved admixtures - 88.5%. The volume of discharged wastes made up 50 m^3, discharge intensity was uniform, the rate of discharge was equal to 1.5 m^3/min. The drilling mud was discharged on the sea surface. The modelling results in the form of horizontal and vertical projections of concentration fields 15 and 30 minutes after the discharge are shown in Fig. 3. Its also shows sampling sites of turbidity and suspended matter concentration measurements. The experiment results revealed significant scattering of observed concentrations, however, averaged observations were in good comparison with calculated concentrations.

Modelling of TSS settling was also fulfilled, the results being tested over the changing grain size of bottom sediments. The grain size of TSS in drilling mud and cutting is finer than the one of the Sakhalin shelf sediments. Thus, in the area of the appraisal well drilled in 1996 the calculated increase of fine fraction content (0.25-0.1 mm) made 0.6-0.8% in the ten-centimeter layer, in the surface two-centimeter layer - 2.0-2.4%. According to the values observed by CSA company the actual increase of fine fraction content in the whole sample made up 0.70-0.85%, in the surface layer - 0.8-1.1%.

Assessments of calculation results carried out for the wells drilled in 1998 revealed that the maximal increase of fine fraction concentrations (0.1-0.25 mm and less) in 10 cm of sediment sample for the first well made up to 5-7%, for the second well - 6-10%. The real percentage revealed by monitoring measurements made 7.5% in the 10 cm layer, for the second well the increase of fine fraction made 60-65%. The latter result was much higher than the calculated numbers. That might be due to a small narrow gully on the sea-floor at the sampling site, where over 5 cm of TSS could have settled.

Verification experiments revealed the sufficient precision of calculations fulfilled for engineering assessments of polluted sea-water and sea-floor zones. The authors believe that the capabilities of the described model are better than the capabilities of OOC model [4], developed for analogous purposes. Further development of the model needs more detailed verification, the effects of "jet sinking" and "flocculation" inclusive.

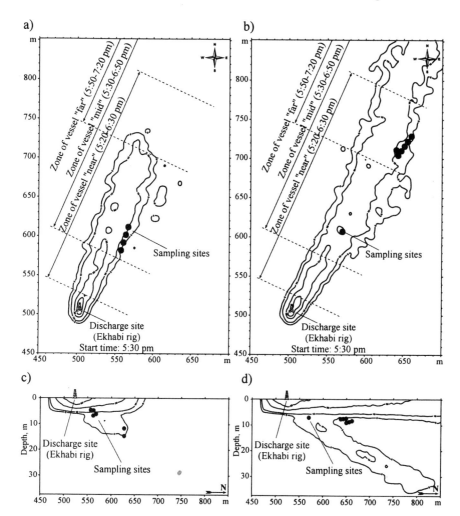

Figure 3: Modelling results and sampling sites. 15 (a,c) and 30 (b,d) minutes after discharge beginning. Horizontal (a,b) and vertical (c,d) projections (concentration isoline gradation is given in 5, 10, 30 and 50 g/m^3).

5 Conclusions

Simulation of drilling waste discharges into marine environment at the various stages of the Sakhalin shelf oil and gas production shall become a component of the environment impact assessment and ecological monitoring procedures. The described approach includes two stages. At the first stage it is the construction of pollution source and environment models (scheme of currents and hydrological conditions) using four types of current scheme presentation. At the second stage the fate of pollutants in marine environment is modelled. The model of pollution transport and settling is constructed for practical use in calculations of potential impact scales on Sakhalin shelf. This model was successfully applied in practical calculations for oil and gas field development projects and ecological monitoring in 1992-2000.

Acknowledgements

The authors express their sincere gratitude to Sakhalinmorneftegas-Shelf Company for the financial support of the fulfilled research and to Dr. V.F. Putov and Dr. V.D. Budaeva for fruitful scientific discussions.

References

[1] Galkin, L.M. *Application of the Monte-Carlo method for diffusive task solutions*, Nauka: Moscow, pp. 1-96, 1975 (in Russian).
[2] Ozmidov, R.V. *Diffusion of contaminants in the ocean*, Hydrometeoizdat: Leningrad, pp. 1–280, 1986 (in Russian).
[3] Kochergin, I.E., & Bogdanovsky, A.A. Transport and turbulence characteristics for the north-eastern Sakhalin shelf conditions. *Proc. of the 2nd PICES Workshop on the Okhotsk Sea and Adjacent Areas*, PICES: Sidney (Canada), pp. 115–121, 1999.
[4] O'Reilly, J.E., Sauer, T.C., Ayers, R.C., Brandsma M.G., & Meek, R. Field verification of the OOC mud discharge model. *Drilling wastes*, Elsevier Applied Science: New-York, pp. 647-665, 1989.
[5] Kurganov, A.M. & Fedorov, N. F. *Reference Book of hydraulic assessments for water supply and canalization systems*, Stroyizdat: Leningrad, pp. 1-408, 1973 (in Russian)
[6] Winterwerp, J.C. A simple model for turbulence induced flocculation of cohesive sediment. *Hydraulic Research*, 36(3), pp. 309-326, 1997.
[7] Budaeva, V.D., & Makarov, V.G. Peculiar water regime of currents in the area of Eastern Sakhalin shelf. *Proc. of the 2nd PICES Workshop on the Okhotsk Sea and Adjacent Areas*, PICES: Sidney (Canada), pp. 131–138, 1999.

Section 4
Oil Slicks and Spills

Toxicity screening, assessment, and reduction of tannery wastewater

M. Cotman[1] & J. Zagorc-Končan[2]
[1]National Institute of Chemistry, Slovenia
[2]University of Ljubljana, Faculty of Chemistry and Chemical Technology, Slovenia

Abstract

Many industrial plants are facing a requirement to conduct toxicity reduction evaluations (TREs). The TRE is an evaluation intended to determine those actions necessary to achieve compliance with water quality based effluent limits (reducing an effluent toxicity or chemical concentrations to acceptable levels).

Wastewater from leather production is known to be heavily loaded with inorganic and organic components. We were faced with wastewater from the largest tannery in Slovenia, which has a system of physico chemical pre-treatment (coagulation and flocculation with Al-sulphate and anionic polyelectrolyte). The wastewater is highly polluted with components, which are specific to the leather industry.

Prior to initiation the TRE program chemical and toxicological analyses were conducted with tannery effluents ten times during the year. An acute toxicity test of whole effluent using an aquatic organism *Daphnia magna* was performed. We identified key toxic components in the pre-treated discharge with fractionation of samples through chemical and physical means (filtration, air stripping, adsorption on activated carbon and zeolite treatment). The goal of each fractionation step was to reduce toxicity due to specific group of chemicals and compare the results to the toxicity present in the unaltered sample.

The inorganic parameters (ammonia and chromium) seem to play a minor role for toxic effects for marine bacteria *Vibrio fischeri*. *Vibrio fischeri* is tolerant against high salt contents and poorly susceptible to metals. Insoluble and suspended material is more toxic for *Daphnia magna*. Toxic effect of ammonia strongly depends on pH value due to concentrations of un-ionized form. Toxicity of ammonia does not have an important effect on toxicity of tannery wastewater when organic load is still present. Due to adsorption of organic load to activated carbon toxicity for *Daphnia magna* and *Vibrio fischeri* decreased. For more complex determination of toxicity in wastewater, containing many unknown compounds, finer fractionation steps are required for toxicity identification purposes.

1 Introduction

Effective environmental management of discharges from industrial wastewater treatment plants must include environmental hazard or impact assessments, as well as technical and economical evaluations of various procedures for pollution reduction, such as dilution, effluent treatment, or implementation of cleaner technologies. The environmental impact is determined by the composition of effluent and the discharged quantities on the one hand and by the dilution and specific characteristic of the receiving water on the other, and both local effects and the possible contributions to a more widespread, general pollution must be taken into account, according to different authors (Nyholm [1]; Nyholm [2]).

1.1 Characterisation of effluent

The characterisation of the harmful properties of effluent is a critical element in any assessment and a key element of strategy. It is usually necessary to rely on a specific chemical characterisation and whole effluent toxicity which both had their advantages and limitations, but which combined may result in an overall level of information sufficient for practical management and control purposes. The environment impact is highly case specific and depends on the composition of effluent, including its temporal variability, and on the discharge volumes.

The Slovenian legislation is based upon fixed emission limits (*Ordinance of Emission of Substances and Temperature at Wastewater Discharging from Pollution Sources*, Official Gazette of Republic Slovenia [3]). This general ordinance is complemented by one for industrial branches (e.g. *Ordinance of Emission of Substances at Wastewater Discharging from Objects and Devices for Leather and Fur Production*, Official Gazette of Republic Slovenia [4]). The legislation prescribes two limits for the discharged wastewater: one for discharge into sewerage system and the second one for discharge into a river.

In our study the effluent was investigated according to the legislation by chemical parameters such as oxygen consuming substances, nutrients and other harmful substances that due to the technology of the plant. The acute toxicity test of whole effluent was also performed with *Daphnia magna*.

1.2 TRE procedure

Many industrial plants are facing a requirement to conduct toxicity reduction evaluations (TREs). The TRE (Ford [5]) is an evaluation intended to determine those actions necessary to achieve compliance with water quality based effluent limits (reducing an effluent toxicity or chemical concentrations to acceptable levels). The U.S. EPA has published a document on toxicity identification procedures for effluents (EPA [6]) but this methodology is adjusted on a case-by case basis depending on effluent characteristic. In our study the following characterisation steps has been used: filtration, air-stripping, adsorption on powdered activated carbon and ion exchange with zeolite.

Filtration is used to remove toxicity associated with particulate. In rare cases, false results have been observed as a resul of purging chemicals as a result of vacuum pressure during the filtration process.

Air-stripping removes volatile and pH extractable organic chemicals as well as some inorganic chemicals (e.g. hydrogen sulfide and ammonia). Samples are vigorously air-stripped for periods of at least two hours at pH 3, pH 11, and original effluent pH. Shifting the pH alters the form of some chemicals, making them less ionic and hence more likely to air-strip. False results in terms of the true mechanism of removal can sometimes be obtained if chemicals are precipitated at high pH and do not return to their original form following pH re-adjustment, or if rearation is oxidising rather then stripping the toxicants.

Powdered activated carbon (PAC) removes chemicals by adsorption, pore entrapment and, ion exchange (in the case of metals). PAC treatments are designed to remove adsorbable chemicals, but remove a wide variety of chemicals due to different mechanisms. PAC treatments are often successful, but relatively non selective.

Zeolite treatment is utilised in conjunction with pH 11 air-stripping and graduated pH tests to confirm or refute the presence of ammonia as the causative toxicant. Although zeolite removes constituents other than ammonia (e.g., some metals) it is useful step for comparison with the results of other tests designed to remove ammonia.

2 Materials and methods

2.1 Source of tannery wastewater

The largest Slovenian tannery in Vrhnika processes 40 to 50 tones of pig skins daily. The leather industry is well known as a high consumer of water, that is 30 to 80 m^3 for 1 ton of processed raw skins (Roš [7]). Wastewater from leather production is heavily loaded with inorganic and organic constituents and its purification is an important task worldwide (Reemtsa [8]). Besides generally high salt contents, the main inorganic contaminants are chromium from chrome-tanning, sulfide used for hair removal, and the liming step and ammonia from deliming. The organic load is also of concern with chemical oxygen demand (COD) in the range of 5000-9000 mgL^{-1} and biological oxygen demands (BOD_5) of 2000-4000 mgL^{-1}. The tannery in Vrhnika has physico-chemical pretreatment system for the wastewater as a whole (precipitation and flocculation with Al-sulfate and anionic polyelectrolyte). This removes 60% of organic load and over 90% of sulfide and chromium. However, the large amounts of sludge produced by flocculation processes give rise to new problems owing to their high organic carbon and chromium contents.

Samples of effluent from the industrial wastewater treatment plant were taken as 24 hours flow proportional samples. Because of varying effluent composition due to different productions, samples were obtained ten times during 1998, that is every month from March to December. The samples were taken at the usual production conditions in the middle of the week.

2.2 The TRE procedure

The fractionated methods applied to samples were filtration, air stripping, adsorption on PAC and zeolite treatment.

2.2.1 Filtration
Samples were filtrated using 0.45μm cellulose acetate filters. Filtrated samples were chemically and toxicologically analysed.

2.2.2 Air Stripping
The 1000 ml samples, previously adjusted to pH 3 and 7 and also an original sample were exposed, during continuous mixing, to 24 hours of aeration with aquarium pumps (Q_{air}= 120L/hour). The samples were chemically analysed followed by neutralisation and toxicity analyses.

2.2 3 Adsorption on PAC
Aliquots of 1L of samples were exposed to 0.1, 0.5, 1.0, 2.0, 5.0, and 10.0 g of activated carbon (GAC, DARCO, 20-40 MESH), during continuous stirring for five hours at 20 °C. The treated samples, after filtering through 0.45μm pore filter were chemically and toxicologically investigated.

2.2.4 Zeolite treatment
Aliquots of 1L of samples were exposed to 1.0, 2.0, 5.0, 10.0, 15.0, and 75.0 g of zeolite clinoptilolite (K-4-B, Zaloška Gorica) during continuous stirring for two hours. The treated samples, after sedimentation, were analysed only for nitrogen compounds and toxicity test with *Vibrio fischeri* due to small volumes of samples.

2.3 Chemical and toxicity analyses

Effluents from tannery were chemically investigated before a toxicity assay. The COD, BOD and oil and grease were performed immediately after sampling. Total P was determined spectrophotometrically after mineralization with persulfate, while sulfide was measured in fixed sample as soon as possible by spectrophotometric methylen blue method. Organic N and ammonia were measured by the macro-Kjeldahl method. Phosphate, sulfate, nitrate and nitrite were determined by chemically suppressed ion chromatography (DIONEX 4000) in the filtrated samples using an 0.2 μm filter. Chromium and aluminium were measured by the ICP-AES (Thermo Jarrell Ash) in whole samples. The 1,5-diphenylcarbazide spectrophotometric method was used for the determination of the most toxic chromium, Cr^{6+}. The toxic potencies of water samples were estimated by using standard nonbenthic bioassay organism as recommended in the *Daphnia magna* - mobility inhibition test and *Vibrio fischeri* - luminescence inhibition. All chemical and toxicological analyses were performed according to standard procedures (*Standard Methods for the Examination of Water and Wastewater* [9]).

3 Results and discussion

The physico-chemical pretreated wastewater from the largest Slovenian tannery was first chemically investigated. Parameters which are prescribed in *"Ordinance of Emission of Substances at Wastewater Discharging from Objects and Devices for Leather and Fur Production"* (Official Gazette of Republic Slovenia [4]) were analysed in 1998. The limit values for the wastewater discharge into the receiving stream are also defined in this regulation. The results of ten samplings during the year 1998 are presented in Table 1and Figure 1.

Table 1. Chemical analyses of pretreated tannery wastewater (10 samplings during the year 1998).

	Tannery effluent (average value)	Limits value
pH	7.4	6.5-9.0
COD, mg/L	3020	300
BOD_5, mg/L	1980	30
Tot. Susp. Solids, mg/L	878	30
S^{2-}, mg/L	1.64	0.5
SO_4^{2-}, mg/L	1347	1500
NH_4^+-N, mg/L	162.6	In summer 15,In winter 30
P_{tot}, mg/L	4.0	2.0
Cr_{tot}, mg/L	3.0	1.0
Al, mg/L	6.0	3.0
Oil and grease, mg/L	315	20

The physico-chemical pretreated tannery wastewater contained very high concentrations of organic and inorganic substances thus almost all prescribed limit values were exceeded. The organic loads (COD, BOD, oil and grease) were very high. Although the concentration of COD and BOD_5 varied between different samplings, the ratio between them remained the same in the range of 0.60 - 0.71, average 0.65. Very high concentrations of ammonia nitrogen were also found in all samplings. It would also be necessary to include in the Ordinance (Official Gazette of Republic Slovenia [4]) the determination of organic nitrogen known from the literature (Reemtsa [8], Fiehn [10]) to be discharged in great quantities from tanneries.

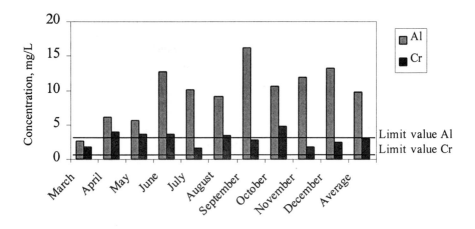

Figure 1. Concentration of aluminium and chromium in the tannery wastewater during the year 1998.

Metals are the substances which cannot be discharged into receiving stream. The physico-chemical pretreatment system at Vrhnika that uses precipitation and flocculation with Al-sulfate and anionic polyelectrolyte is unsatisfactory. The addition of Al-sulfate is not optimal as excess aluminium, whose harmful properties had been previously investigated was detected in treated effluent. With this system chromium is also insufficiently removed and at all samplings total chromium was higher than the limits set for the discharge into the river. The most dangerous form of chromium, Cr^{6+}, was not present due to a reductive matrix (Milačič [11]). There were also high concentrations of sulfide and sulfate.

Chemical analyses of tannery wastewater were complemented with toxicological analyses. The results of toxicity test are presented in Figure 2. The comparison of the 24 and 48 hour toxicity tests demonstrated that tannery wastewater was more toxic in the 48 hour test. Due to different sensitivity of organisms to harmful substances it would be better to prescribe in legislation a set of organisms for toxicity testing. According to our experience and literature data (Fiehn [10], Reemtsa [8]) *Vibrio fischeri* seems to be suitable organism, tolerant to high salt contents but poorly susceptible against heavy metals.

Although both aspects give rise for criticisms against the luminescence inhibition test, it is advantageous for investigating the toxicity of organic compounds in tannery wastewater.

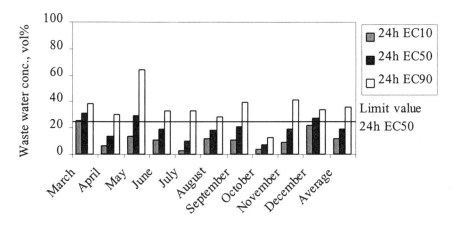

Figure 2. Toxicity of tannery wastewater to *Daphnia magna* during the year 1998.

We identified key toxic components in the tannery discharge with fractionation of samples through chemical and physical means (filtration, air stripping, adsorption on PAC and zeolite treatment). The goal of each fractionation step was to reduce toxicity due to specific group of chemicals and compare the results to the toxicity present in the unaltered sample. The most effective procedures for removal of pollution were air stripping at pH 11, adsorption on PAC at concentration 10 g/L and zeolite treatment at concentration 75 mg/L. Results of those procedures are presented in Table 2 for chemical analyses and in Figure 3 and Table 3 for toxicological analyses.

With the filtration of tannery wastewater the suspended solids were removed so the treated sample contained less chromium and aluminium. Metals are more toxic for *Daphnia magna* than for *Vibrio fischeri*. The quantity of aluminium and chromium could be decreased by improvement of existent treatment of wastewater with coagulation and floculation. The air-stripping at pH 11 removes ammonia from the tannery waste. The toxicity of ammonia depends on pH because for the aquatic organisms the un-ionized ammonia is more toxic, especially for *Daphnia magna*. Adsorbtion on PAC was the most effective procedure for toxicity reduction of tannery wastewater. 80% of organic pollution was removed. This pollution seems to be the main reason for toxicity of wastewater. It would be obvious that the tannery at Vrhnika upgrades from chemical treatment to biological treatment.

Table 2. Chemical analyses of raw and treated tannery wastewater (4 procedures).

	Raw	Filtration	Air stripping	Adsorption on PAC	Zeolite treatment
pH	7.4	7.2	11.0	-	-
COD, mg/L	2227	1392	1999	504	-
BOD_5, mg/L	1800	850	1175	-	-
Al, mg/L	11.9	<0.2	7.7	7.8	-
Cr_{tot}, mg/L	1.9	0.13	1.1	1.7	-
$N_{Kjeldhal}$	262,2	221.1	148.6	144.8	149.6
NH_4^+-N, mg/L	136,5	136.7	22.2	120.7	64.4
SO_4^{2-}, mg/L	1210	1210	1040	1400	1280
NO_3^-, mg/L	2.1	2.1	4.9	-	1.4
PO_4^{3-}, mg/L	4.0	4.0	2.8	4.1	4.4
Cl^-, mg/L	3430	3430	3017	3780	4060

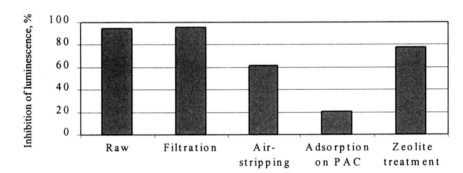

Figure 3. Toxicity of raw and treated tannery waste water (4 procedures) to *Vibrio fischeri*.

Table 3. Acute toxicity test with *Daphnia magna* of raw and treated tannery wastewater (3 procedures).

	Raw	Filtration	Air-stripping	Adsorption on PAC
24h EC 10, vol %	18.3	46.6	17.5	60.0
24h EC 50, vol %	28.3	50.4	34.6	77.4
24h EC 90, vol %	43.6	54.4	68.3	98.9

In our study zeolite treatment was not very suitable procedure for the ammonia removal. Tannery wastewater contains large concentration of salts, so the sites of exchange on the natural zeolite - clinoptilolite changed with other cations (Na^+ and K^+) and so reduced the exchange capacity for ammonia.

The final solution of the problem, as indicated in our study, is upgrading from mechanical treatment to biological stage including nitrification (ammonia removal)/denitrification (nitrogen removal) process which would result in decrease of both effluent toxicity and organic pollution. Such a treatment process is necessary to achieve compliance with water quality based effluent limits, a prerequisite for fulfillment of the goal to minimize the toxic effects that discharges may cause in aquatic environs.

5 References

[1]   Nyholm, N., Bach, H., Birklund, J., Jensen, T. L., Kusk, K. O., Schleicher, O., &   Schrodre, H. Environmental studies of a marine wastewater discharge from a sulphite pulp mill - example of a general study approach for marina industrial discharges. *Water. Sci. Techol.*, **23**, (3), pp.151-161, 1991.

[2]   Nyholm, N. Environmental impact assessment and control of marine industrial wastewater discharges. *Water. Sci. Techol.*, **25**, (11), pp. 449-456, 1992.

[3]   Official Gazette of Republic Slovenia. Ordinance of emission of substances and temperature at wastewater discharging from pollution sources. No. 35 , pp. 2953-2960, 1996.

[4]   Official Gazette of Republic Slovenia. Ordinance of emission of substances at wastewater discharging from objects and devices for leather and fur production. No. 35, pp. 2972-2973, 1996.

[5]   Ford D.L., *Toxicity   Reduction   Evaluation   and   Control*   Techomic Publishing Company, Lancaster, 1992.

[6]   EPA, *Generalized   Methodology   for   Conducting   Industrial   Toxicity Reduction Evaluations (TREs)*, EPA. Ohio, 1989.

[7]   Roš, M. &   Gantar, A. Possibilities of reduction of recipient loading of tannery wastewater in Slovenia. *Water Sci. Technol.*, **37**, (8), pp.145-152, 1998.

[8]   Reemtsma, T., & Jekel, M. Dissolved organic in tannery waste waters and their alteration by a combined anaerobic and aerobic treatment. *Water. Res.*, **31**, pp. 1035-1046, 1997.

[9]   *Standard Methods for the Examination of Water and Wastewater.* 19th edn., American Public Health Association/ American Water Works Association/ Water Environmental Federation, Washington, DC, 1995.

[10] Fiehn, O., Vigeahn, L., Kalnowski, G., Reemtssma, T., & Jekel, M. Toxicity-directed fractionatation of tannery wastewater using solid phase extraction and luminescence inhibition in microtiter plates. *Acta hydrochim. hydrobiol.*, **25**, pp.11-16, 1997.

[11] Milačič, R., Štupar, J., Kožuh N. & Korošin, J. Critical evaluation of three analytical techniques for the determination of chromium (VI) in soil extracts. *Analyst*, **117**, pp. 125-130, 1992.

Biological treatment-nitrification/denitrification seems to be the only way to achieve feasible ammonia removal and beyond this nitrogen removal.

4 Conclusions

Tannery wastewater is a complex mixture of biogenic matter of the hides and a large variety of organic chemicals added during the tanning process. The physico-chemical pretreated tannery wastewater contained very high concentrations of organic and inorganic substances. Beside organic pollution (COD, BOD, oil and grease) very high concentrations of ammonia nitrogen was present in all samples. With the existing treatment system chromium was not satisfactory removed from wastewater, so the concentration of total chromium was always higher than the limits set for the discharging into the river. The most toxic form of chromium, Cr^{6+}, was not present due to the reductive matrix of the tannery wastewater. There were high concentrations of sulfide and sulfate. The results of toxicity test with *Daphnia magna* concurred with the results of chemical analyses. The most toxic were samples that contained the greatest organic pollution load, with the highest concentrations of COD, BOD_5, oil and grease and the highest concentrations of chromium. These substances besides the ammonia and organic nitrogen were the probable reason for the toxicity of tannery wastewater.

TRE procedure included filtration, air stripping, adsorption on PAC and zeolite treatment. The goal of each fractionation step was to reduce toxicity due to specific group of chemicals and compare the results to the toxicity present in the unaltered sample. The inorganic parameters (ammonia and chromium) seem to play a minor role for toxic effects for marine bacteria *Vibrio fischeri*. *Vibrio fischeri* is tolerant against high salt contents and poorly susceptible to metals. Toxic effect of ammonia strongly depends on pH value due to concentrations of un-ionized form. Toxicity of ammonia does not have an important effect on toxicity of tannery wastewater when organic load is still present. Due to adsorption of organic load to activated carbon toxicity for *Daphnia magna* and *Vibrio fischeri* decreased significantly. For more complex determination of toxicity of wastewater, containing many unknown compounds, finer fractionation steps are required for toxicity identification purposes.

The results of our study clearly demonstrate that the physico-chemical pretreatment system in the tannery at Vrhnika for wastewater as a whole (precipitation and flocculation with Al-sulfate and anionic polyelectrolyte) is unsatisfactory. The findings of TRE procedure indicate the predominant value of dissolved organic pollution in exerting toxicity as opposed to ammonia and chromium.

Optimisation of existing physico-chemical treatment by using a simple and cost effective jar-test procedure can immediately result in decrease in Al and Cr concentrations, while the change from whole effluent treatment to treatment of individual (single) tanning process wastewater stream would considerably reduce the quantity of dangerous sludge. Each treatment step must be optimized individually to improve the efficiency of the entire treatment process.

Numerical modelling of tidal propagation in the Severn Estuary using a depth-adapted non-orthogonal grid

R.W. Barber[1] & L.J. Scott[2]

[1]Computational Engineering Group, Daresbury Laboratory, U.K.
[2]Advantage CFD, Reynard Motorsport Ltd., U.K.

Abstract

With greater use being made of computational hydrodynamic and pollution transport models for water quality assessments of coastal and estuarine areas, increasing emphasis needs to be focused on providing an accurate treatment of the complex boundary configurations which are typical of coastal regions. This paper presents details of a depth-averaged two-dimensional numerical model which makes use of a curvilinear or boundary-fitted non-orthogonal coordinate system. The curvilinear approach provides an accurate representation of the complex shape of natural flow domains and avoids the approximate 'staircase' representation associated with models based upon regular Cartesian finite-difference grids. Special attention is given to the simulation of the exposure and re-submergence of the inter-tidal zones which cause the flow perimeter to change shape as the tide ebbs and flows. In addition, grid adaptation to deep water is used to improve the resolution of navigation channels. The versatility of the model is demonstrated by simulating the upper reaches of the Severn Estuary on the west coast of England.

1 Introduction

One of the major difficulties in solving the motion of estuarine and near-coastal waters is the problem of resolving the complex shape of the shoreline contour. The most commonly used approach is to employ a finite-difference grid which approximates the boundary of the flow domain by a series of line segments parallel to the Cartesian coordinate axes (Leendertse & Gritton [1]). A drawback of such an approach is that the solution may poorly represent the flow conditions close to the coastline.

Boundary-fitted grids are designed to overcome the problem of boundary non-alignment, by allowing the use of a finite-difference scheme which accurately follows the curved perimeter. While the use of non-orthogonal boundary-fitted meshes is widespread in computational fluid dynamics [2], the technique has not been extensively applied to computational hydraulics. Johnson & Thompson [3] were the first authors to present the non-orthogonal curvilinear versions of the shallow water equations. Johnson [4] later solved the equations using a semi-implicit finite-difference formulation and tested the scheme on an idealised estuary. Johnson employed a staggered grid for the solution variables, with Cartesian depth-averaged velocity components, U and V, specified at cell corners and water surface elevation, ζ, at cell centres but was only partially successful at achieving a stable solution due to troublesome 'decoupling effects' in the velocity vectors. Häuser et al. [5] have also utilised the same layout of variables to solve the *linearised* non-orthogonal shallow water equations but again experienced decoupling problems. More recently, Borthwick & Barber [6] developed a numerical scheme to solve the non-linear curvilinear shallow water equations using a Leendertse 'C-grid' in order to overcome the instability encountered by earlier researchers. The present paper describes an improved numerical scheme employing a depth-adaptation process which can be used to attract the mesh to areas of deep water in order to improve the resolution of navigation channels.

The simulation of shallow estuaries is further complicated by the existence of inter-tidal zones which are exposed and re-submerged each tidal cycle, altering the shape of the flow domain. It is not uncommon for the nature of an estuary to change from a two-dimensional open sea area at high tide to a complicated arrangement of interconnecting channels at low tide. To allow for the changing shape of the flow domain, a 'flooding and drying' scheme has been developed whereby hydrodynamic cells are added to or removed from the computations by checking the local water surface elevation.

2 Non-orthogonal grid-generation

Following Thompson et al. [7], the boundary-fitted mesh was generated by solving a pair of Poisson equations:

$$\left.\begin{aligned} \xi_{xx} + \xi_{yy} &= P(\xi,\eta) \\ \eta_{xx} + \eta_{yy} &= Q(\xi,\eta) \end{aligned}\right\} \tag{1}$$

relating the physical (x,y) coordinates to the transformed (ξ,η) coordinates. (Here, the subscripts denote the usual shorthand notation for partial differentiation). Interchanging the dependent and independent variables in equation (1) yields a quasi-linear elliptic system:

$$\left.\begin{array}{l} \alpha x_{\xi\xi} - 2\beta x_{\xi\eta} + \gamma x_{\eta\eta} + J^2(Px_\xi + Qx_\eta) = 0 \\[2mm] \alpha y_{\xi\xi} - 2\beta y_{\xi\eta} + \gamma y_{\eta\eta} + J^2(Py_\xi + Qy_\eta) = 0 \end{array}\right\} \qquad (2)$$

where

$$\alpha = x_\eta^2 + y_\eta^2 \quad , \quad \beta = x_\xi x_\eta + y_\xi y_\eta \quad , \quad \gamma = x_\xi^2 + y_\xi^2$$

and J is the Jacobian of the transformation, given by $J = x_\xi y_\eta - x_\eta y_\xi$.

To construct a reasonably smooth grid, a suitable shape of transformed mesh is chosen which can be moulded to the shape of the flow domain without excessive stretching. Dirichlet boundary conditions are imposed on the grid generation equations by specifying the (x,y) coordinates of the coastline in the physical domain. The mapping expressions shown in (2) are rewritten as finite-differences and solved using a successive-over-relaxation procedure to find a bijection between the rectilinear transformed plane and the curvilinear physical plane.

3 Depth-dependent grid adaptation

The aim of grid adaptation is to improve the accuracy of the flow solution at low tide by increasing the number of cells which are computationally active throughout the tidal cycle. This is achieved by making the control functions, P and Q, in the grid generation equations dependent on a weight function, W, which in turn depends on the depth of the estuary bed. An *equidistribution* of weight function and cell size results in control functions:

$$\left.\begin{array}{l} P = \dfrac{1}{WJ^2}(\alpha W_\xi - \beta W_\eta) \\[4mm] Q = \dfrac{1}{WJ^2}(\gamma W_\eta - \beta W_\xi) \end{array}\right\} \qquad (3)$$

which give small cells where the weight function is large and vice versa. The weight function is thus chosen to depend on the depth of the bed, h, below a fixed datum. Initially, a linear variation, $W = 1 + c_1 h$ was tested, where c_1 is a parameter which controls the overall strength of the attraction. However, it was found that a more elaborate relationship was necessary to provide an adequate depth-adapted grid. Numerical experimentation indicates that a suitable weight function for estuarine bathymetries is given by

$$W = 1 + c_1\left(1 + \tanh\left(\frac{h - c_2}{c_3}\right)\right) \qquad (4)$$

where h is the depth of the bed below a fixed datum and c_1, c_2 and c_3 are parameters which control the overall strength of the attraction.

4 Governing hydrodynamic equations

The two-dimensional vertically-averaged shallow water equations can be written in the Cartesian (x,y) coordinate system as:

$$\frac{\partial \zeta}{\partial t} + \frac{\partial (UD)}{\partial x} + \frac{\partial (VD)}{\partial y} = 0 \tag{5}$$

$$\frac{\partial U}{\partial t} + U\frac{\partial U}{\partial x} + V\frac{\partial U}{\partial y} = f_c V - g\frac{\partial \zeta}{\partial x} + \frac{\tau_{wx}}{\rho D} - \frac{\tau_{bx}}{\rho D} + v_t\left[2\frac{\partial^2 U}{\partial x^2} + \frac{\partial^2 V}{\partial x \partial y} + \frac{\partial^2 U}{\partial y^2}\right] \tag{6}$$

$$\frac{\partial V}{\partial t} + U\frac{\partial V}{\partial x} + V\frac{\partial V}{\partial y} = -f_c U - g\frac{\partial \zeta}{\partial y} + \frac{\tau_{wy}}{\rho D} - \frac{\tau_{by}}{\rho D} + v_t\left[\frac{\partial^2 V}{\partial x^2} + \frac{\partial^2 U}{\partial x \partial y} + 2\frac{\partial^2 V}{\partial y^2}\right] \tag{7}$$

where U and V are the depth-averaged velocity components in the x- and y-direction, ζ is the surface elevation above a fixed horizontal datum, g is the acceleration due to gravity, f_c is the Coriolis parameter, $D=h+\zeta$ is the total water depth, where h is the distance between the bed and the datum, ρ is the fluid density, τ_w and τ_b are the wind and bed stresses respectively, and v_t is the coefficient of eddy viscosity. The wind and bed stresses are expressed using quadratic drag laws:

$$\tau_{wx} = \rho_a C_a W_x\sqrt{W_x^2 + W_y^2} \qquad \tau_{wy} = \rho_a C_a W_y\sqrt{W_x^2 + W_y^2} \tag{8}$$

and

$$\tau_{bx} = \rho C_f U\sqrt{U^2 + V^2} \qquad \tau_{by} = \rho C_f V\sqrt{U^2 + V^2} \tag{9}$$

where W_x and W_y are the wind speeds in the x- and y-direction, ρ_a is the density of air, and C_a and C_f are the drag coefficients at the surface and bed respectively. In the present study, the bed friction coefficient, C_f, is evaluated using Manning's equation:

$$C_f = \frac{n^2 g}{D^{1/3}} \tag{10}$$

where n is the Manning roughness factor.

The above equations combine with boundary conditions specified at the perimeter of the physical (x,y) domain to describe the hydrodynamics of the estuary. However, it is preferable to recast the equations so that all numerical computations are performed in the transformed (ξ,η) domain which has a rectilinear boundary. The transformation of the Cartesian shallow water equations (5, 6 & 7) is accomplished using the derivative relationships presented by Thompson et al. [7]:

$$
\left.
\begin{aligned}
\frac{\partial f}{\partial x} &= \frac{1}{J}(y_\eta \frac{\partial f}{\partial \xi} - y_\xi \frac{\partial f}{\partial \eta}) \\[2mm]
\frac{\partial f}{\partial y} &= \frac{1}{J}(x_\xi \frac{\partial f}{\partial \eta} - x_\eta \frac{\partial f}{\partial \xi})
\end{aligned}
\right\}
\tag{11}
$$

where f denotes an arbitrary function of x and y. Substituting for all derivatives of x and y in the governing equations (5, 6 & 7) yields the transformed shallow water expressions:

$$
\frac{\partial \zeta}{\partial t} + \frac{1}{J}\left[y_\eta \frac{\partial(UD)}{\partial \xi} - y_\xi \frac{\partial(UD)}{\partial \eta} + x_\xi \frac{\partial(VD)}{\partial \eta} - x_\eta \frac{\partial(VD)}{\partial \xi} \right] = 0
\tag{12}
$$

$$
\frac{\partial U}{\partial t} + \frac{1}{J}\left[y_\eta U \frac{\partial U}{\partial \xi} - y_\xi U \frac{\partial U}{\partial \eta} + x_\xi V \frac{\partial U}{\partial \eta} - x_\eta V \frac{\partial U}{\partial \xi} \right] = f_c V - \frac{g}{J}\left[y_\eta \frac{\partial \zeta}{\partial \xi} - y_\xi \frac{\partial \zeta}{\partial \eta} \right]
$$
$$
+ \frac{\tau_{wx}}{\rho D} - \frac{\tau_{bx}}{\rho D} + \text{DIFFUSIVE TERMS}
\tag{13}
$$

$$
\frac{\partial V}{\partial t} + \frac{1}{J}\left[y_\eta U \frac{\partial V}{\partial \xi} - y_\xi U \frac{\partial V}{\partial \eta} + x_\xi V \frac{\partial V}{\partial \eta} - x_\eta V \frac{\partial V}{\partial \xi} \right] = -f_c U - \frac{g}{J}\left[x_\xi \frac{\partial \zeta}{\partial \eta} - x_\eta \frac{\partial \zeta}{\partial \xi} \right]
$$
$$
+ \frac{\tau_{wy}}{\rho D} - \frac{\tau_{by}}{\rho D} + \text{DIFFUSIVE TERMS} .
\tag{14}
$$

The transformed diffusive terms in the momentum equations are too long to be presented here but are fully expanded by Borthwick & Barber [6] and Scott [8]. The problem is thus changed from solving a set of p.d.e.'s in the physical plane with a complex boundary configuration to solving more complex p.d.e.'s in the computational plane with a rectilinear boundary.

5 Numerical solution procedure

The transformed governing equations (12, 13 & 14) were discretised on a space-staggered (ξ, η) grid and solved using an explicit finite-difference procedure. In contrast to earlier non-orthogonal schemes where the U- and V-velocity components were stored at cell corners [4,5], the velocities in the present scheme were located at the middle of cell faces. This reduces the amount of spatial averaging and prevents the 'decoupling effects' reported in earlier studies. To allow for the changing shape of the estuary as the tide ebbs and flows, a 'flooding and drying' scheme was employed, whereby grid cells were added to, or removed from, the computational domain depending upon the local water depth.

The flooding and drying algorithm was based upon that developed by Falconer & Chen [9], who introduced parameters so that a cell *may* dry once its depth has fallen below a shallow height, SH, and *must* dry once its depth has fallen below a roughness height, RH. Each timestep all cells currently dry were checked to see if they should become wet and vice versa. Specifically a cell was allowed to dry if <u>any</u> of the following conditions were satisfied:

(a) The cell centre depth was less than the roughness height, RH.

(b) The depths at all four faces of the cell were less than RH.

(c) The cell centre depth was less than the shallow height, SH, and the overall flow was out of the cell towards surrounding wet cells.

A dry cell was allowed to reflood if <u>both</u> of the following conditions were satisfied:

(d) At least one of the neighbouring cells was wet, and

(e) The overall flow was into the cell.

The flooding and drying rules prevent undue surface elevation noise in the vicinity of the coastline and help stabilise the numerical model.

6 Model application

The aim of the research was to provide a storm surge prediction model for the fluvio-tidal reaches of the Severn Estuary on the west coast of England. Present work has concentrated on the wider lower portion of the estuary where the flow patterns are essentially two-dimensional and tidal effects dominate. The estuary presents a very stringent test of model capabilities because of its extreme tidal conditions. With a mean spring tide of 12.2m and spring tides regularly in excess of 14m, the Severn Estuary has the second largest tidal range in the world. The estuary is also characterised by very steep gradients in bathymetry.

Figure 1 shows the extent of the modelled area, the upstream and downstream limits and the location of intermediate tidal stations used in model validation, whilst Figure 2 illustrates the transformed and physical hydrodynamic meshes employed in the present study. To drive the model, tidal curves of surface elevation against time were specified at the seaward and landward open boundaries. The model was run for four tidal cycles from a stationary initial condition with the resulting elevations and velocities being recorded on the fourth cycle. The Manning roughness coefficient, n, was calibrated by comparing computed tidal profiles at intermediate sites along the estuary against harmonic tidal predictions; the closest agreement was obtained for a Manning's n of 0.02. In the present study, the flooding and drying parameters were specified as RH=0.08m and SH=0.16m: these values minimised the artificial disturbances in surface elevation created by the flooding and drying procedure.

Figure 3 illustrates predicted surface elevation profiles for four intermediate sites along the estuary. The predicted tidal elevations at each site were compared with harmonic tidal predictions and all sites showed good correlation, as illustrated in Figure 4 which details the comparison of surface elevations at Beachley, approximately 12km from the seaward boundary. Although there is a small discrepancy at low water, it can be seen that the model successfully replicates the lengthening ebb and shortening flood tides which are generated as the tidal wave propagates up the estuary. The accuracy of the velocity predictions is more difficult to assess in the absence of accurate flow measurements for the interior of the model area. Some very approximate tidal stream data are given on Admiralty Chart No. 1166 for a point near Avonmouth. Figure 5 compares this observed velocity data against computed results and shows that the numerical model successfully predicts the phase of the velocity variations but indicates the correlation is less satisfactory for the magnitude. The inaccuracy in predicting the magnitude of the velocity can be partly attributed to the difficulty in obtaining accurate and up-to-date bathymetry data for the estuary due to the mobile nature of the deep water channels and inter-tidal mudflats. In addition, the extreme gradients in bathymetry at the edges of the deep water channels may invalidate the 'nearly-horizontal' flow assumption of the depth-averaged shallow water equations. This latter difficulty can only be overcome by employing a three-dimensional hydrodynamic model.

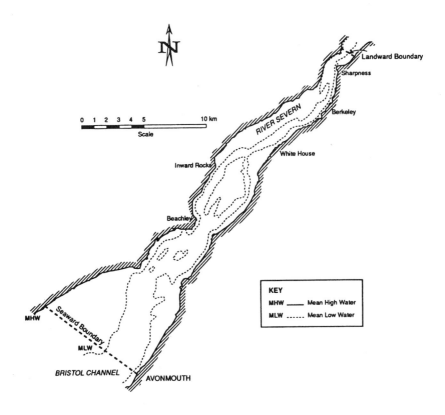

Figure 1: Map of the Severn Estuary

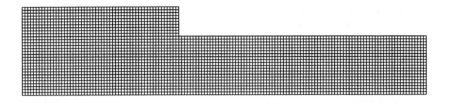

(a) Transformed/computational (ξ,η) mesh

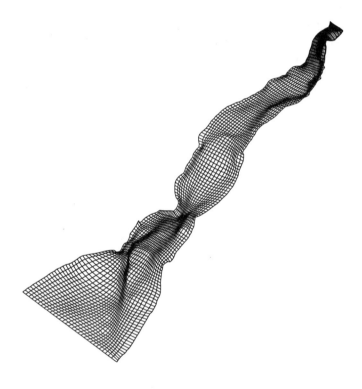

(b) Depth-adapted physical (x,y) mesh

Figure 2: Boundary-fitted grid representative of the Severn Estuary

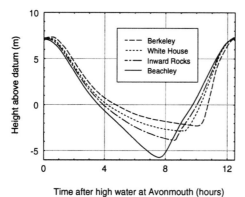

Figure 3: Surface elevation predictions for a mean spring tide

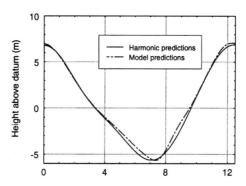

Figure 4: Comparison of surface elevation predictions at Beachley

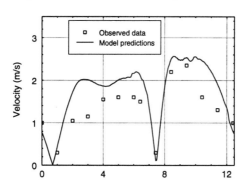

Figure 5: Comparison of velocity predictions at Avonmouth

7 Conclusions

A technique for solving the vertically-averaged shallow water equations in irregularly shaped estuaries has been presented. The procedure employs a non-orthogonal boundary-fitted coordinate system and uses an explicit finite-difference scheme to discretise the governing curvilinear hydrodynamic equations. Grid adaptation to the deep water channels is achieved using a control function, equidistributional approach with the aim of improving the resolution of the navigation channels, whilst a 'flooding and drying' scheme has been employed to allow for the changing shape of the flow domain as the tide ebbs and flows. The results demonstrate that the depth-adapted boundary-fitted methodology has considerable promise for simulating the hydrodynamics of complex estuaries.

References

1. Leendertse, J.J. & Gritton, E.C. *A water-quality simulation model for well-mixed estuaries and coastal seas: Vol 2, Computation Procedures*, The Rand Corporation, R-708-NYC, 1971.
2. Thompson, J.F., Warsi, Z.U.A. & Mastin, C.W. Boundary-fitted coordinate systems for numerical solution of partial differential equations- a review, *J. Comp. Phys.*, **47**, 1-108, 1982.
3. Johnson, B.H. & Thompson, J.F. *A discussion of boundary-fitted coordinate systems and their applicability to the numerical modelling of hydraulic problems*, U.S. Army Engineer Waterways Experiment Station Hydraulics Laboratory, Vicksburg, USA, 1978.
4. Johnson, B.H. *VAHM - A vertically averaged hydrodynamic model using boundary-fitted coordinates*, U.S. Army Engineer Waterways Experiment Station Hydraulics Laboratory, Vicksburg, USA, 1980.
5. Häuser, J., Paap, H.G., Eppel, D. & Sengupta, S. Boundary conformed coordinate systems for selected two-dimensional fluid flow problems. Part 2: Application of the BFG method, *Int. J. Num. Methods in Fluids*, **6**, 529-539, 1986.
6. Borthwick, A.G.L. & Barber, R.W. River and reservoir flow modelling using the transformed shallow water equations, *Int. J. Num. Methods in Fluids*, **14**, 1193-1217, 1992.
7. Thompson, J.F., Thames, F.C. & Mastin, C.W. Automatic numerical generation of body-fitted curvilinear coordinate system for field containing any number of arbitrary two-dimensional bodies, *J. Comp. Phys.*, **15**, 299-319, 1974.
8. Scott, L.J. Numerical modelling of tidal propagation in the Severn Estuary using a boundary-fitted coordinate system, *Ph.D. Thesis, University of Salford*, U.K. , 1996.
9. Falconer, R.A. & Chen, Y. An improved representation of flooding and drying and wind stress effects in a two-dimensional tidal numerical model, *Proc. Instn. Civ. Engrs.*, **91**, 659-678, 1991.

Prediction of storm impacts and shoreline retreat induced by hypothetical storms on open coastlines

Ó. Ferreira & J. Alveirinho Dias
UCTRA/CIACOMAR, Universidade do Algarve, Portugal

Abstract

Storms are the main mechanism responsible for coastal erosion on open coastlines. In the last few years an effort has been made to apply and validate simple models determining coastal retreat and dune erosion due to storms on highly energetic coasts. In this study, the convolution method was applied to determine the impact of two hypothetical storms called the "mean storm" (return period of 0.7 years) and the "century storm" (return period of 100 years) on the coast between Aveiro and Cape Mondego (Northwest Portugal). The application of the model to different sites resulted in a prediction of beach erosion volumes between 70 and 110m^3/m for the "mean storm", without showing dune retreat at any of the analysed sites. According to calculations, the "century storm" would be expected to cause erosion volumes between 390 and 680m^3/m with a mean computed dune retreat of about 26m. Predicted erosion volumes and dune retreat for this storm are able to completely destroy the dune ridge at most of the tested beaches, resulting in severe overwashes and damage.

The use of the convolution model as a predictor of potential storm erosion on open coastlines seems to be of great interest, allowing the identification of potential hazards associated with future storms. It is also potentially an important tool for beach managers, defining set back lines and beach nourishment design.

1 Introduction

Under present conditions of rising sea level and reduced sediment supply, extreme storms are the most significant factor affecting short-term shoreline movement [1] and the episodic erosion history of a coastline [2]. Storm dominance on controlling beach erosion is especially important on open coastlines where up-drift erosion is the primary source of sand for the natural beach replenishment [1]. Storms are also responsible for the largest and most rapid changes in beach profiles [3] [4], their effects being devastating, inducing loss of lives and extensive damages. The study of beach profile response to storms is of great interest both for the understanding of the fundamental processes taking place on beaches during storms, and for the prediction of shoreline evolution. This knowledge constitutes an important tool for effective coastal zone management.

The most effective way for the correct determination of storm effects on open sandy shores is constant and intense field monitoring. Such monitoring should include, on a daily basis, both morphological surveys (topography and bathymetry), and hydrodynamic data (waves, currents and storm surge). The acquisition of these data sets for a significant number of storms is too expensive and almost impossible for most of the scientific groups presently working on this subject. To minimise this problem and to improve the prediction of storm action, several numerical models have been developed during the past decades. These models are based on different concepts and solutions; thus it is difficult to select which one should be used on a given coastal stretch. However, almost all authors mention which fundamental parameters and steps a model should include. Those can be generally synthesised on the ones proposed by Birkemeier *et al.* [5]: i) offshore wave height and wave period; ii) water level, including wave set-up; iii) representative grain size or fall velocity for the profile; iv) representative initial shape of the profile, including the dune; v) onshore boundary conditions for dune erosion; vi) calculation procedure for the cross-shore sediment transport rate or, alternatively, direct calculation of the morphologic change of the profile based on a given idealised form; vii) calculation procedure for computing dune erosion, containing a temporal dependence; viii) and verification with field data over the range of conditions for which the model will de applied. Additionally, several authors agree that the storm surge duration is also fundamental for a correct volumetric prediction of coastal erosion induced by storms [6] [7] [8] [9] [10]. These authors concluded that the magnitude of the erosion should be directly dependent on the actuation time of the forcing mechanisms. After a detailed analyses of different models and their application, it can be concluded that it is virtually impossible to find a universally valid theory for cross-shore sediment transport [11] and also not possible to define conclusively which model(s) is(are) superior to the others [10]. According to Schooness and Theron [10] each model may be the best for a specific case (purpose) or under specific conditions. Therefore, the basic principles for the selection of one model to be applied to a case study is to apply a model that integrates the fundamental parameters giving good results for the studied area.

In this study the main goal will be the prediction of potential morphological impacts on an open sandy shore (Aveiro – Cape Mondego), located on the West Coast of Portugal (Figure 1). The definition of these impacts will be achieved through the use of a simple model (the convolution model) after validation for the study area.

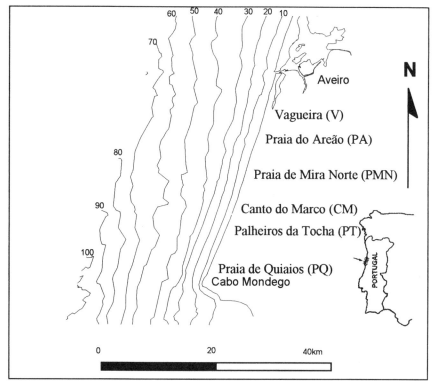

Figure 1: Location of the studied area.

2 The convolution model

2.1 Model description

The convolution model was developed by Kriebel and Dean [7], and consists of simple analytical solutions to predict the time-dependent beach profile response to severe coastal storms. The model tries to answer the need for simple methods of analysing beach erosion or accretion due to variable wave and water-level conditions [7]. Based on field observations, it assumes that beaches subjected to steady-state erosion forcing conditions respond toward a stable or equilibrium form in approximately an exponential manner. The forcing mechanisms considered are storm-induced variations in water level, with the magnitude of the erosion response being determined by two parameters: i) the maximum potential response (R_∞), that would occur if the beach was allowed to respond completely to a new equilibrium; ii) the characteristic erosion time scale (T_s), that governs the exponential rate at which the profile responds toward the new equilibrium.

The initial beach profiles are split into two parts, the submerged part below mean sea level and the upper part of the profile above mean sea level. For the upper part of the profile a schematic profile can be used, based on actual profiles.

To do this simplified profile parameters like berm elevation above mean sea level, berm width, dune height and beach face slope are needed. For the submerged portion the model uses the equilibrium profile defined by Dean [12]. To compute the maximum potential beach retreat (R_∞) and the associated maximum potential volume erosion (V_∞), a series of closed-form solutions was developed. These computations are dependent on storm characteristics and the initial beach profile form: i) square-berm profile with infinite foreshore slope; ii) equilibrium profile with linear foreshore slope; iii) equilibrium profile with dune with no backshore, and iv) equilibrium profile with dune and wide backshore. Due to the relatively slow morphological response typical of natural beaches only a fraction of the maximum potential erosion and retreat is realised before recovery is initiated [7]. Kriebel and Dean developed a solution for the time-dependent erosion response in order to obtain R_{max} (maximum retreat) and V_{max} (maximum erosion volume above mean sea level) as a function of R_∞ and the storm duration. The solution for R_{max} and V_{max} is obtained by shifting the equilibrium profile vertically through a distance S (maximum sea level, including both tide and storm surge) above mean sea level, and then landward through a distance R_{max} such that a mass balance is achieved between the eroded and deposited sand volumes. Thus, the solution for the equilibrium beach recession is obtained by equating the volume eroded from the beach face with the volume deposited offshore.

A detailed model description including the different formulations can be found in Kriebel and Dean [7], while examples of the model application are shown in the Coastal Engineering Manual [13].

2.2 Model validation

The convolution model has already been used and applied to the Portuguese West Coast by Ferreira *et al.* [14], Ferreira and Dias [15], and Gama [16]. However, model validation was not made prior to those studies. For the present study and in order to achieve a good understanding of model limitations and errors, a validation was performed. This validation was based on the comparison of beach profiles obtained with the model and those observed after two storms with different hydrodynamic characteristics (storms of 24-27 February 1989 and 16 December 1992). The storm characteristics needed as input data for the model are shown in Table 1.

Table 1. Storm data used on the model validation for the storms of February 89 and December 92. H_{so}=deep water significant wave height; S=elevation above mean sea level; T_D=storm duration.

	H_{so} (m)	Return period (years)	S (m)	T_D (h)
February 89	8.83	3	1.53	62
December 92	6.6-6.9	<1	1.37	11.7

For the February 1989 storm morphological data from four beaches (Vagueira, Barra de Mira, Canto do Marco and Praia de Quiaios) were used.

These beaches are located along a 40 km stretch of the Portuguese West Coast, between Aveiro and Cape Mondego (Figure 1). The storm induced an average beach erosion volume above mean sea level of 180m^3/m, with minimum and maximum erosion values of 139 and 210m^3/m, respectively. The results obtained with the model showed an average volume underestimation of about 10.8%, with a maximum deviation of 23.6% (49m^3/m). Figure 2 shows the best and worst fits between measured and predicted beach profiles for the February 1989 storm.

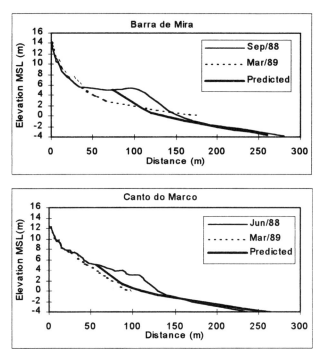

Figure 2. Comparison of predicted and observed post-storm profiles for the February 89 storm. Barra de Mira: worst fit; Canto do Marco: best fit.

For the December 1992 storm five profiles from the same beach (Palheiros da Tocha) were used, distributed along 100m in length. The measured beach erosion volumes above mean sea level induced by this small storm ranged from 14.4 to 27.8m^3/m, with an average of 20.1m^3/m. The observed average deviation between model predictions and field results was an overestimation of 23.4%, with a maximum deviation of 58.6% (8.5m^3/m). Figure 3 shows the best and the worst beach profile fits for the December 1992 storm. Since the erosion volume was small it can be seen that profiles generated by the model are generally well fitted with the actual ones, even though the volume deviation was relatively greater compared to the February 1989 storm.

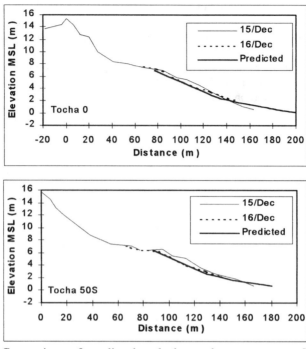

Figure 3. Comparison of predicted and observed post-storm profiles for the December 92 storm. Tocha 0: worst fit; Tocha 50S: best fit.

Erosion volumes computed with the convolution method and measured in the field showed a reasonable agreement. This agreement is better for average computations along a coastal stretch and slightly worse for individual applications. Differences between predicted and observed values are within the range of observed along-shore variations in actual erosion induced by the studied storms. Therefore, the time-dependent and morphological assumptions established by the model seem to be adequate for the study area. Beach profiles represented by the model are also seen to be generally well fitted with actual post-storm profiles, with a few exceptions, namely for the storm with higher energy and inducing stronger erosion (storm of February 1989). After these tests of the convolution model it can be concluded that the method can be used on the exposed sandy beaches of the Portuguese West Coast to predict coastal erosion induced by storms, giving acceptable accuracy.

3 Model application to hypothetical storms

The prediction of beach behaviour exposed to a defined storm with imposed conditions (hypothetical storm), through the use of a previously validated model, can be useful in defining beach and foredune robustness along a coastal stretch or at single sites. For this main goal, two hypothetical storms were defined: the "mean storm" and the "century storm". These storms are intended to represent the hydrodynamic conditions associated with an average event and an extreme event with one hundred years return period. Schematic beach profiles were defined for 4 selected sites belonging to the coastal stretch from Aveiro to Cape Mondego, representing both erosional (Vagueira Sul and Praia do Areão) and stable areas (Praia de Mira Norte and Palheiros da Tocha). The schematic beach profile represents the average behaviour of each site, being based on the analysis of topographic data obtained from 5 surveys carried out between September 1992 and June 1993 [17].

3.1 "Mean storm"

3.1.1 Storm characteristics
Costa [18] presented a complete list of storms recorded at Figueira da Foz wave buoy (placed offshore Cape Mondego) between 1981 and 1990. This list was used to define the characteristics of an average storm at the study area. The average offshore significant wave height (H_{so}) of all events was 6.2m, with an associated significant period of 14.0s. According to Ferreira [19], storms with H_{so}=6.2m have a return period of less than one year (approximately 0.7 years). Using the generic formulation of Komar and Gaughan [20], the significant breaking wave height was estimated at 7.6m. Data published by Costa [18] also allowed the computation of an average storm duration, which is around 36 hours. To determine the elevated sea level at which the storm would act it was assumed that the hypothetical storm would occur during average spring tides (high tide of about 1.4m above mean sea level, MSL) with a superimposed significant storm surge (0.31m at Aveiro according to Gama *et al.* [21]). The sum of these values gives a sea level value of about 1.71m above MSL, which was used for the simulation.

3.1.2 Model application
Using the average beach profile and the defined mean storm hydrodynamic conditions, the convolution model was applied to the selected sites, thus determining R_{max} (maximum retreat), V_{max} (maximum erosion volume above mean sea level) (Table 2), and the post-storm profiles (Figure 4).

Table 2. R_{max} and V_{max} values, computed by the convolution model for a hypothetic "mean storm", at the selected sites.

Site	R_{max} (m)	V_{max} (m^3/m)
PMN	18.4	85.7
PT	17.3	90.6
VS	13.2	71.2
PA	17.4	77.3
Average	16.6	81.2

According to the computed values, no dune retreat was estimated for such a storm at all profiles, meaning the erosion was confined to the berm. However, there is a significant difference between beaches under erosion (Vagueira Sul and Praia do Areão) and the other two sites. At Vagueira Sul and Praia do Areão the berm is expected to be reduced to only 5m width, while at Praia de Mira Norte and Palheiros da Tocha the predicted post-storm profile shows a berm with no less than 35m width. In a general analysis, it can be said that the study area has berms with an adequate volume to withstand storms with intermediate characteristics. However, Vagueira Sul and Praia do Areão would not withstand a consecutive storm without significant dune retreat. The occurrence of storm sequences in association with successive meteorological fronts is relatively frequent on the Portuguese West Coast.

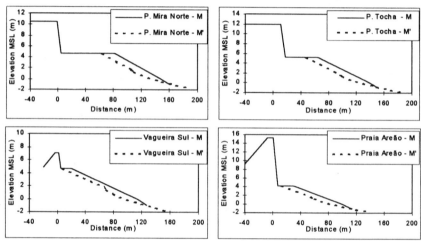

Figure 4. Predicted post-storm profiles for the "mean storm". M – initial profile, M' – post-storm profile.

3.2 "Century storm"

3.2.1 Storm characteristics
Since there are no storm records for periods as long as one hundred years, assumptions had to be made to obtain the century storm parameters. Based on the papers of Pires and Pessanha [22] [23] and Carvalho [24] a value of 13.9m was defined as the offshore significant wave height associated with a return period of 100 years. The analysis of the studies developed by Capitão [25] and Costa [18] show that the maximum recorded significant period for the study area is about 18s. The estimated significant breaking wave height associated with these values was 16.0m. The time scale for the storm duration was defined as 14 days (336h), corresponding to the maximum value so far recorded in Portugal. The elevated sea level was determined by the sum of the sea level reached at high tide of average spring tides (1.4m above MSL) with the maximum storm surge recorded for Aveiro (approximately 0.8m [21]), giving a total value of 2.2 m above mean sea level.

3.2.2 Model application
The results obtained with the use of the convolution model on the selected sites for the conditions related with the "century storm" are expressed in Table 3. The R_{max} value is only relative to dune erosion, while V_{max} represents the sum of predicted erosion volumes for berm and dune. Dune erosion is expected to occur at all sites as shown in Figure 5.

Table 3. R_{max} and V_{max} values, computed by the convolution model for a hypothetic "century storm", at the selected sites.

Site	R_{max} (m)	V_{max} (m³/m)
PMN	4.9	407.4
PT	25.6	548.1
VS	45.0	415.3
PA	37.1	663.2
Average	28.2	508.5

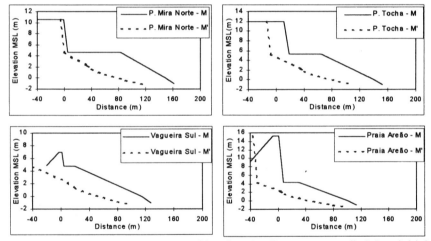

Figure 5. Predicted post-storm profiles for the "century storm". M – initial profile, M' – post-storm profile.

According to the predictions the dune ridge at Vagueira Sul and Praia do Areão would be completely destroyed leading to the existence of overwashes and extensive coastal floods. For Praia de Mira Norte and Palheiros da Tocha the average berm volume, the dune ridge width and the dune height would be sufficient to avoid these consequences.

4 Conclusions

This study used a validated beach and dune erosion model to predict the expected consequences of two hypothetical storms. The adopted method (the convolution model) is highly dependent on the sea level used and on the response time of the beach profile to each storm. The erosion predicted for the "mean storm" was not enough to induce dune erosion at any of the studied sites. However, for beaches that have undergone historical retreat and with smaller sand stock an extreme berm width reduction was foreseen. This can lead to a situation of fragility if there is no time for berm recovery between storms. To avoid dune retreat predicted for the "century storm", it would be necessary to have an average available volume of about 500m³/m above MSL at all sites. Such a volume is only observed at a few beaches in the studied area, and then only during summer. Thus, an event with the characteristics of the "century storm" would cause extensive dune ridge erosion and its total destruction in places where there has been historical erosion. In these areas coastal floods would be expected.

Predictions like the ones obtained in this work can easily be applied to this and other coastal areas, after a model validation, to compute set-back lines, to define limits for coastal occupation and to predict volumes needed for beach renourishment in order to avoid dune retreat. However, it is important to stress that the above predictions were developed for single storms. The effects of consecutive storms were not analysed and should be further developed in future studies. Moreover, the model used (or in fact any other model) should be constantly evaluated and tested against field data to correctly define the conditions in which it can be considered a useful tool for coastal management.

Acknowledgements

This work was partially funded by projects INDIA (CEC, Directorate-General for Science, Research and Development, MAST contract MAS3-CT97-0106) and RIMAR (PRAXIS XXI n.2/2.1/MAR/1743/95). The authors wish to say "muito obrigado" to Brad Morris that transformed the text into proper English.

References

[1] Morton, R.A., Gibeaut, J.C. & Paine, J.G. Meso-scale transfer of sand during and after storms: implications for prediction of shoreline movement. *Marine Geology*, 126, 161-179, 1995.
[2] Russel, P. Mechanisms for beach erosion during storms. *Continental Shelf Research*, 13 (11), 1243-1265, 1993.
[3] Birkemeier, W.A. The effects of the 19 December 1977 coastal storm on beaches in North Carolina and New Jersey. *Shore and Beach*, 47, 7-15, 1979.
[4] Birkemeier, W.A. Time scales of nearshore profile changes. Coastal Engineering '84, ASCE, 1507-1521, 1984.
[5] Birkemeier, W.A., Kraus, N.C., Scheffner, N.W. & Knowles, S.C. Feasibility study of quantitative erosion models for use by the Federal Emergency Management Agency in prediction of coastal flooding. *TR87-8*, CERC, 1987.
[6] Balsillie, J. Beach and coast erosion due to extreme event impact. *Shore and Beach*, 54 (4), 22-37, 1986.

[7] Kriebel, D.L. & Dean, R.G. Convolution method for time-dependent beach-profile response. *Journal of Waterway, Port, Coastal, and Ocean Engineering*, 119 (2), 204-226, 1993.

[8] Wise, R.A. & Kraus, N.C. Simulation of beach fill response to multiple storms, Ocean City, Maryland. *Beach Nourishment Eng. and Management Considerations*, eds. D.K. Stauble & N.C. Kraus, ASCE, 133-147, 1993.

[9] Van de Graff, J. Coastal and dune erosion under extreme conditions. *Journal of Coastal Research, Special Issue n° 12*, ed. C.W. Finkl, 253-262, 1994.

[10] Schoones, J.S. & Theron, A.K. Evaluation of 10 cross-shore sediment transport/morphological models. *Coastal Engineering*, 25, 1-41, 1995.

[11] Lee, C.E., Kim, M.H. & Edge, B. Numerical model for on-offshore sediment transport with moving boundaries. *Journal of Waterway, Port, Coastal, and Ocean Engineering*, 122 (2), 84-92, 1996.

[12] Dean, R.G. Equilibrium beach profiles: U. S. Atlantic and Gulf coasts. *Ocean Engineering Report n° 12*, University of Delaware, 1977.

[13] C.E.R.C. *Coastal Engineering Manual, Part III, Chapter 3, Cross-shore sediment transport processes.* U.S. Army Corps of Engineers, 1998.

[14] Ferreira, Ó., Dias, J.A., Gama, C. & Taborda, R., 1995. Quantification of beach erosion caused by storms on the Portuguese coast. *Directions in European Coastal Management*, eds. M.G. Healy & J.P. Doody, 267-273.

[15] Ferreira, Ó. & Dias, J.A. Avaliação da acção de temporais na costa Oeste portuguesa (sector Aveiro – Cabo Mondego). *Colectânea de Ideias sobre a Zona Costeira de Portugal*, EUROCOAST Portugal, 429-447, 1997.

[16] Gama, C. *Caracterização do fenómeno da sobreelevação do nível do mar de origem meteorológica em Portugal Continental.* MSc thesis. Universidade de Lisboa, 1997.

[17] Ferreira, Ó. *Morfodinâmica de praias expostas: aplicação ao sector costeiro Aveiro–Cabo Mondego.* PhD thesis, Universidade do Algarve, 1998.

[18] Costa, M. Inventário de dados de agitação marítima. Medições efectuadas pelo Instituto Hidrográfico, volume 2. *PO-WAVES 5/93-A*, IH/LNEC, 1993.

[19] Ferreira, Ó. *Caracterização dos principais factores condicionantes do balanço sedimentar e da evolução da linha de costa entre Aveiro e o Cabo Mondego.* MSc thesis, Universidade de Lisboa, 1993.

[20] Komar, P. & Gaughan, M. K. Airy wave theory and breaker height prediction. *Coastal Engineering '72*, 405-418, 1972.

[21] Gama, C., Dias, J.A., Ferreira Ó. & Taborda, R. Analysis of storm surge in Portugal, between June 1986 and May 1988. *Littoral'94*, 381-387, 1994.

[22] Pires, O. & Pessanha, L.E. Wave power climate of Portugal. *Hydrodynamics of Ocean Wave-energy Utilization*, eds. D. Evans & A. Falcão, 157-167, 1986.

[23] Pires, O. & Pessanha, L.E. Estima da distribuição de probabilidades dos valores extremos, utilizando séries climatológicas curtas. *Revista do Instituto Nacional de Meteorologia e Geofísica,*, 1986.

[24] Carvalho, M.M. Valores extremos da agitação marítima, Figueira da Foz – tratamento local dos dados existentes. *PO-WAVES 13/92-A*, IH/LNEC, 1992.

[25] Capitão, R.P. Síntese de dados de temporais ocorridos em Portugal Continental. *PO-WAVES 14/92-A*, IH/LNEC, 1992.

Numerical simulations of an oil spill accident in Guanabara Bay, Rio de Janeiro, Brazil

R.F. Cantão[1], R.F. De Oliveira[2] & J.F.C.A. Meyer[1]
[1]*Department of Applied Mathematics, State University at Campinas, Brazil*
[2]*Department of Mathematics, Federal Rural University at Rio de Janeiro, Brazil*

Abstract

In this work we present a classical evolution equation for the movement of oil slicks in marine water in its second phase (Fay, 1969), and, setting it in its variational formulation, discretize it with a view towards the use of the Finite Element Method (using first degree approximations for oil concentration). For simulating marine currents special upwinding techniques are adopted so as to eliminate main oscillations caused by numerical options. These currents are given by the solution of Stokes' equation using second order finite elements, and boundary conditions are obtained from data furnished by the Brazilian National Hydrographic Directory. Resulting currents were then used for a qualitative simulation so as to verify the model comparing it to information obtained after the accident and during clean-up operations. The model is also used to show the presence of oil in an environmentally protected area (called Guapimirim), an area of mangroves, essential for marine life in the bay. Program results are compared to circulated news on the oil spill movements and effects.

1 Introduction

In the early hours of January the 18th, 2000, something like 1292 tons of oil leaked for four hours from the Brazilian National Petroleum Company (Petrobrás) facilities inside Guanabara Bay. This accident was due to a leak in a broken underwater pipeline, located quite far from the marine entrance to the Bay, where circulation caused by currents is very much smaller than wind-dominated movements. The oil slick initially moved because of currents induced by a wind from southwest, touching the shoreline along the northern tip of the bay, and subsequently moved down driven by a current induced by wind from the northwest, reaching the isle of Paquetá, as well as some beaches on the western shore of the Governador Island. This oil slick has severely affected the fishing communities in this region (mainly from the town of Magé) as well as the preservation area of wetlands inside Guanabara Bay. Media coverage of clean-up activities brought the oil spill into main news, and drew the attention of many sectors in society to the need of severe law-enforcement measures in this area. Press releases did not coincide completely with scientific facts.

In order to test both mathematical modelling techniques as well as numerical schemes for the simulation of oil spill movements, the approximate data relative to this accident were used in order to be able to compare simulation results with those obtained by media information, satellite images and local photographs.

2 Mathematical model

Several authors have worked with oil spill movements using mathematical models based upon the classical Partial Differential Equation of the Conservation of Mass ([1], [3], [4], [7], [8], and [9]). These models take into account the second phase of an oil spill, as defined by Fay [5] the so-called diffusive-advective phase, which describes the behaviour of an oil spill a few hours (4-6) after it occurs until about a week (something between 10 and 15 days) [6]. Diffusion was considered as being constant, and a single term was used as a linear approximation to several processes of degradation. Coherent boundary conditions, for this level of approximation, were chosen much in the line of Meyer, Cantão and Poffo (1998). The studied domain, identified as Ω, is a discretization of Guanabara Bay, the triangles of which are obtained with the use of convenient software. Its boundary is given by $\partial\Omega$, and outer normal unit vector of this border is given by η. The chosen model is therefore given in its classical formulation by:

Obtain $u = u(x,y;t)$ for $(x,y)\in\Omega\subset\mathbf{R}^2$, $t\in(0,T]\subset\mathbf{R}$, such that

$$\frac{\partial u}{\partial t} + \operatorname{div}(-\alpha\nabla u) + \operatorname{div}(\mathbf{V}u) + \sigma u = f, \quad \text{for}$$

$$u(x,y;0) = u_o(x,y), \quad (x,y)\in\Omega\subset\mathbf{R}^2, \quad \text{with} \tag{1}$$

$$u,f:\Omega\times[0,T]\to\mathbf{R}, \quad \mathbf{V}:\Omega\to\mathbf{R}^2 \quad \text{and} \quad \sigma:[0,T]\to\mathbf{R}.$$

Boundary conditions used for this case study use the fact that very little oil that washed up on beach shores was retained upon those shores (this part of the boundary will be identified as Γ_1), and that no oil reached the southern part of the bay (this part will be named Γ_0), so that the adopted boundary conditions will be, for $\partial\Omega = \Gamma_0 \cap \Gamma_1$, of mixed type, Dirichlet and von Neumann homogeneous:

$$\left.\partial u \middle/ \partial n\right|_{\Gamma_1} = 0 \text{ and } u|_{\Gamma_0} = 0. \tag{2}$$

For constant diffusivity, an approach that is both practical and coherent for Fay's second phase of an oil spill, this classical formulation becomes:

$$\frac{\partial u}{\partial t} - \alpha\Delta u + \mathrm{div}(\mathbf{V}u) + \sigma u = f, \tag{1'}$$

and we can now make the option for a variational formulation, seeking to use whatever benefits we may from weaker conditions imposed upon parameters, solutions and information as well as along the boundaries. We will now consider the space for test functions as

$$H = \{\, v \in H^1(\Omega) \;:\; v|_{\Gamma_0} = 0 \,\} \subset H^1(\Omega), \tag{3}$$

besides the space in which we, theoretically at least, seek to obtain the solution $u(x,y;t)$ as:

$$W = \{\, v \in L^2((0,T], H^1(\Omega)) \;:\; \forall t \in (0,T], \frac{\partial v}{\partial t} \in L^2(\Omega) \text{ and } v|_{\Gamma_0} = 0 \,\}. \tag{4}$$

Formulation (1) becomes the following equation (in which we have used Green's Theorem), where classical notation is adopted:

$$(\frac{\partial u}{\partial t} \mid v) + \alpha(\nabla u \parallel \nabla v) + (\mathbf{V} \bullet \nabla u \mid v) + \sigma(u \mid v) = (f \mid v), \forall v \in H. \tag{5}$$

The boundary conditions have been made part of this formulation, and the only condition that was not previously mentioned was that the velocity field is conservative, that is, $\mathrm{div}(\mathbf{V}) = 0$. This expression permits the choice of an approximation via Galerkin's Method by choosing a subspace V_h of H, which is finite-dimensional and generated by chosen finite elements φ_j. This choice separates space and time variables, by the use of

$$u_h = \sum_{j=1}^{n} c_j(t)\varphi_j(x,y), \quad \forall\, \varphi_j \text{ from the basis } \beta \text{ of } V_h. \tag{6}$$

This will transform equation (5) into a system of Ordinary Differential Equations.

The initial condition, for this type of second-phase model is given by the satellite image presented in daily papers (*Jornal o Globo*, January 23rd, 2000) 12 hours after the period in which the leak occurred. The resulting oil slick reached from well beyond western boundaries of the refinery region, touched the Mauá and Anil beaches of Magé county and was (at that time) limited on the eastern side by the point on the continent north of Paquetá Island, as indicated in the first chart of figure 3.

3 Approximate solutions

Besides the choice for (6), we will also make an option for a second-order approximation method in time: Crank-Nicolson, and we will, therefore, define

$$u_h(x_j, y_j; t_n) \cong \sum_{j=1}^{n} c_j(t_n)\varphi_j(x, y) \cong \sum c_j^{(n)} . \varphi_j(x, y), \quad \varphi_j \in \beta \subset V_h. \quad (7)$$

The use of this expression in the weak formulation given by (4), besides the choice for $c^{(0)}_j = u_0(x_j, y_j)$, leads us to a linear system of equations in the unknowns $c_1^{(n+1)}, c_2^{(n+1)}, c_3^{(n+1)}, ..., c_N^{(n+1)}$:

$$\sum_{j=1}^{n} c_j^{(n+1)}\left[\left(1 + \sigma\frac{\Delta t}{2}\right)(\varphi_j|\varphi_i)_\Omega + \alpha\frac{\Delta t}{2}(\nabla\varphi_j \| \nabla\varphi_i)_\Omega + \right.$$

$$\left. + \frac{\Delta t}{2}\left(V_x\frac{\partial\varphi_j}{\partial x}|\varphi_i\right)_\Omega + \frac{\Delta t}{2}\left(V_y\frac{\partial\varphi_j}{\partial y}|\varphi_i\right)_\Omega \right] =$$

$$= \sum_{j=1}^{n} c_j^{(n)}\left[\left(1 - \sigma\frac{\Delta t}{2}\right)(\varphi_j|\varphi_i)_\Omega - \alpha\frac{\Delta t}{2}(\nabla\varphi_j \| \nabla\varphi_i)_\Omega - \right.$$

$$\left. - \frac{\Delta t}{2}\left(V_x\frac{\partial\varphi_j}{\partial x}|\varphi_i\right)_\Omega - \frac{\Delta t}{2}\left(V_y\frac{\partial\varphi_j}{\partial y}|\varphi_i\right)_\Omega \right] + \Delta t(f^{(n+\frac{1}{2})}|\varphi_i)_\Omega$$

$$(8)$$

with $i = 1, ..., N$ and for given $c^{(0)} = (c^{(0)}_1, c^{(0)}_2, ..., c^{(0)}_N)$.

In order to obtain values for V_x and of V_y on each element, the numerical solution of Stokes' equation using second-order finite elements is used on the Guanabara Bay discretized domain, and the obtained results are added to wind-induced currents:

$$-\text{div}(\nabla V) + \nabla p = f, \text{ on } \Omega \subset R^2$$
$$\text{div}(V) = 0, \text{ on } \Omega \quad (9)$$
$$V = V_0, \text{ on } \partial\Omega$$

The boundary conditions for Stokes' equation were determined using data provided by the National Hydrographic Directory (DHN), a service maintained by the Brazilian Navy. The time series of measures for surface currents were treated with the Matlab "Rose" function to obtain set patterns of behaviours. These were tested against historical data before adding wind terms. The obtained velocity fields are given in figures 1 (without wind) and 2 (circulation plus wind velocity field).

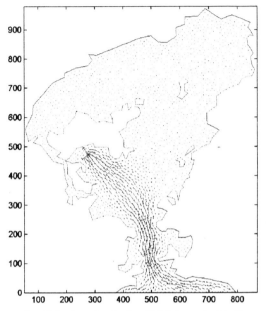

Figure 1 – Calculated velocity field (no wind influence).

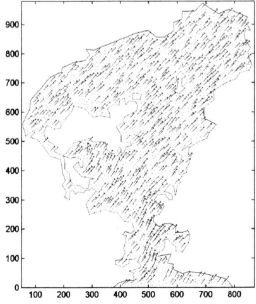

Figure 2 – Calculated velocity field (with wind influence)

With the output of this numerical approximation of Stokes' equation as input for the main program, and making use of upwinding techniques [2] in order to eliminate numerical oscillations which plague the numerical solution of this kind of system, we obtained the following results for the oil spill movements during the first period. During this period, wind came in from the Southwest causing the oil slick to spread along the northern shores of the bay touching on lands of the Refinery (REDUC), and reaching beaches of the Magé region, as well as beginning to enter the Guapimirim mangroves, as indicated in figure 3. The used upwinding technique consisted of adding a fixed term in the direction from which the velocity field originates, and subtracting this same term from the finite element when it is placed downwind, introducing a discontinuity along boundaries separating the triangles which make up the grid mesh.

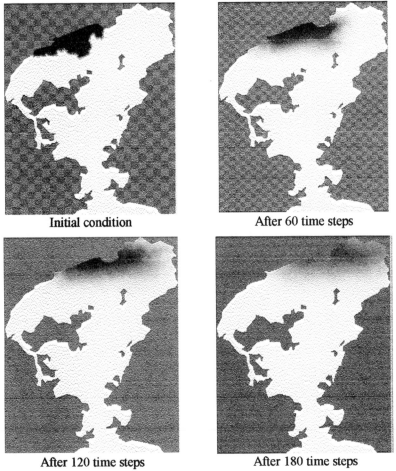

| Initial condition | After 60 time steps |
| After 120 time steps | After 180 time steps |

Figure 3

After 6 hours, a change in the wind direction occurred, with wind blowing from the northwest. Using this new situation in the generation of wind components, the numerical code obtained the following final results:

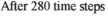

After 280 time steps   After 400 time steps

Figure 4

What we can now identify is that, quite like what effectively happened, the oil slick reaches Paquetá Island, while simultaneously remaining in the northeast corner of the Bay. In spite of the decrease in concentration, the continued presence of the oil in this region is of major concern in almost all environmental aspects. Figure 1 correctly describes the reduced circulation of bay waters in this region, while local characteristics of vegetation and fauna may very well amplify the negative medium and long-term effects for all present communities.

Conclusions

The main improvement upon previous work is the inclusion of Stokes' equations to model water circulation when, on the one hand, some information is available, and that information is of good quality credibility-wise, as well as when no information can be used and viable scenarios are to be simulated. The improvement upon cases (as in [7]), in which the velocity fields are created using only qualitative information and circulation is obtained by interpolating these data, is the time necessary for this to be developed.

Simulations in the present case show a very good degree in agreement with satellite photographs and descriptions of the evolutionary behaviour of the spill, showing that these coupled models (Stokes *cum* Diffusion-advection) represent a very valuable tool for decision-makers and emergency planners. Of course, a Navier-Stokes approximation for the water circulation can be used, but the main objective in this case was to consider a quasi-stationary situation, as described by technicians present at the moment, due to the geographical location of the oil spill at the furthermost end of Guanabara Bay, where, as mentioned above, circulation is mainly provoked by the predominant winds.

Acknowledgements

This work was partially sponsored by the São Paulo State Science Foundation: FAPESP

References

1. Benqué, J. P., Hauguel, A. & Viollet, P. L. in *Engineering Applications of Computational Hydraulics II*, Pitman Advanced Publ. Programme, 1980.
2. Brooks, A. & Hughes, T. J. R. Streamline Upwind/Petrov-Galerkin Formulations for Convection Dominated Flows with Particular Emphasis on the Incompressible Navier-Stokes Equations, *Computer Methods in Applied Mathematics and Engineering*, 1982, **32**, 199-259.
3. Carbonel, C. Modelling of Coastal Ocean Response Induced by Wind Fields, *Laboratório Nacional de Computação Científica*, 1995.
4. Cuesta, I., Grau, F. X. & Giralt, F. Numerical Simulation of Oil Spills in a Generalised Domain, *Oil & Chemical Pollution*, 1990, 7, 143-159.
5. Fay, J. A. The Spread of Oil Slicks on a Calm Sea. *Oil in the Sea*, 1969, Plenum Press, 53-63.
6. Meyer, J. F. C. A. Derrames de Petróleo em Águas Costeiras: Modelagem Matemática e Simulação Numérica, *in III° Simpósio de Ecossistemas da Costa Brasileira*, pp. 238-247, Serra Negra, São Paulo, Brasil, 1993.
7. Meyer, J. F. C. A., Cantão, R. F. & Poffo, I. R. F.: (1998), Oil Spill Movement in Coastal Seas: Modelling and Numerical Simulations, in Brebbia, C.A. (ed.), Oil Spill 98, Comp. Mech. Publ., Southampton, 1998, pp. 76-87.
8. Stolzenbach, K. D., Madsen, O. S., Adams, E. E., Pollack, A. M. & Cooper, C.K. A Review and Evaluation of Basic Techniques for Predicting the Behaviour of Surface Oil Slicks. M.I.T. *Sea Grant Report*, 1977, **222**, Mass. Inst. of Technol.
9. Valencia, M.J.: East Asian Seas: Hypothetical Oil Spill Trajectories. *Mar. Pollut. Bull.*, **23**, 1991.

Section 5
Erosion and Drift

Effect of turbulence on tidal suspended transport

B.A. O'Connor
Department of Civil Engineering, University of Liverpool, UK

Abstract

The effect of enhanced turbulence upon tidal suspended transport has been investigated using a combination of a one-dimensional (1DV) computer model and a turbulence-modified bulk transport model. An existing 1DV model, O'Connor and Nicholson [2], has been modified by including a random distribution of horizontal flow velocities into the model, which are also allowed to change mixing coefficients and their distribution. Operation of the model at turbulent space and time steps for a typical field situation with a fine sand bed shows that transport rates for normal turbulence are enhanced by some 4-25% at all times during the tidal cycle compared to calculations performed neglecting turbulence effects. A doubling of turbulence levels was found to increase transports by some 19-72% although a very large increase, albeit small in absolute terms, occurred near critical entrainment times due to earlier erosion. For higher enhanced turbulence levels, a similar turbulence velocity modification was done to a bulk transport predictor of the Van Rijn [4] type and calibrated against 1DV model ouput. Results for turbulence enhancement by up to an order of magnitude of normal values showed large increases in transport rates and large potential scour depth immediately downstream of structures. A series of parameterised equations provide preliminary guidance on the inclusion of enhanced turbulence effects in computations. Limited checking of the IDV model output against field data shows realistic results but further field data is needed to check the equations produced.

Introduction

The effect of turbulence in modifying suspended sediment transport rates is generally ignored in engineering calculations on the grounds that such effects are small in comparison with transport by the turbulent-mean flow field. Such calculations also often assume the occurrence of equilibrium (steady flow) conditions, even for tidal situations. However, for situations with fine sands, the majority of sediment transport occurs within the water column as suspended load and significant non-equilibrium lag effects can occur, MacDowell and O'Connor [1]. If structures are present, they will also add significantly to turbulence levels, particularly close to them so that enhanced suspended transport and bed scour will also result.

    The difficulty with quantifying the effect of turbulence on suspended load is the lack of suitable instrumentation to measure insitu suspended sediment concentrations at many points within the water column without disturbing the flow itself. Even modern acoustic sediment probes have difficulties of both flow disturbance and calibration when a range of particle sizes are present in the water column. In order to study the effect of turbulence on suspended transport rates, it is proposed, therefore, to make use of an existing one-dimensional (1DV) tidal suspended sediment computer model, O'Connor and Nicholson [2], and to modify it so as to include turbulence effects. The work forms part of a multi-institute ongoing EU-sponsored MAST III Research Project, SCARCOST.

1   Methodology

The 1DV suspended sediment concentration equation for tidal conditions is given by the equation:-

$$\frac{\partial c}{\partial t} = \varepsilon_z \frac{\partial^2 c}{\partial z^2} + \left(\frac{\partial \varepsilon_z}{\partial z} - w + w_f \right) \frac{\partial c}{\partial z} \tag{1}$$

where c is the suspended sediment concentration; t is time; ε_z is the vertical sediment diffusion coefficient; and $\partial \varepsilon_z / \partial z$ is its vertical gradient; ω_f is the fall velocity of the sediment particles; w is the vertical fluid velocity, which is generally ignored in relation to ω_f; and z is a vertical co-ordinate measured positive upwards from the bed. Because of the presence of large concentration gradients near the sediment bed, equation (1) is usually solved by numerical methods that use a transformed grid system over the flow depth, see O'Connor and Nicholson [2].

    For equilibrium turbulent-mean suspended transport, $\partial c/\partial t = 0$, and the solution to equation (1) depends on the specification of ε_z, $\partial \varepsilon_z/\partial z$ and ω_f. For a parabolic variation of ε_z w.r.t. z, see equation (2), the solution is the classical Rouse form, equation (3), that is:-

$$\varepsilon_z = \beta \kappa u_* h \eta (1 - \eta) \tag{2}$$

where β is a correction factor linking sediment diffusion to fluid flow diffusion often taken as unity; κ is Von Karman's constant (= 0.40 for clear fluid); u_* is the flow shear velocity; h is the water depth; η is a non-dimensional elevation (= z/h).

$$c = c_a \left[(1/\eta - 1) / (1/\eta_a - 1) \right]^Z \tag{3}$$

where c_a is a turbulent-mean reference concentration at elevation η_a (=a/h), taken equal to half the bedform height; Z is Rouse's exponent ($= \omega_f / (\beta \kappa u_*)$).

For equilibrium conditions, the suspended load transport is given by the equation:-

$$T_\infty = h \int_{\eta_a}^{1} u c \, d\eta \tag{4}$$

where u is the horizontal flow velocity at elevation η. For a steady, fully-rough turbulent-mean flow, u has a logarithmic variation given by:-

$$u = 2.5 u_* \ell n (\eta / \eta_o) \tag{5}$$

where η_o is the effective non-dimensional bed roughness height (=z_o/h, z_o is the roughness height of the bed, RR/30, and RR is the Nikuradse roughness height), and it is assumed that $\kappa = 0.40$.

For tidal conditions in the absence of vertical density gradients, the effect of flow acceleration and deceleration is usually small and hence equation (2) can be used to represent tidal mixing provided u_* is allowed to vary with time, that is:-

$$u_* = U / c_* \; ; \; c_* = C / \sqrt{g} \tag{6}$$

where C is Chezy's roughness coefficient; g is the acceleration due to gravity; and U is the tidal depth-mean velocity, which for an M2 tide is given by the equation:

$$U = U_m \operatorname{Sin}(\omega t) \tag{7}$$

where ω is the tidal wave frequency (= $2\pi/T$), T is tidal period; and U_m is the maximum depth-mean tidal velocity. The water depth will also vary tidally as given by the equation:-

$$h = h_m \pm R \operatorname{Cos}(\omega t) / 2 \tag{8}$$

where h_m is the tidal-mean depth at a site; and R is the tidal range.

   The effect of turbulence is to cause high frequency (1-10Hz) random variations in flow and mixing parameters, u, w, ε_z, $\partial \varepsilon_z/\partial z$, relative to turbulent mean values, which implies spatial and temporal scales of (0.1 - 1m, 0.1 - 1s) for field scale maximum tidal velocities of 1m/s. By using numerical grids with logarithmically increasing spacing from the bed (typically 32 grids where used which gives spacing of 4.53mm near the bed varying to 1.75m near the surface) and time steps of 0.1s, equation (1) has been solved numerically, see O'Connor and Nicholson [2], through a tidal cycle to obtain turbulence-affected suspended sediment concentrations.

   The effect of high-frequency variations upon the flow and mixing parameters was introduced into the 1DV model using a random number generator and the assumption of a gaussian variation of fluctuating velocities, that is:-

$$u = \overline{u} + u' \; ; \; u' = 3\sigma_u (2Rd(1) - 1) \tag{9}$$

where $\overline{u}$ is the turbulent-average velocity given by equation (5) with the turbulent-average u∗; Rd(1) is a random number in the range 0-1; and σ_u is the standard deviation of the horizontal velocity fluctuations $\left(\sqrt{u'^2} \right)$. The variation of σ_u with elevation was based on laboratories evidence, see for example, Savelle [3] and Van Rijn [4], that is:-

$$\sigma_u = 3n u_* (\exp(-0.80\eta)) \tag{10}$$

where n has a value of about 1 for normal rough-turbulent flow. The effect of turbulence on w was neglected.

   Turbulence was introduced into the mixing coefficient ε_z via equation (6) by the introduction of a depth-average velocity variation, U', that is:-

$$U' = 3 <\sigma_u> (2Rd(1) - 1) \tag{11}$$

where $<\sigma_u>$ is the depth-average value of equation (10).

Finally, the bed boundary condition, c_a, in the 1DV model was modified by introduction of equation (11) into Van Rijn's formula for c_a, Van Rijn [4], that is:-

$$c_a = 0.015 \rho_s d_{50} T^{1.5} / (d_*^{0.30} \eta_a h) \tag{12a}$$

where
$$T = (u_*' / u_{*_c})^2 - 1 \tag{12b}$$

$$u_*' = (U + U') C' / \sqrt{g} \tag{12c}$$

$$C' = 5.75 \sqrt{g} \log(4h/d_{90}) \qquad (12d)$$

$$d_* = d_{50} (\Delta g / v^2)^{1/3} \qquad (12e)$$

and Δ is the submerged relative density $\rho_s/\rho - 1$, ρ_s, ρ are the sediment grain and fluid densities, respectively); v is the kinematic eddy viscosity of the fluid; d_{50}, d_{90} are % finer values of the bed sediment grading curve; and u_{*c} is the critical shear stress for sediment motion (a function of d_*, see Van Rijn [4]).

In order to take into account the varying response of different sediment grain-sizes to turbulence, the 1DV model was used with the bed grading curve divided into twelve grain-size fractions with mean sizes between 65-385µm. The concentration profiles for the "instantaneous" c values, together with the "instantaneous" u velocity profiles of equations (5) and (9) where used in equation (4) to determine the "instantaneous" suspended sediment transport rate. These "instantaneous" values were than averaged over each 3000 consecutive time steps to determine the combined turbulent-average and turbulent-fluctuation transport rate. The contributions from each grain-size fraction were then combined together according to the % occurrence in the bed grading curve so as to produce a composite time-averaged suspended sediment transport rate every five minutes throughout the tidal simulation.

2   Results

The 1DV model was run firstly for normal turbulence conditions for a typical unstratified, fine sediment coastal site (see also [2]) with a tidal-mean water depth of $h_m = 10m$; a tidal range R = 3m; a maximum tidal velocity of $U_m = 1m/s$; d_{50}, d_{90} values of 150µ, 300µ, respectively; and a water temperature of 10^0C ($v = 1.39 \times 10^{-6}$ m²/s) for which suspended load transport rates were already available, see O'Connor and Nicholson [2]: all time varying bedform sizes (η_a) and friction factors (c_*) were fixed at previous values.

The tidal results for transport at any tidal state (T_t) were expressed in terms of "velocity-average" ($\overline{T_t}$) values and deviations (DT) from the velocity-average, that is:-

$$T_t = \overline{T_t} (1 \pm DT / \overline{T_t}) \qquad (13)$$

so that the effect of turbulence upon both "velocity-average" and deviation (lag) quantities could be estimated.

Table 1 shows the results scaled to the transport at maximum tidal velocities (T_m @ U_m) for the flood phase of the tidal cycle for varying tidal depth-mean velocity (U) scaled to the maximum tidal velocity (U_m), that is X = U/U_m. It is clear that enhanced transport occurs for all velocities even below critical values (U_c = 0.4076m/s) and that maximum enhancement near critical is some 25% of mean values. A simple model of the results with m = 0.25 is the equation:-

$$RA\,(\%) = 100\,(1 + m\,[(1 - X)/(1 - X_c)])^{1.2} \text{ for } X > X_c \qquad (14a)$$

$$RA\,(\%) = 100\,(1 + m) \text{ for } X \le X_c \qquad (14b)$$

The effect of turbulence upon tidal deviation transport is shown in Table 2.

It is clear that turbulent-mean transport rates can vary by almost a factor of two at critical velocities ($X_c = 0.4076$) but that only minor influences are felt on deviation quantities: on average there is only some 5% change over all velocity values.

Comparison of the model results with field data is difficult at present because of lack of information. Results by Soulsby et al [5] for a flood tide using a sediment impact probe and electromagnetic current meter with a response frequency of up to 5Hz above the crest of a sandwave of height 750mm and wavelength 25m in the Taw Estuary, UK with sediment of $d_{50} = 165\mu m$ suggested that the turbulent flux at maximum tidal velocity at two points above the bed (130mm, 330mm for $h_m = 2.7m$) was some 1% of turbulent-mean values, which is a little lower (4%) than the values in Table 1, but clearly of the correct order. Lower field values may be due to suppression of turbulence by flow acceleration towards the bed form crest: larger values would be expected in the lee.

The 1DV model was next run for enhanced turbulence ($n = 2$) by assuming that the turbulence profile over the depth was enhanced by the same factor (2) at all levels. The results are shown in Table 3.

A similar effect is found as for normal turbulence but with a larger transport factor. Equation (14) provides a realistic model with $m = 0.72$, which suggests a non-linear enhancement factor with increasing turbulence level. The enhancement of maximum transport rates (T_{mt}) is seen from Table 2 to have increased by some 20% (18.92%). Unfortunately, there is no data available to

Table 1: Effect of normal turbulence ($n = 1$) on velocity-average suspended load.

$X(\%)$	$\overline{T}_t / \overline{T}_{mt}\,(\%)$	$\overline{T} / \overline{T}_m\,(\%)$	Ratio(RA%)	Eq(14)
12.05	4.6041E-2	3.8421E-2	119.83	125
23.93	1.6814E-2	1.4081E-2	119.41	125
35.46	5.1788E-1	4.2693E-1	121.30	125
46.47	1.5165	1.2163	124.67	122.1
56.81	4.1362	3.5311	117.14	117.1
66.31	9.7248	8.6752	112.10	112.7
74.85	19.244	17.795	108.14	108.9
82.30	32.799	31.298	104.80	105.9
88.55	49.813	48.402	102.92	103.5
93.50	67.344	66.959	100.58	101.8
97.10	84.274	83.847	100.51	100.7
99.27	95.828	95.720	100.11	100.1
100	100	100	100	100

$\overline{T}_{mt} = 1.0999\,kg/m/s : \overline{T}_m = 1.0562\,kg/m/s : \overline{T}_{mt}/\overline{T}_m = 1.0413$

test these results.

The effect on deviation quantities is shown in Table 4. Again a similar small effect is found with deviation quantities reducing by some 11% (a factor of 0.89) on average across all velocities. Again data to confirm these results in lacking.

The 1DV model can clearly be run for different enhancement factors. Unfortunately, model run time is large taking some 12 hours per individual grain size computation on a 486PC. To examine the possible effect of larger enhancement values, an alternative approach has been used.

Table 2: Effect of normal turbulence (n = 1) on deviation transport.

X(%)	$DT_t / \overline{T}_t$ (%)	$DT / \overline{T}_t$ (%)	RB (%)	Eq. 19
12.05	100	100	100	100
23.93	100	100	100	100
35.46	99.20	100	99.20	100
46.47	86.81	94.07	92.28	94.3
56.81	67.42	72.35	93.19	94.3
66.31	50.57	53.57	94.40	94.3
74.85	38.05	39.95	95.24	94.3
82.30	29.14	30.03	97.04	94.3
88.55	22.37	22.29	100.36	94.3
93.50	15.21	15.89	95.72	94.3
97.10	10.23	10.26	99.71	94.3
99.27	4.67	5.04	92.66	94.3
100	0	0	0	0

Table 3: Effect of enhanced turbulence (n = 2) on velocity-average suspended load.

X(%)	$\overline{T}_t / \overline{T}_{mt}$ (%)	$\overline{T} / \overline{T}_m$ (%)	Ratio(RA%)	Eqn (14)
12.05	5.4791E-2	3.8421E-2	142.61	172
23.93	2.0224E-2	1.4081E-2	143.63	172
35.46	6.6189E-1	4.2693E-1	155.03	172
46.47	2.0129	1.2163	165.49	163.8
56.81	5.3015	3.5311	150.14	149.3
66.31	11.826	8.6752	136.31	136.6
74.85	22.251	17.795	125.04	125.8
82.30	36.081	31.298	115.28	116.9
88.55	52.935	48.402	109.37	110.0
93.50	69.029	66.959	103.09	105.1
97.10	85.115	83.847	101.51	101.9
99.27	95.777	95.720	100.05	100.4
100	100	100	100	1

$\overline{T}_{mt} = 1.2561 \text{kg} / \text{m} / \text{s}$: $\overline{T}_m = 1.0562 \text{kg} / \text{m} / \text{s}$: $\overline{T}_{mt} / \overline{T}_m = 1.1892$

Table 4: Effect of enhanced turbulence (n=2) on deviation transports

X(%)	$DT_t / \overline{T}(\%)$	$DT / \overline{T}(\%)$	Ratio(%)	Eq. 19
12.05	100	100	100	100
23.93	100	100	100	100
35.46	95.84	100	95.84	93.4
46.47	78.49	94.07	83.44	89.4
56.81	59.30	72.35	81.96	89.4
66.31	43.84	53.57	81.84	89.4
74.85	32.68	39.95	81.80	89.4
82.30	25.46	30.03	84.78	89.4
88.55	20.13	22.29	90.31	89.4
93.50	13.16	15.89	82.82	89.4
97.10	9.38	10.26	91.42	89.4
99.27	4.16	5.04	82.54	89.4
100	0	0	0	0

Table 5. Enhanced suspended loads at maximum tidal flows.

Enhan. (n)	$\overline{T}_{mt}$ (Eq. 15)	$\overline{T}_m$ (Eq. 16)	$\overline{T}_{mt} / \overline{T}_m$
0	1.056	1.056	1.0
1	1.105	1.106	1.047
2	1.257	1.256	1.289
3	1.502	1.517	1.436
4	1.865	1.889	1.788
5	2.354	2.385	2.258
6	2.985	3.026	2.865
7	3.779	3.834	3.630
8	4.757	4.837	4.580
9	5.946	6.064	5.741
10	7.371	7.547	7.145

Van Rijn [4] has developed a widely used suspended load formula for use in river, coastal and estuarine environments, that is:-

$$T_{\infty} = a \rho_s \, d_{50} \, U \, [(U - U_c) / \sqrt{(\Delta g d_{50})}]^b / d_*^{\,c} \qquad (15)$$

where a = 0.012; b = 2.4; and c = 0.60.

Application of equation (15) to the earlier data of O'Connor and Nicholson [2] for equilibrium tidal conditions at maximum flow gives a = 0.0607; b = 2.9463; c = 3.10 and shows a larger reduction in transport for larger grains than Van Rijn's river conditions, but has been used herein to provide comparison with earlier work [2].

Turbulence effects have been included into equation (15) to give $T_{\infty t}$ by replacing U by $U + \beta_0 U'$, where β_0 is a scale parameter. The maximum tidal turbulent transport (T_{mt}) is then related to $T_{\infty t}$ by a second scale factor $\gamma (= T_{mt} / T_{\infty t})$ which earlier work [2] suggests will be in the range 0.6-1. The turbulence modified equation (15) was then used for 3000 trials for enhanced ($n = 2$) turbulence conditions with U' varying as a gaussian distribution about mean values. Comparisons with the 1DV model gave scale factors of $\beta_o = 0.5399$; $\gamma = 0.9098$. For normal turbulence, the turbulent-average form of equation (15) with the fitted β_o, γ values gives $\overline{T}_{mt} = 1.106$ kg / m / s which is within 0.6% of the 1DV values, see Table 1.

The calibrated turbulence-modified form of equation (15) was then run for a range of enhanced turbulence values up to a factor of 10 times normal values. These results were then used to modify equation (15) so as to include enhanced turbulence effects in a simple engineering level formula. The result is the equation:-

$$\overline{T}_{mt} = \gamma a \rho_s\, d_{50}\, U[(U - U_c + \alpha\{\sqrt{k}\}) / \sqrt{(\Delta g d_{50})}]^b / d_*^c \qquad (16)$$

with $\gamma = 0.9098$; $b = 2.9463$; $c = 3.10$; $\{\sqrt{k}\}$ represents the depth – averaged turbulent kinetic energy per unit mass; and α is a parameter which varies with enhanced turbulence level (n), that is:-

$$\alpha = a_o + a_1 K + a_2 K^2 + a_3 K^3 + a_4 K^4 \qquad (17)$$

where $K = \ell n(n)$; $a_0 = 8.140606 \times 10^{-2}$; $a_1 = 5.94318 \times 10^{-2}$; $a_2 = 8.053073 \times 10^{-2}$; $a_3 = 1.587392 \times 10^{-2}$; $a_4 = 1.36214 \times 10^{-3}$.

Using equation (16), the enhanced suspended transport at minimum flow can be related to the "non-turbulent" value T_m by the equation:-

$$\overline{T}_{mt} = \overline{T}_m\, [1 + \alpha\{\sqrt{k}\} / (U_m - U_c)]^b \qquad (18)$$

which indicates a 100% increase is $\overline{T}_m$ for $n = 4.8$, see Table 5.

Equation (16) has also been used to extend equation (14) for large enhancement values. The result is an equation for m w.r.t n, the enhancement ratio, that is:-

$$m = a'_1 n + a'_2 n^2 + a'_3 n^3 + a'_4 n^4 \qquad (19)$$

where $a'_1 = 1.05046111 \times 10^{-1}$; $a'_2 = 1.6391761 \times 10^{-1}$; $a'_3 = -1.9707111 \times 10^{-2}$; $a'_4 = 7.43389 \times 10^{-4}$

The effect of higher turbulence levels upon DT/$\overline{T}$ values requires use of the 1DV model. A possible functional form based on the $n = 1,2$ values of Tables 2, 4, is as follows with RB = $(DT_t / \overline{T}_t) / (DT / \overline{T})$, that is

$$RB = p = \exp(-0.0583 n^{0.943}) \quad \text{for} \quad X = X_c \qquad (20a)$$

where

$$RB = p + (1-p)(X_c - X)/(X_c - X'_c) \quad \text{for} \quad X_c' < X < X_c \qquad (20b)$$

and $\quad X'_c = X_c(1 + 6.2n/c_*)$ $\qquad (20c)$

The fit of equation (19) is shown in Tables 2, 4.

Table 6. Net tidal cycle scour depth variation with n.

Factor (n)	Scour Depth (m)
1	0
2	0.11
3	0.29
4	0.55
5	0.90
6	1.35
7	1.92
8	2.62
9	3.49
10	4.53

In order to illustrate the effect of enhanced turbulence upon scour depths, equation (16) has been combined with a bed level change equation (20) to product scour depths over a half tidal cycle:

$$\frac{\partial T_t}{\partial x} + \rho_s(1 - n_*)\frac{\partial z_b}{\partial t} = 0 \qquad (21)$$

where T_t is the total bed and suspended load transport during the tide, n_* is the bed porosity and z_b is the bed level change relative to a horizontal datum. If it is assumed that $\overline{T_t}$ varies linearly to maximum transport rates $\overline{T}_{mt}$ during the tide and DT_t and bed load are small compared with $\overline{T}_t$; that flow streamlines remain parallel (1m apart) but a structure adds enhanced turbulence over a horizontal distance (Δx) equal to the flow depth ($h_m = 10m$); and the bed consists of fine sand as tested in the IDV model with $U_m = 1m/s$, the value of z_b over half a tidal cycle (no turbulence added over the reversed part of the cycle since it is now upstream of the structure) is found to vary from 0.1 - 4.5m with enhanced turbulence level, see Table 6.

3 Discussion and Conclusions

The IDV model results are clearly of the same order as field observations while Table 2 suggests that the neglect of turbulent effects leads to errors in transport of some 4-25% over the tidal cycle. The influence on tidal lag is much less although transport starts at velocities below the average value (X_c' compared with X_c, see equation 19). The turbulent fluxes are equally as important in relative terms as the turbulent-mean fluxes near critical conditions although absolute transport values are very small (<1% of maximum tidal values, see Table 1). The effect of neglecting turbulence on scour depths near structures is also small for normal turbulence levels, see Table 6.

Operation of the 1DV model at prototype spatial and temporal scales leads to long computer run times if a realistic number of grain fractions are used. An alternative approach, based on a modified Van Rijn engineering formula has enabled results to be obtained for high turbulence levels after using the IDV results for calibration purposes. The model results show an order of magnitude increase in transport rates at maximum tidal flows for an order of magnitude increase in turbulence levels, see Table 5, and a corresponding change in local scour depths, see Table 6 which points to the need for bed protection in such situations.

If information is available on velocity-average and deviation transport rates for "zero turbulence (n = 0)" calculations, see also [2], equations 14, 18 provide possible enhancement factors for velocity-average effects while equation (19) provides similar, but less precise, factors for deviation quantities. In the absence of such information equations (16) and (17) provides enhanced transport rates accurate to a factor of two or so using the present constants (γ, a, b, c) or those of Van Rijn [4], that is (1, 0.012, 2.4, 0.60).

Clearly, the suggested formulae and the various modelling approaches need further testing in order to refine estimates. The recent development of multi-frequency field acoustic sediment probes is a welcome step in this direction.

4   Acknowledgements

The work was partially supported by the Commission of the European Communities Directorate General for Science and Education, Research and Development under contract number MAS3-CT97-0097. The author is grateful to Mrs J. Price and Ms E. Hooton for preparation of the manuscript.

References

[1] MacDowell, D.M., O'Connor, B.A. *Hydraulic behaviour of estuaries,* MacMillan, 1977.

[2] O'Connor, B.A. and Nicholson, J. Tidal sediment transport, *Computer Modelling of the Seas and Coastal Regions III,* eds. J.R. Acinas and C.A. Brebbia, Computational Mechanics Publications, Southampton, pp. 367-380, 1997.

[3] Savell, I.A. PhD Thesis, Department of Engineering, University of Manchester, 1986.

[4] Van Rijn, L.C. *Principles of Sediment Transport in Rivers, Estuaries and Coastal Seas,* Aqua Publications, 1993.

[5] Soulsby, R.L., Salkfield, A.P., Haine, R.A. and Wainwright, B. Observations of the turbulent fluxes of suspended sand near the seabed, *Euromech 192, Transport of Suspended Solids in Open Channels, Neubiberg, 11-15th June 1985,* pp. 183-186, 1985.

Simulations to examine the effect of enhanced turbulence levels on bed load transport

C.P. Rose & B.A. O'Connor
University of Liverpool, Dept. of Civil Engineering, Brownlow St., Liverpool, L69 3GQ, UK

Abstract

The effect of enhanced turbulence levels on the magnitude of bed load transport rates is examined by computer simulations for both hydraulically smooth and rough beds. A method is demonstrated to estimate the root mean squared (RMS) bed shear stress for hydraulically rough beds. This method as well as an existing numerical integration method to calculate bed shear stresses in turbulent smooth bed conditions are used to simulate high frequency temporal shear stresses and bed load transport rates. The simulation method is shown to give similar temporally-averaged results to an existing approach of van Rijn [3] for a smooth bed. The present work suggests that reduced or enhanced shear stress fluctuations can significantly affect the magnitude of the temporally-averaged bed load transport rates for both the smooth and the rough bed cases. The results also demonstrate that the enhancement effects are of greatest significance near the threshold of sediment motion. The turbulent shear stress characteristics developed in the present work for a hydraulically rough bed were added into van Rijn's [3] wave-current bed load transport model and used to study enhanced transport rates. The results indicate that the random turbulent shear stress fluctuations have a significant effect on the magnitude of bed load transport rates for enhanced turbulence levels as might be experienced in the presence of structures.

1 Introduction

There are many situations of engineering significance where an ability to predict the effect of enhanced turbulence levels on bed load transport is important. For example, the presence of man-made structures in a river or coastal environment leads to flow disturbances in the immediate vicinity of the structure, which may in turn enhance local sediment transport inducing bed scour. Similarly, the flow separation zone which forms on the leeward side of steep bedforms is associated with enhanced levels of turbulence which is an important aspect in the morphodynamic predictions of bedforms (de Ruiter [1]). Further, in wave-current conditions, the effect of enhanced steady flow type turbulent fluctuations on bed load transport is also an important area requiring further study.

The bed shear stress is a key hydrodynamic parameter used to predict bed load transport. In the scour and bedform examples, the bed shear stress is enhanced due to the presence of advective acceleration effects, a decreased flow cross-section (in the case of the scour problem) and enhanced turbulence levels. The former two points can be successfully represented with numerical models, however the latter point remains an area requiring further research, particularly for hydraulically rough beds. The enhancements discussed above may affect both the mean and fluctuating bed shear stress components. In the present work attention is focused upon the effect of enhanced shear stress fluctuations on bed load transport. In section 2.1 of this paper, methods to predict shear stress fluctuations for smooth and rough beds are examined and in section 2.2 the effect of a given modification to the RMS bed shear stress on the temporally averaged bed load transport rate is quantified. In section 2.3, the level of enhancement to the random shear stress fluctuations due to wave-current interaction is predicted and the enhancement in the bed load transport rates examined.

2 Results

2.1 Bed Shear Stress Characteristics

In order to examine the effect of enhanced temporal shear stress variations on bed load transport, it is first necessary to quantify bed shear stress characteristics for "normal" conditions. Defining the Cartesian co-ordinates (x,y,z) in the stream-wise, cross-stream and vertical directions, respectively, corresponding to instantaneous turbulent velocities (u,v,w), respectively, the instantaneous shear stress in the x-z plane is given for a uniform smooth turbulent flow as

$$\tau_{xz}(z,t) = -\rho u w + \rho \upsilon D(z)\frac{\partial u}{\partial z} \qquad (1)$$

where $D(z)$ is a viscous damping function, ρ is the density of the water and υ is the kinematic viscosity. For a uniform open channel flow the v and w velocity components are only composed of mean zero velocity fluctuations ($v = v'$, $w = w'$) while the u component is the sum of a temporal mean component and mean zero velocity fluctuations ($u = U + u'$). For the above situation, at the bed level, u and w are zero and $D(0) = 1$, thus the RMS of the mean zero bed shear stress fluctuations, $\tau_{xz}{'}_{rms}(0)$ is given as

$$\tau_{xz}'{}_{rms}(0) = \rho\upsilon \frac{\partial u'{}_{rms}(0)}{\partial z} \qquad (2)$$

where $u'{}_{rms}$ is the RMS of the mean zero stream-wise velocity fluctuations. From eqn (2) and experimental measurements (Grass [2]), the following non-dimensional parameter has been quantified in smooth turbulent conditions

$$\frac{\tau_{xz}'{}_{rms}(0)}{\bar{\tau}_{xz}(0)} = 0.4 \qquad (3)$$

where $\bar{\tau}_{xz}(0)$ is the temporal mean total bed shear stress and for sediment transport calculations, the probability distribution function (PDF) of the bed shear stress is commonly assumed to be Gaussian for simplicity (de Ruiter [1]; van Rijn [3]), though a skewed function was obtained from data summarised in Grass [2]. The Gaussian distribution can be defined using

$$N(\mu,\sigma) = \frac{1}{\sqrt{2\pi}\sigma}\exp\left(-\frac{(X-\mu)^2}{2\sigma^2}\right) \quad \text{for} \quad -\infty < X < \infty \qquad (4)$$

see figure (1a), where X is a dummy variable and $N(\mu,\sigma)$ denotes a Gaussian PDF with a mean of μ and a standard deviation of σ.

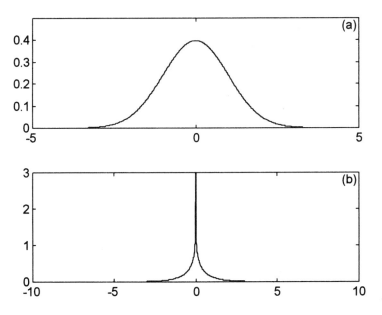

Figure 1: (a) Gaussian PDF, $N(0,1)$ and (b) PDF formed from the product of 2 uncorrelated Gaussian signals.

For a hydraulically rough bed, the presence of the roughness elements disrupts the viscous shear layer and the instantaneous bed shear stress is given solely by the temporal Reynolds stresses. The Reynolds stresses corresponding to the *x-z* and *y-z* planes are given as follows:

$$\left.\begin{array}{l} \tau_{xz} = -\rho u'w' \\ \tau_{yz} = -\rho v'w' \end{array}\right\} \qquad (5)$$

as *v* and *w* are mean zero signals for the 1D case. It is noted that the Reynolds stress terms in eqn (5) is different in structure to the viscous term in eqn (2), thus it is reasonable to expect that the temporal shear stress characteristics for a rough bed are different to the smooth bed case. This can be illustrated by considering the simplified case when *u* and *w* are uncorrelated. As the velocity fluctuations can be well represented by Gaussian PDFs, the shear stress will not be Gaussian but instead will follow a symmetric PDF with a kurtosis of 9 instead of the Gaussian value of 3, see figure (1b). This may be a good approximation for the *y-z* plane Reynolds shear stress as the correlation between the velocity signals may indeed be zero. However, in reality, the *u* and *w* components are correlated leading to a skewed PDF function with a mean component. Thus the *x-z* plane Reynolds stress is composed of two components, namely

$$\tau_{xz} = \overline{\tau}_{xz} + \tau_{xz}' \qquad (6)$$

where $\overline{\tau}_{xz}$ is the temporal mean component and τ_{xz}' is the mean zero fluctuating component. For simplicity, in the present approach the mean shear stress component is included though the skewness of the PDF is neglected. Therefore it is assumed that the mean zero shear stress fluctuations follow the PDF shown in figure (1b). This model therefore neglects the skewness caused by the correlation between the velocity fluctuations. A full examination of the assumptions in the present model is an area of on-going work. Using the present assumptions, the rough bed $\tau'_{rms}(0)$ values for the two cases can be estimated using

$$\left.\begin{array}{l} \tau_{xz}'_{rms}(0) = \rho u'_{rms}(0)w'_{rms}(0) \\ \tau_{yz}'_{rms}(0) = \rho v'_{rms}(0)w'_{rms}(0) \end{array}\right\} \qquad (7)$$

where v'_{rms} is the RMS of the cross-stream velocity fluctuations, w'_{rms} is the RMS of the vertical velocity fluctuations and the PDF of the shear stress fluctuations can be estimated using the curve shown in figure (1b).

Based upon the measurements of Nezu and Nakagawa [4] and specifying in the present case a rough bed where near-bed viscous damping effects are not present, the parameters to solve eqn (7) can be given as

$$\left.\begin{array}{l} u'_{rms}(0) = 2.30u_* \\ v'_{rms}(0) = 1.63u_* \\ w'_{rms}(0) = 1.27u_* \end{array}\right\} \qquad (8)$$

where u_* is the friction velocity which is related to the mean bed shear stress by $\overline{\tau}_{xz}(0) = \rho u_*^2$. Eqn (8) can be substituted into (7) to give the following results

$$\left. \begin{array}{l} \dfrac{\tau_{xz\,'rms}\,(0)}{\overline{\tau}_{xz}\,(0)} = 2.92 \\[4mm] \dfrac{\tau_{yz\,'rms}\,(0)}{\overline{\tau}_{xz}\,(0)} = 2.07 \end{array} \right\}$$ (9)

Thus the present method suggests that the normalised x-z plane RMS bed shear stress for a rough bed is a factor of 7.3 times larger than the case for a smooth bed. Further the PDF functions are also different.

2.2 Bed Load Transport Simulations For Steady Flow Conditions

To simulate temporal bed load transport, use will be made of the van Rijn [3] bed load transport rate expression per unit width, which is given as

$$Q_{bl} = \alpha(s-1)^{0.5} g^{0.5} D_{50}^{1.5} D_*^{-0.3} T_m^{2.1}$$ (10)

where $\alpha = 0.1$ for the smooth bed case,

$$D_* = \left[\frac{(s-1)g}{\upsilon^2} \right]^{1/3} D_{50}$$ (11)

and T_m is an instantaneous excess shear stress parameter given as

$$T_m = \frac{(\tau'_b - \tau_{cr})}{\tau_{cr}}$$ (12)

where $\tau'_b = \mu\overline{\tau}(0)$ is the instantaneous effective bed shear stress and μ (= 1) is the efficiency factor, $\tau_{cr} = 1.5\overline{\tau}_{cr}$ (De Ruiter [1]) is the instantaneous critical bed shear stress and $\overline{\tau}_{cr}$ is the mean critical bed shear stress which can be obtained from the van Rijn [5] parameterisation of the Shields curve.

It is noted that particle inertia effects may reduce the magnitude of high frequency bed load events and introduce phase lags between the shear stress and the bed load fluctuations. A method to include such effects is outlined in Fredsoe and Rasmussen [6] for oscillatory flows. However, for bed load dominated conditions, Fredsoe and Rasmussen [4] show that the reduction in the bed load transport rate is typically in the range of 0% to 5% compared to the original magnitude without lag effects. An initial examination of this method by the present authors for steady flow situations suggest that this effect is also small and is therefore not included in the present simulations.

Van Rijn [3] derived numerical solutions for temporally averaged smooth bed Q_{bl} values based upon a Gaussian PDF to represent τ'_b and numerical integration of the portions of the curve above the positive and negative instantaneous critical shear stress values. In the present work, time series of Q_{bl} will be generated to also examine the temporal characteristics and to compare the two solutions for the temporally averaged values. In order to generate temporal values of Q_{bl} from eqn (10) in the present work, use will be made of eqns (3) and (4) to generate temporal bed shear stress values. To generate instantaneous bed load transport rates, three instantaneous conditions are used: (a) when the magnitude of the shear stress is less than critical the corresponding bed load is set at zero, (b) when the shear stress is positive and greater than critical, eqns (10) to (12) are used directly to calculate the corresponding bed load transport rate and (c) when the shear stress is negative and less than the negative critical condition, the magnitude of the shear stress is taken for calculation of eqn (11) and a negative value is applied to eqn (10). These operations correspond to the conditions applied in the van Rijn [3] numerical integration method and point (c) assumes that a negative shear stress corresponds to transport against the main flow direction, following the van Rijn [3] approach.

A typical example of a small section of a computer generated shear stress and bed load transport rate temporal record for normal turbulence levels is shown in figure (2a) and (2b) respectively, corresponding to a mean excess shear stress of 2. The horizontal lines on the plots denote the temporal mean values. This plot highlights the large difference between the mean and peak shear stress and bed load transport rates, further the plot indicates that negative instantaneous bed load transport rates are relatively rare for normal turbulence levels. However, these events do become more frequent for enhanced turbulence conditions. At the present state-of-the-art there is no recognised method to measure high frequency sandy bed load transport in the laboratory or the field. Thus it is not possible to compare the characteristics of figure (2b) with data at present.

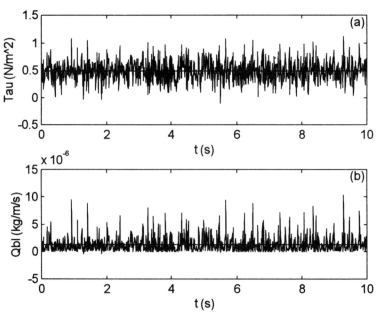

Figure 2: Smooth bed temporal simulations of (a) the bed shear stress and (b) bed load transport rate.

Having established that the ratio of the RMS to the mean bed shear stress is approximately 0.4 for smooth beds and estimated at 2.92 for rough beds in section 2.1, numerical experiments are performed to systematically vary this ratio, in the range of half to three times these values, corresponding to reduced through to enhanced RMS bed shear stresses, to examine the enhancement or reduction in the temporally-averaged bed load transport rates compared to the value for normal turbulence levels. In all the calculations the excess mean shear stress was set at 2. Figure (3) shows the results from the smooth bed time series simulations (o) compared to the smooth bed results using the van Rijn [3] numerical solution involving integration over the relavent parts of the Gaussian distribution (solid line). Figure (3) also shows the results for the rough bed simulations using the time series approach (x). For the smooth bed case, the figure confirms that both the numerical and time series approach give very similar results and an increase by a factor of 3 in the RMS shear stress corresponds to approximately a factor of 3.5 increase in the bed load ratio. For the rough bed case the enhancement and reduction factors are slightly greater than for the smooth bed ratio.

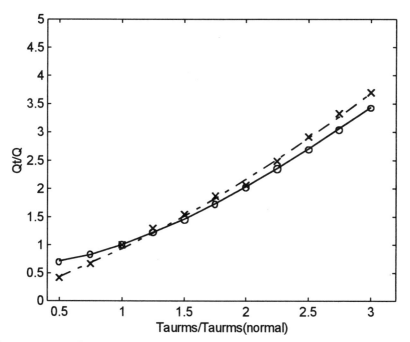

Figure 3: Non-dimensional bed load enhancement as a function of the non-dimensional RMS shear stress; (o) and (-) hydraulically smooth beds, (x) and (dot dashed line) hydraulically rough beds.

The present rough bed results can be parameterised using the equation

$$\frac{Q_{enhanced}}{Q} = 0.07 + 0.78n + 0.15n^2 \qquad (13)$$

where $n = \tau_{xz}{}'_{rms(enhanced)} / \tau_{xz}{}'_{rms}$. This is shown as the dash-dotted line in figure (3). For rough bed situations, the bed shear stress is proportional to the bed turbulent kinetic energy (TKE) per unit mass, $k(0)$, therefore the non-dimensional RMS shear stress ratio in eqn (13) is equal to $k_{enhanced}(0) / k(0)$ when the product of each of the three numbers squared in eqn (8) remains the same in enhanced turbulence conditions. However, this assumption may not generally be true for scour or bedform problems. In situations where these assumptions are approximately valid, the non-dimensional TKE parameter is of more direct use in engineering calculations.

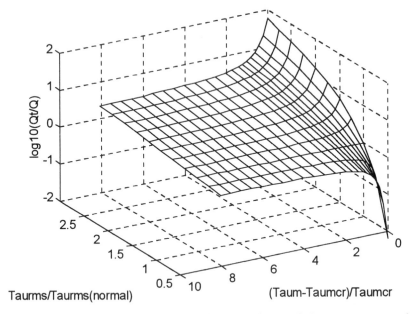

Figure 4: A 3D plot demonstrating the dependence of the mean excess shear stress and the non-dimensional RMS shear stress on the non-dimensional bed load transport rate.

Methods to calculate the bed load transport rates in rough bed conditions with enhanced turbulence levels can now be described. A full Reynolds stress closure could be applied in 2 or 3D depending on the problem, to calculate the bed RMS velocity fluctuations. Eqn (7) could then be used to estimate the bed RMS shear stresses. Finally eqn (13) could be used with the RMS bed shear stress values for normal turbulence levels, which can be obtained from empirical eqns such as (9) together with a standard bed load transport rate expression to finally estimate the bed load transport in enhanced turbulence conditions. An alternative, more approximate method of solution would be to apply a k or $k - \varepsilon$ model to the problem and replace the non-dimensional bed RMS shear stress by the non-dimensional bed TKE ratio. Further work is required to examine if the skewness in the RMS shear stress PDF significantly effects the results shown in figure (3).

The combined effect of the excess mean shear stress parameter and the non-dimensional RMS shear stress on the non-dimensional bed load transport rate is shown in figure (4). The results were obtained from smooth bed simulations using the van Rijn [3] integration method. The figure clearly demonstrates the greatest enhancement in the bed load transport occurs near critical conditions for sediment motion. When the excess mean shear stress parameter is zero, the enhancement is up to 60 times normal levels. These results indicate the importance of enhanced turbulence on bed load transport near critical conditions.

2.3 Bed Load Transport Simulations For Wave-Current Flow Conditions

In coastal environments, the presence of waves combined with currents enhances the bed shear stress and thus the transportation of sediment compared to the equivalent current alone situation. The van Rijn [3] bed load model TRANSPOR includes the enhancements in the mean bed shear stress due to the presence of the waves and additionally includes the temporal oscillatory-induced shear stresses and resulting bed load transport. The model calculates the instantaneous wave-current direction and allows for asymmetry induced in the oscillatory shear stress caused by the presence of the currents. In the present work the rough bed wave-current enhanced random shear stress fluctuations are added to the model for both the bed shear stress and the bed load transport calculations. To generate the RMS random shear stress component the mean wave-current shear velocity is used with the x-z and y-z RMS shear stress components in eqn (9). At each instant the instantaneous wave-current bed shear stress direction, $\phi_{wc}(t)$, is used to generate the RMS value of the random shear stress component in that direction by solving the following eqn of an ellipse

$$\tau_{\phi}{}'_{rms} = \frac{1}{\sqrt{\dfrac{1}{\tau_{xz}{}'_{rms}}\dfrac{1}{1+\tan^2(\phi_{wc})} + \dfrac{1}{\tau_{yz}{}'_{rms}}\dfrac{\tan^2(\phi_{wc})}{1+\tan^2(\phi_{wc})}}} \quad (14)$$

Eqn (14) is used together with the PDF shown in figure (1b) to generate the random bed shear stress fluctuations at 10 Hz over 5000 wave cycles. Using the times series approach for solution, plots corresponding to figure (2) were obtained for wave-current conditions, where the instantaneous shear stresses were composed of a mean, oscillatory and random component and the resulting instantaneous bed load transport rates were calculated using the method already described for steady flows using an eqn modified from eqn (10) and resolved in the mean flow direction as described in van Rijn [3].

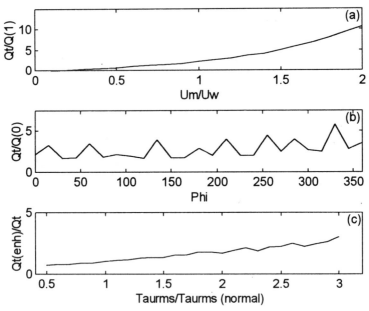

Figure 5: Wave-current plots of dimensionless bed load transport rate as a function of (a) Um/Uw, (b) ϕ_{wc} and (c) RMS bed shear stress divided by the RMS bed shear stress for normal conditions.

In wave-current conditions there are a large number of parameters which influence the bed load transport rates. Thus for the present work a default set of values for these parameters are set. These are depth-averaged mean velocity, $Um = 1$ m/s, peak orbital velocity at top of wave boundary layer, $Uw = 1$ m/s, $\phi_{wc} = 0$, $D_{50} = 0.3$ mm, the near bed wave period, $T = 8$ s, the ripple height was set at 0.05 m and the ripple wavelength = 0.4 m. Only the bed load transport component is examined in the present work. The parameters Um/Uw and ϕ_{wc} were individually varied to examine the corresponding dependency of the non-dimensional bed load transport rates. Further the magnitude of the bed RMS shear stress was also varied to obtain results corresponding to figure (3). The results of the simulation are shown in figure (5). Figure (5a) shows the total mean bed load transport (with random fluctuations) normalised by the mean bed load transport rate for default conditions without random fluctuations, as a function of Um/Uw, (5b) uses the same bed load normalisation to show the effect of changes in ϕ_{wc} and (5c) shows the mean total bed load transport rate normalised by the mean total transport rate for normal turbulence levels as a function of the RMS shear stress divided by the RMS shear stress for normal turbulence levels. Figure (5a) demonstrates that the total bed load transport is sensitive to the ratio Um/Uw, while (5b) shows that the dependence on ϕ_{wc} is less strong. Figure (5b) also more clearly demonstrates that the inclusion of the random shear stress fluctuations significantly increases the magnitude of the mean bed load transport rates. Finally figure (5c) demonstrates that the effect of enhancing the RMS bed shear stress ratio is similar to the case in rough bed steady flows shown in figure (3) with a factor of 3 increase in the RMS shear stress corresponding to approximately a factor of 3 increase in the non-dimensional bed load transport rate. Second order regression through the data in figure (5c) suggest the following relationship under these particular conditions

$$\frac{Q_{enhanced}}{Q} = 0.37 + 0.56n + 0.08n^2 \qquad (15)$$

Conclusions

A method is demonstrated to estimate the root mean squared (RMS) bed shear stresses for hydraulically rough beds. The present work suggests that reduced or enhanced shear stress fluctuations can significantly affect the magnitude of the temporally averaged bed load transport rates for both hydraulically smooth and rough beds, particularly near initiation of sediment motion. The addition of random turbulent shear stress fluctuations in an existing wave-current model has a significant effect on the magnitude of bed load transport rates.

Acknowledgements

The work was funded as part of the EU MAST III project SCARCOST, by the Commission of the European Communities Directorate General for Science and Education, Research and Development under contract number MAS3-CT97-0097. The authors thank R.G. Tickell for advice on random series analysis.

References

1. de Ruiter, J.C.C. The mechanism of sediment transport on bed forms. *Euromech* 156, 137-142, 1982.

2. Grass, A.J. Initial instability of fine bed sand. *Journal of the Hydraulic Division Proc. Of the American Soc. Of Civil Eng.*, HY 3, 619-632, 1970.

3. van Rijn, L.C. Principles of sediment transport in rivers, estuaries and coastal seas. *Aqua Publications*, 1993.

4. Nezu, I. and Nakagawa, H. Turbulence in open-channel flows. *A.A. Balkema, Rotterdam, Netherlands*, 1993.

5. van Rijn, L.C. Sediment transport: part I: bed load transport. *Journal of Hydraulic Engineering*, 110(10), 1431-1456, 1984.

6. Fredsoe, J. and Rasmussen, P. Inertia of bed load in oscillatory flow. *ISVA Progress Report* 51, 29-35, 1980.

Liquefaction analysis using a three-dimensional numerical model for the determination of pore pressure in a seabed

C. O'Donnell, B.A. O'Connor & K.H.M. Ali
Department of Civil Engineering, University of Liverpool, UK

Abstract

A new flexible three-dimensional numerical pore pressure model is developed and used to assess the liquefaction potential at a field site susceptible to scour. The model is tested against a more complex analytical solution and found to give good results. The field test shows that partial saturation has a significant effect on liquefaction potential and that failure depths are similar to erosion depths reported at the site.

1 Introduction

Determination of pore pressure under wave conditions is an increasingly important aspect of seabed stability analysis in offshore engineering. Phenomena such as wave-induced liquefaction should be accounted for, especially where enhanced wave conditions exist as in the vicinity of marine structures. Various theoretical methods exist ranging from simple two-dimensional steady-flow approaches to complex analytical three-dimensional models. The aim of this paper is to present a simple general three-dimensional numerical model that can be adapted for a range of coastal situations without the need for excessive numerical manipulation.

2 Numerical Theory

Evaluation of wave-induced pore pressure is usually based upon some assumptions regarding the compressibility of the soil skeleton and pore water. Theories by Putnam [1] and Sleath [2] assumed that both the bed and pore water are incompressible and, using Darcy's law, presented two-dimensional solutions of the Laplace equation for steady flow through an isotropic bed:-

$$\frac{\partial^2 p}{\partial x^2} + \frac{\partial^2 p}{\partial z^2} = 0 \qquad (1)$$

where p is the pore pressure and x and z are the two co-ordinate directions.

Terzaghi [3] accounted for bed skeleton compressibility in his classic one-dimensional consolidation theory.

$$\frac{k_z}{\gamma_w} \frac{\partial^2 p}{\partial z^2} = m_v \frac{\partial p}{\partial t} \qquad (2)$$

where k_z is the vertical permeability, γ_w is the unit weight of water, m_v is the coefficient of volumetric compressibility and t is time. O'Connor and Clarke [4] obtained accurate pore pressure results from a two-dimensional expansion of eqn(2) which was based on the original one-dimensional assumptions of Terzaghi's original theory.

The flow theory of Jacob [5] also allows consideration of material compressibility. His governing equation for one-dimensional flow is similar to that of Terzaghi:-

$$\frac{k_z}{\gamma_w} \frac{\partial^2 p}{\partial z^2} = \left(m_v + \eta\beta \right) \frac{\partial p}{\partial t} \qquad (3)$$

with η is the porosity and β is the compressibility of the pore water.

A realistic three-dimensional expansion of Terzaghi's original one-dimensional consolidation theory was presented by Biot [6]. His general theory allows the compressibility of the pore water and the bed skeleton to be taken into account. The relevant governing equation for isotropic conditions can be written (Verruijt [7]):-

$$\frac{k_z}{\gamma_w} \left(\frac{\partial^2 p}{\partial x^2} + \frac{\partial^2 p}{\partial y^2} + \frac{\partial^2 p}{\partial z^2} \right) = \frac{\partial e}{\partial t} + \eta\beta \frac{\partial p}{\partial t} \qquad (4)$$

where e is the volumetric strain. This method provides consistency between the flow of water and the stress response of the bed skeleton in three dimensions.

It will be assumed in the present study that the soil skeleton and pore water are compressible and hence that the flow into a volume of soil is not necessarily the same as the flow out of the same volume. The flow theory of Jacob [5] will be used and the model accuracy checked with the analytical solution of eqn(4) presented by Hsu et al [8].

2.1 Model Formulation

A finite-difference scheme has been employed to solve the governing differential equation. Central-difference approximations of the second-order spatial derivatives are used:-

$$\frac{\partial^2 p}{\partial x^2} = \frac{p_{i-1,j,k}^n - 2p_{i,j,k}^n + p_{i+1,j,k}^n}{\Delta x^2} \tag{5}$$

$$\frac{\partial^2 p}{\partial y^2} = \frac{p_{i,j-1,k}^n - 2p_{i,j,k}^n + p_{i,j+1,k}^n}{\Delta y^2} \tag{6}$$

$$\frac{\partial^2 p}{\partial z^2} = \frac{p_{i,j,k-1}^n - 2p_{i,j,k}^n + p_{i,j,k+1}^n}{\Delta z^2} \tag{7}$$

While a forward-difference scheme will be used for the time derivative:-

$$\frac{\partial p}{\partial t} = \frac{1}{\Delta t}\left(p_{i,j,k}^{n+1} - p_{i,j,k}^n\right) \tag{8}$$

where *i, j* and *k* represent nodes along the x, y and z axes respectively, *n* is the current time-step and *n+1* is the future time-step.

These approximations can be substituted into the three-dimensional expansion of eqn (3) which, for non-isotropic conditions, can be written:-

$$k_x \frac{\partial^2 p}{\partial x^2} + k_y \frac{\partial^2 p}{\partial y^2} + k_z \frac{\partial^2 p}{\partial z^2} = \gamma_w\left(m_v + \eta\beta\right)\frac{\partial p}{\partial t} \tag{9}$$

where k_x and k_y represent the permeability in the x and y directions respectively.

The finite-difference approximations described in eqn(5) to eqn(8) can be substituted into eqn(9) which can be rewritten in the form:-

$$p_{i,j,k}^{n+1} = p_{i,j,k}^n + \frac{1}{s_s}\left\{ r_x k_x (p_{i-1,j,k}^n - 2p_{i,j,k}^n + p_{i+1,j,k}^n)\right.$$

$$+ r_y k_y (p_{i,j-1,k}^n - 2p_{i,j,k}^n + p_{i,j+1,k}^n)$$

$$\left. + r_z k_z (p_{i,j,k-1}^n - 2p_{i,j,k}^n + p_{i,j,k+1}^n)\right\} \tag{10}$$

where $s_s = \gamma_w(m_v + \eta\beta)$ and $r_i = \dfrac{\Delta t}{\Delta i^2}$ (i = x, y or z).

Hence, if values of pore pressure are available at time-step n then these, together with suitable boundary conditions, can be used to determine values at time-step $n+1$. The solution domain is of the form of figure 1. Therefore six boundary conditions are required.

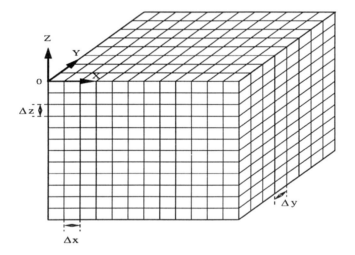

Figure 1: Solution domain

The finite-difference method allows some flexibility in the choice of these boundary conditions with the model being adaptable to the particular situation under examination. The solution domain provides a physical three-dimensional representation of the bed under examination allowing physical boundaries, such as the base of a laboratory flume, to be included.

3 Case Study – Sooyung Seawall

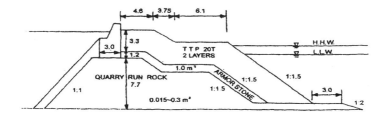

Figure 2: The Sooyung seawall (dimensions are metres)

In order to utilise the model a real-life erosion problem will be simulated. The Sooyung Seawall is located on the coast of Korea. It is predominantly a rubble-mound structure and a serious seabed scour problem has been observed in this area. The numerical model developed above will be used to assess the potential for wave-induced liquefaction at this location. A cross-section of the seawall is shown in Figure 2. The site itself is wave-dominant with a mean Spring tidal range of 0.5m. The seabed around the structure consists mainly of sand with some silt. The mean grain diameter is approximately $300\mu m$. Table 1 shows the wave and bed conditions used in the present study, these being largely based upon previous studies carried out at this location (see Kim et al [9]).

Table 1. Model Parameters

Wave Length L (m)	Wave Period T (s)	Wave Height H (m)	Water Depth d_w (m)	Shear Modulus G (N/m^2)	Bed Thickness d_b (m)	Permeability k_z (m/s)
85	9.9	5.0	8.5	1×10^8	20	9×10^{-4}

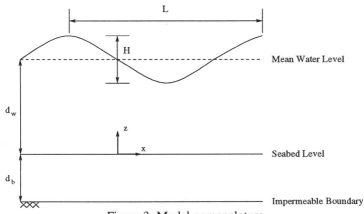

Figure 3: Model nomenclature

3.1 Solution Domain and Boundary Conditions

A solution domain of 61 x 61 x 31 nodes was employed with spatial step distances in all three co-ordinate directions being equal to the bed thickness divided by the number of vertical spatial steps, thirty in this case. Wave reflection was also taken into account at the seabed surface. Hsu et al [8] defined the wave-induced pore pressure on the seabed for a three-dimensional short crested wave system as:-

$$p_o = \frac{\gamma_w H}{2\cosh kd}\cos nky \cos(mkx - \omega t) \tag{11}$$

where p_o is the wave-induced pore pressure at the bed, H is the wave height, k is the wave number ($= \frac{2\pi}{L}$, where L is the wavelength), d is the water depth, x and y are distances in the xy-plane, ω is the angular frequency ($= \frac{2\pi}{T}$, where T is the wave period) and m and n are factors related to the angle of obliquety of the wave, where:-

$$n = \cos\theta \quad \text{and} \quad m = \sin\theta \tag{12}$$

where θ is the angle between the wave-orthogonal and the normal to the wall, as shown in figure 4.

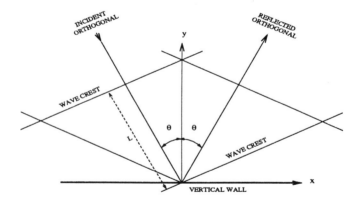

Figure 4: Short-crested wave system

The attenuation of pore pressure at the base of the solution domain will be assumed to be negligible and the following boundary condition is applied:-

$$\frac{\partial p}{\partial z} = 0 \quad \text{at} \quad z = -d_b \tag{13}$$

Due to the sinusoidal nature of the wave loading a continuation-of-gradient method will be used for the remaining lateral boundaries, where:-

$$\frac{\partial^3 p}{\partial x^3} = \frac{\partial^3 p}{\partial y^3} = 0 \tag{14}$$

The bed is assumed to be isotropic with the permeability being equal in all three co-ordinate directions.   The remaining significant unknown is the compressibility of the pore water. The influence of this on bed stability will be examined herein. Liquefaction will be assessed using the three-dimensional expansion of the excess pore pressure criterion of Zen and Yamazaki [10] where liquefaction is assumed to have occurred when the excess pore pressure is greater than the effective stress of the bed. This was presented by Hsu et al [11], as:-

$$-\frac{1}{3}\left(1 + 2K_o\right)\left(\gamma_s - \gamma_w\right)z + \left(p_0 - p\right) \leq 0 \tag{15}$$

where K_o represents the earth coefficient at rest, γ_s is the specific unit weight of the sediment, z is the vertical distance measured positively upwards from the seabed surface and $-(p_b - p)$ is the 'excess' pore pressure. The earth coefficient at rest represents the ratio of the principal effective stresses σ_1' and σ_3' in a soil undisturbed by any forces other than its own weight and is related to Poisson's ratio μ by:-

$$K_0 = \frac{\mu}{1-\mu} \tag{16}$$

4 Results

The model was run for a range of saturation values. Figure 5 shows the variation in pore pressure profiles measured at the centre of the seabed layer for three conditions. It is clear that the attenuation of pore pressure, and hence the bed stability, is significantly influenced by the degree of saturation.

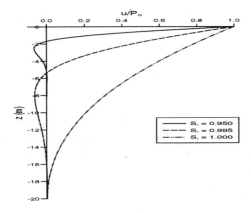

Figure 5: Attenuation of pore pressure for various degrees of saturation using the numerical model.

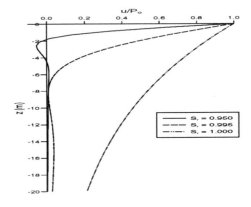

Figure 6: Attenuation of pore pressure for various degrees of saturation using the Biot model for an infinite bed (Hsu et al [8]).

Comparison of the two solution methods shows that they both produce similar results, especially near the seabed surface where scour will initially occur. The presence of as little as 0.5% gas in the pore water leads to significant phase lag. As this lag increases so does the 'excess' pore pressure under a wave trough, leading to a greater potential for liquefaction. The two sets of results for full saturation become more divergent with depth but this is possibly a consequence of the boundary condition selection at the bottom of the solution domain with the analytical solution assuming an infinite depth of deposit. In order to assess the liquefaction potential the effective stress was determined throughout a wave period for the case of $s_r = 0.950$ using eqn(13). Results are shown in figure 7.

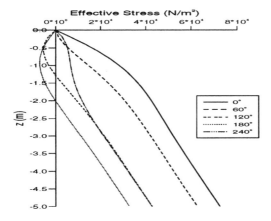

Figure 7: Variation in effective stress over a wave period.
$0°$ represents the wave crest.

Under the wave trough, when the pore water is flowing *out* of the bed, the effective stress falls to a value below zero, indicating that the bed has failed and the sediment has become liquefied. Laboratory tests (not shown here) by the authors as part of the present research programme indicate enhanced bedload transport as critical conditions are approached. The maximum theoretical depth for this failure is approximately 2m, similar to the reported erosion depth at the site (see Kim et al [9]).

5 Conclusions

The numerical model developed in this paper has been used to examine the liquefaction potential of the seabed in the vicinity of the Sooyung seawall off the Korean coast. Results from this model would suggest that the existence of gas within the pore water has a significant effect on the pore pressure distribution and subsequent effective stress profiles. Seabed liquefaction may be of significance in the erosion process where this partial saturation exists.

The accuracy of the model has been assessed by comparison with an analytical three-dimensional solution of Biot's general consolidation theory (Biot [6]) from Hsu et al [8]. Results are similar, especially for partially saturated conditions, indicating that the numerical model can adequately describe the behaviour of a non-saturated seabed under the effect of wave action. The numerical nature of this new model means, however, that it can be used for a much wider range of conditions through the use of suitable boundary conditions.

Acknowledgements

This research has been funded by a grant provided by the EPSRC and this study is partially supported by the Commision of the European Communities, Directorate-General XII for Science, Research and Development Program Marine Science and Technology (MAST III) Contract No. MAST3-CT97-0097, Scour Around Coastal Structures (SCARCOST).

References

[1] Putnam, J.A. Loss of wave energy due to percolation in a permeable sea bottom, *Trans. Amer. Geophys. Union*, **30(3)**, pp. 349-456, 1949

[2] Sleath, J.F.A. Wave-induced pressures in beds of sand, *Journal of the Hydraulics Div.*, ASCE, **96(HY2)**, pp. 367-378, 1970

[3] Terzaghi, K. *Theoretical Soil Mechanics*, John Wiley and Sons, Inc. 1943

[4] O'Connor, B.A. and Clarke, C.S.J. The dishpan scour problem, *3rd Indian Conf. Offshore Engng*, 2, pp. 1-12, 1986

[5] Jacob, C. On the flow of water in an elastic artesian aquifer, *Trans. Amer. Geophys. Union*, **21**, pp. 574-586, 1940

[6] Biot, M.A. General theory of three-dimensional consolidation, *Journal of Applied Physics*, **12**, pp. 155-164, 1941

[7] Verruijt, A. Elastic storage of aquifers (Chapter 8). *Flow Through Porous Media*, ed. R.J.M. De Wiest, Academic Press: New York and London, pp. 331-376, 1969

[8] Hsu, J.R.C, Jeng, D.S. and Tsai, C.P. Short-crested wave-induced soil response in a porous seabed of infinite thickness, *International Journal for Numerical and Analytical Methods in Geomechanics*, **17**, pp. 553-576, 1993

[9] Kim, H., O'Connor, B.A., Hwang, K-N, Lee, T.H and Kim, T.H. Modelling wave-induced scouring at a seawall front, Sooyung Yacht Centre, Korea, *Proc. of Littoral '98*, Barcelona, Spain, pp. 457-466, 1998

[10] Zen, K., and Yamazaki, H. Oscillatory pore pressure and liquefaction in seabed induced by ocean waves, *Soils and Foundations, Japanese Society of Soil Mechanics and Foundation Engineering*, **30(4)**, pp. 147-161, 1990

[11] Hsu, J.R.C., Jeng, D.S. and Lee, C.P. Oscillatory soil response and liquefaction in an unsaturated layered seabed, *International Journal for Numerical and Analytical Methods in Geomechanics*, **19**, pp. 825-849, 1995

Combatting nutrient spillage in the Archipelago Sea—a model system for coastal management support

H. Lauri[1], H. Ylinen[1], J. Koponen[1], H. Helminen[2] & P. Laihonen[2]
[1] *Environmental Impact Assessment Centre of Finland Ltd, Finland*
[2] *South-West Finland Regional Environment Centre, Finland*

Abstract

The Archipelago Sea is a sea area in the Baltic Sea between Åland and the Finnish mainland strongly influenced by the Baltic Sea forcing. Archipelago Sea waters are relatively clean but threatened from the southern and eastern sides by more nutrient rich waters from the Gulf of Finland and the Baltic Proper, and also from local nutrient loads from Finnish mainland and fish farming. In order to find out the relative impact of different nutrient sources a hydrodynamic model was applied to the area. Modelling the Archipelago Sea required taking into account large scale variations of the Baltic Sea as well as the small scale topography created by the numerous islands, therefore a nested three-dimensional grid with three grid refinement levels was used. Flows in the Archipelago Sea are dominantly wind-induced but affected by salinity and temperature gradients.

Computed flows were compared to flow measurements performed in the Archipelago Sea during the open water periods of 1993 and 1994. These simulated flows were then used to model transport in the area to investigate the effect of local nutrient sources and water exchange between the Archipelago Sea and neighbouring sea areas. Transport computations were further verified using salinity measurements.

In order to make the model more usable a model management system was constructed. The system simplifies model input data handling, model computation, and result visualisation. As a basis of the user interaction a map-based view of the modelled area is shown, including a set of interactive symbols that allow geographic location-based data access and model parameter manipulation.

Introduction

Archipelago Sea is one of the most complicated archipelago areas in the world, numerous small and big islands and rocks make the archipelago highly fractal in character. Large open water area and main channels form the central part of the archipelago that is strongly influenced by the Baltic Sea forcing, while intermediate and inner archipelago areas surround the central part on the Åland Island and Finnish mainland sides. Inner archipelago areas are to a large extent isolated from the main Baltic Sea influence, storm events, however, can be important for the exchange of the inner archipelago waters.

Figure 1. Archipelago Sea

The Archipelago Sea is affected by nutrient transport from the surrounding seas, mainly Baltic Proper, and nutrient loads from land runoff and point sources, and internal bottom load from sediments. Some areas of the Archipelago Sea suffer from excessive eutrofication demonstrating itself as sliming, extensive growth of filamentous algae, oxygen depletion and algae blooms.

In order to better understand the origin nutrients causing the eutrofication, the influence of the surrounding seas to the Archipelago sea was investigated by computing the water flows in the area and water exchange with the neighbouring seas. The goal was to achieve information on the relative importance of local nutrient sources compared to long distance transport (Helminen 1998). The flow computations were verified using set of flow measurements performed during open water period in several locations on years 1993 and 1994.

In addition to computation results a modelling support system was built to enable generation and investigation of computation results by the end users, that is, the local environmental office. The complete modelling system consist of a hydrodynamic and water quality models, and a model management system helping users in the complex task of input data management, model running and result visualisation (Lauri 1998).

Computational model

The hydrodynamic model used is a 3D primitive equation, z-coordinate, free surface model, with constant eddy viscosity coefficients (Koponen 1992, Simons 1980). The model grid covers the whole Baltic Sea with two nestings that locally increase the model resolution. The grid box sizes used are 24, 6 and 1.5 km, where the finest grid covers the Archipelago Sea. Vertically there are ten layers with thickness of one meter at the surface and increasing downwards.

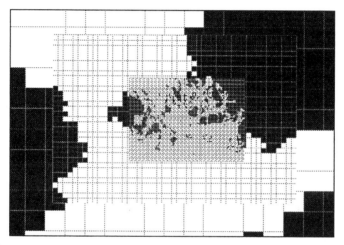

Figure 2. Part of the model grid showing the two finest grid nestings.

As the whole Baltic Sea was modelled there are no open boundaries. The water exchange through the Danish straits and fresh water river inputs were either neglected or taken into account as constant concentrations, and in the model initial state. As the tide in the Baltic Sea is negligible, the main current driving forcing in is wind.

Comparison of computed and measured flows

To investigate model reliability model results were compared to measured flows performed with recording current meters at ten-minute intervals. There were three point measurements from summer −93 and four from summer −94. The measurement points were mostly located on straits on the Archipelago Sea. The comparison between computed and measured flows were performed using along the strait flow component, typically the cross-strait flow components were significantly smaller in magnitude.

Goodness of fit of computed flow to measured flow was estimated using goodness of fit (R^2) coefficient. For the two whole measurement periods the R^2 varied from negative values to 0.51 at different points. Below a figure showing comparisons performed using a moving one-week time range for year –94 data, from which periods of good and bad model fit can be identified. As a reference the lighter colored line in the figures identifies the values of R^2 achieved when using a constant zero as the computed flow.

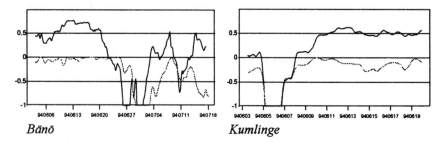

Bänö   *Kumlinge*

Figure 3. Time-dependent R2 coefficient between computed and measured flows

The model performance goes in some cases below the reference line. However, in this kind of comparison a slight phase error may cause poor values for the R^2 coefficient. Also this comparison was done with measured point flow data, while the model always computes average flow of 1.5 x 1.5 km grid box. Generally the modelled results compared to measured values rather well, and in some cases the model fit could be called good, taking into account the above limitations.

Water exchange

Water exchange was computed using an initial state where Baltic, Gulf of Finland, Bothnian Bay and Archipelago Sea were marked individually. The movement of these waters was then followed for eight months in different years. The water exchange varied greatly between different years, mainly due to strong winds occurring most often during spring and autumn. A difference between outer (left) and inner archipelago is demonstrated in the figure below.

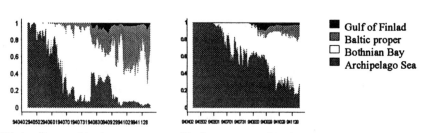

Kihti, outer archipelago   *Turku, inner archipelago*

Figure 4. Water exchange timeseries for eight-month simulation.

Nutrient dispersion

To find out the relative importance of local nutrient sources in different areas of Archipelago Sea the local nutrient sources were mapped and divided into three main categories: 1. loads from Finnish shoreline including river discharges as well as municipal and industrial sources, 2. Loads in eastern archipelago including mostly fish farming, and 3. Loads in Ahvenanmaa archipelago. In the figure below the load locations are shown on the map. Distribution of nutrients was then estimated by letting the released tracker spread with the computed flow and plotting timeseries of tracker concentrations on given locations for the eight month simulation period. A clear distinction can be seen between the inner archipelago, where local source are significant due to smaller water exchange and large loads from shore, and the outer archipelago, where local loading disperse quickly.

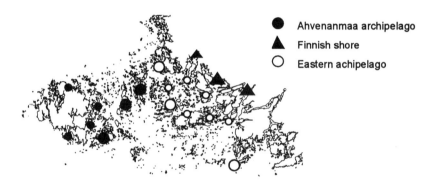

Point loads on the Finnish shore, Eastern and Ahvenanmaa archipelago

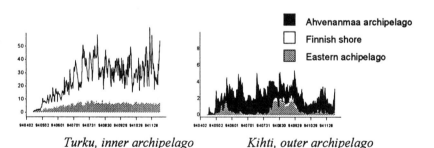

Turku, inner archipelago   *Kihti, outer archipelago*

Figure 5. Simulated relative active phosphorous fractions from different load areas in two points on Archipelago Sea during eight-month simulation.

Model management system

The Archipelago Sea model management system to integrates a set of existing modelling and data processing software tools into a modelling system, and provides a basis for further development of modelling tools. The integration was done by providing a common base for starting individual tools and sharing data. Different user needs from model result viewing to model application building can be supported by providing different tool sets for different modelling tasks.

A common software tool launching system was implemented by using a standard operating system file browser and file system. The user sees a list of model related files, where double clicking a file will start a corresponding modelling tool. The tool launch environment consist thus a set of file type - software tool associations, which can be modified according the user needs. Additionally the model related files are associated to a specific model run, for example, information about the model parameters used computation is available to visualisation tool. A database server for holding model run information is being developed, which will further increase the possibilities to share model information between different modelling tools.

Below a figure of the system structure and a list of modelling tasks and related software tools.

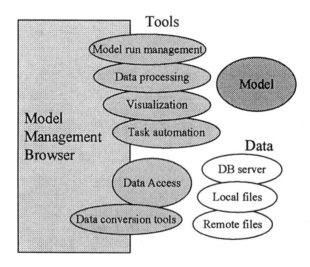

Figure 6. Model management system structure

System management
• Model data browser (operating system file browser used)
Input data handling
• Model grid generation tool
• Timeseries import tool

• Data conversion tools
Model run management:
• Model run management tool
• Computational model
Data visualisation & reporting
• Flow and scalar field visualisation tool
• Timeseries data visualisation tool
• Report generator

The above set of tools consists mostly of custom software implemented using a GUI development system. Currently the system works a stand-alone version on a workstation.

Model run management and 2d field visualisation tools

The model run management tool is used to manage model input data and parameters. It is based on following ideas :

• The main window contains a view of the model grid, a toolbar and a menu.
• The window menu contains the following menu items: File, View, Source data, Computation, and Results.
• Model simulation parameter sets can loaded and stored to files similarly to text documents. Each parameter set contains all model input parameters and access information of the timeseries input data and result files.
• Model input data and suitable model parameters are shown on the model grid as symbols, where each symbol can be used to access corresponding information. Same data is also accessible through menu selections.
• Model parameters are set using dialog boxes accessible through the computation menu or map symbols. There are automated routines and help available for setting the parameters.
• Timeseries measurement data exists separately, and does not depend on the model cases. Model can use this data, for example, for input boundary conditions.
• Result data, such as timeseries and 2d fields, is accessible within the model run manager can be easily visualised and exported to documents.

Below in a figure 7 the main window of the water quality model with a zoomed view into the Archipelago Sea grid.

A 2d-field visualisation tool is used to inspect 2d and slices of 3d fields. Figure 8 shows the main window of this tool. Similarly to GIS programs a concept of data layers is used. The main window consists of a picture area for displaying data layers and a list of layers in a box on the left part of the window, through which layers can be added, removed and selected independently. Cursor coordinates and corresponding data values are displayed in the toolbar. Other miscellaneous data such as animation time, depth layer and length scale is shown in the left part of the window.

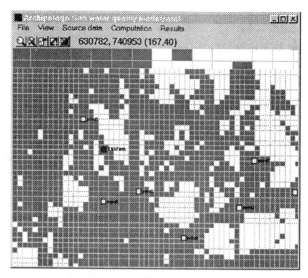

Figure 7. Water quality model main window

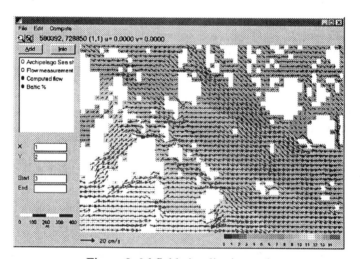

Figure 8. 2d field visualisation tool

Typical tasks possible using the 2d visualisation tool include
- Displaying map information from the model area.
- Displaying measurement data from the model area.
- Displaying selected model results, for example, as timeseries, 2d fields, or 2d animations.
- Comparing modelled and measured data using visual comparison and statistical analyses.

- Computing statistical information from the model and measurements data.
- Computation of differences between two fields.
- Exporting modelled and measured data as numbers or figures for further processing.

Conclusions

A hydrodynamic model was applied with a 1.5 km resolution to the Archipelago Sea area. Comparisons with measured currents were used to calibrate the model and also to verify the model results. Calibrated currents were then applied to water exchange and nutrient dispersion computations. Further use of the model will include applications using more refined scale in specific parts of the Archipelago Sea.

The model results were transferred to end users in the form of an interactive modelling system. This way the computation results could be investigated for any given time period and location with given wind and boundary condition data. The model is being taken into use but for now not much data is available about the usability of the system. Further development of the system will include streamlining the system user interface, further integration with databases and a toolkit usable though a web browser.

Acknowledgements

The work here was partly funded by the European Commission under the DGXII MAST Programme (MAS3-CT97-0089).

References

[1] Helminen, H., Juntura E., Koponen J. and Ylinen H. Assessing of Long Distance Background Nutrient Loading to the Archipelago Sea, Northern Baltic with a Hydrodynamic Model. *Environmental Modelling & Software* 13, pp 511-518. Elsevier, 1998.

[2] Koponen, J., Alasaarela, E., Lehtinen, K., Sarkkula, J., Simbierowicz, P., Vepsä, H., Virtanen, M., Modelling the dynamics of a large sea area. *Publications of the Water and Environment Research Institute, 7*. National Board of Waters and Environment: Helsinki, Finland, 1991..

[3] Simons, T.J., Circulation models of lakes and inland seas. *Can. Bull. Fish. Aquat. Sci.* 203, 145, 1980.

[4] Lauri H., Koponen J., Virtanen M., Alasaarela E., Integrated water quality management supported by graphical user interface, *Proc. of the 3rd Int. Conf. On Hydroinformatics*, eds. V.Babovic, L.C.Larsen, Balkema: Rotterdam, pp. 309-314, 1998.

Section 6
Sediment Transport

Three-dimensional model of dispersion of mercury in marine environment

D. Žagar[1], R. Rajar[1], A. Širca[1], M. Horvat[2] & M. Četina[1]
[1] University of Ljubljana, Faculty of Civil and Geodetic Engineering,
Hajdrihova 28, Ljubljana, Slovenia
[2] Josef Stefan Institute, Jamova 2, Ljubljana, Slovenia

Abstract

The Gulf of Trieste is subject to mercury pollution from the River Soča which drains contaminated sediments from the region of a former mercury mine in Idrija, Slovenia. This results in elevated mercury levels in some marine organisms. Due to concern for human health, a study has been undertaken to predict mercury contamination trends through the use of a field program and mathematical modelling. An existing three-dimensional baroclinic mathematical model PCFLOW3D was upgraded to simulate the transport of mercury in its dissolved and particulate form. Hydrodynamics due to wind, tidal forcing and river inflow momentum can be simulated and stratified conditions can be taken into account. A large amount of data on temperature and salinity fields and winds in the Gulf of Trieste, River Soča discharges, suspended sediment concentrations and temperatures was collected and interpreted. This data, together with measurements of mercury concentrations in water, suspended sediment, bottom sediment and pore waters in the River Soča and the Gulf of Trieste are used as input for the model. A scenario for long-term simulations on the basis of seasonally averaged parameters and a few shorter inserts of strong wind and high discharges of the River Soča was developed as a substitute for simulations based on annual averaged input data. Measurements and observation data from 1995, 1996 and 1997 were applied to verify and calibrate the PCFLOW3D model. Although some complex mercury transformation processes are not well known and were therefore not taken into account in the simulations, an acceptable qualitative agreement of results and measurements was achieved. Whenever a quantitative comparison was possible, an accordance of measured and computed results within a factor of two was attained.

1 Introduction

The Gulf of Trieste is situated in the eastern part of the Northern Adriatic Sea. It covers an area of about 25 x 30 km (Fig. 1). The average depth is about 16 m and reaches 25 m in the central part. Recent measurements in the Gulf have shown greatly increased mercury concentrations in the water, sediment and some marine organisms. Concentrations in the water and the biota were as much as an order of magnitude higher, and concentrations in the bottom sediment, even as much as two orders higher than the corresponding natural background values. Occasional anoxia at the bottom of the Gulf may increase the methylation of mercury; thus, there is a potential impact on the humans living near the Gulf. Recent studies have shown that the former Idrija Mercury Mine, where mining was active for about 500 years, is the main source of the mercury pollution in the Gulf of Trieste. The suspended and bottom sediment of both rivers, the River Idrijca and the River Soča is highly contaminated with mercury, which is carried away to the Gulf of Trieste.

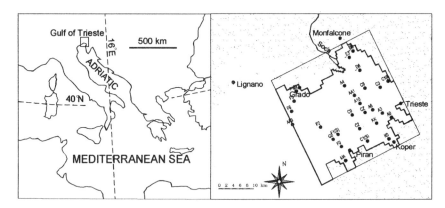

Figure 1: The Gulf of Trieste: location (left) and the extent of the computational domain and measuring points (right).

Even 10 years after the closure of the Idrija Mercury Mine, concentrations in river sediments, water and sediment at the bottom of the Gulf do not show a significant decrease (Horvat et. al. [1], Horvat et. al. [2]). Therefore, extensive research on mercury cycling in the Gulf is in progress. Besides the measurements of physical, chemical and biological parameters, mathematical modelling was also used to simulate mercury cycling in the Gulf of Trieste. A two-dimensional (2D) steady-state model STATRIM was developed first for the simulation of hydrodynamic circulation and mercury transport and fate. Some of the results of the 2D simulations using annually averaged input data are described in Rajar et al. [3] and in Širca and Rajar [4]. The simulation of vertical distribution of parameters was not possible with the 2D model; therefore an existing three-dimensional (3D) unsteady state model PCFLOW3D was upgraded and used to simulate the transport and dispersion of mercury in the Gulf of Trieste. As most of the mercury, flowing to the Gulf, is bound to suspended sediment particles, a new sediment transport module was first developed and included into the model. The basic model, the new sediment transport module and some simulations of transport of suspended sediment are described in detail in Rajar et. al. [5], Rajar et. al. [6] and Rajar and Četina [7].

2 Description of the three-dimensional model

The PCFLOW3D hydrodynamic and transport-dispersion model was developed at the Faculty of Civil and Geodetic Engineering of the University of Ljubljana. It has already been applied to many practical hydrodynamic and pollutant dispersion problems in Slovenia and abroad. It is a non-linear baroclinic model, composed of three modules: a hydrodynamic (HD) module, a transport-dispersion (TD) module, and a recently developed sediment-transport (ST) module. A short description of the modules is given below.

2.1 Hydrodynamic and transport-dispersion modules

The HD and TD modules are both based on the finite volume method; the system of differential equations is solved using a hybrid (central-upwind) implicit scheme. In the horizontal plane, the eddy viscosity and diffusivity are constant, while in the vertical direction the simplified one-equation turbulence model of Koutitas is included. The simulation of some biochemical processes has also been included.

The TD module, which is solved coupled with the HD module, simulates temperature, salinity or any contaminant which can influence water density and, at the same time, the velocity field. Therefore, stratified conditions during the warmer half of the year, as well as density-driven flow can be simulated using the model. The simulation of transport and the dispersion of heat from heat sources and from the atmosphere has also recently been included to enable the simulation of thermal pollution in surface waters. There are two methods of solving the transport equation in the model, an Eulerian finite difference method (FDM) and a Lagrangean particle tracking method (PTM). Each of them has its benefits as well as its deficiencies. The FDM was used for the mercury transport simulations.

Mercury transformation equations have not yet been included in the 3D model. In the present state, the transport of dissolved and particle-bound mercury in both non-methylated and methylated forms was simulated.

2.2 The sediment-transport module

The sediment transport module is based on the equations of van Rijn [8]. Non-cohesive sediment material can be simulated. The module basically resolves the advection-diffusion equation for suspended sediment concentration, where the empirical equation for the sedimentation velocity (described by van Rijn [8]) of the particles is accounted for. As the bottom boundary condition, resuspension or settling of suspended sediment which depends on the bottom shear stress caused by current velocities (result of the HD module) and wave parameters is calculated. The mass conservation equation for the sediment is used to calculate erosion/deposition thickness at the bottom.

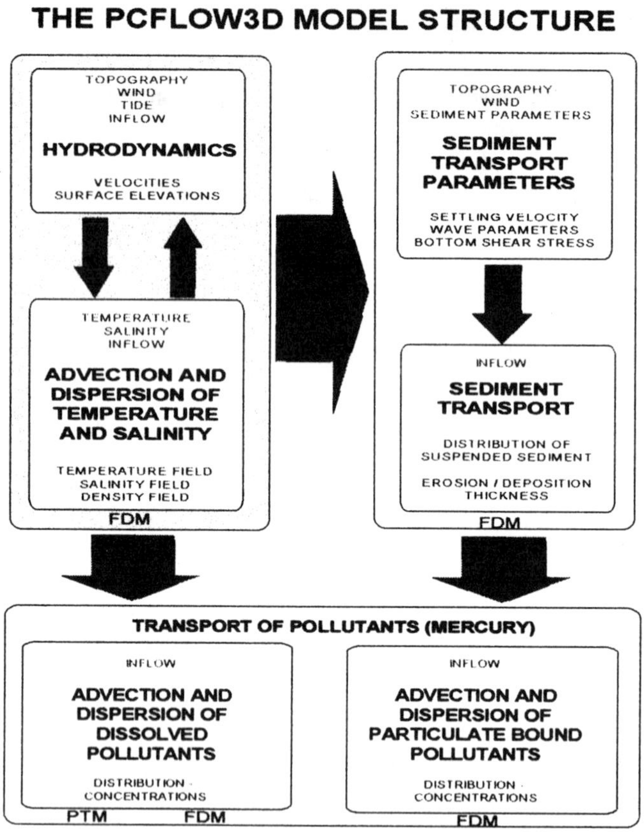

THE PCFLOW3D MODEL STRUCTURE

Figure 2: Flow-chart of the PCFLOW3D model.

3 Long-term simulations

3.1 The basic principle

The unsteady state 3D model needs a very large amount of input data to work properly. In a relatively large area, such as the Gulf of Trieste, it is very difficult to measure all the parameters simultaneously in enough sampling points.

Usually steady state simulations with time-averaged input data are sufficient for the modelling of long-term processes, while real-time simulations over the short time periods of the measurements must be performed to calibrate and verify the model. However, mercury transport and dispersion was found to be a highly unsteady state process. It is known that over 90 % of the annual inflow of suspended sediment and mercury are flushed into the Gulf with two flood waves of the River Soča, usually during spring and autumn rains. It is also known that strong wind is the most important cause of the transport processes in the Gulf. This latest phenomenon mostly occurs during the winter months between November and February.

Despite numerous measurements of different parameters, it was not possible to perform fully unsteady state simulations over several months; therefore, a new approach was used. Annually averaged input data were replaced with seasonally averaged input data. Four main seasons, more or less identical to the calendar seasons, were accounted for. A few inserts of strong wind and the River Soča flood-peaks, which statistically occur with high probability at approximately the same time every year, were added to the main seasons. These inserts are of great importance, as they represent a significant step forward in comparison with the previously performed steady state modelling.

Furthermore, real-time modelling was replaced by the quasi-steady state principle. Here, an unsteady state simulation is performed for a certain period of time after input parameters have been changed significantly. Afterwards, when the hydrodynamic parameters and temperature and salinity fields are stabilised, they are treated as fixed, and only the transport of the pollutant is further calculated. In this way the computational time is also essentially reduced, as about 80 % of total computational time needed is used for the hydrodynamics computation when density driven flow and stratified conditions are taken into account.

Finally, a typical year was partitioned into 12 sequences (Fig. 3), which were simulated successively. With the shorter (up to a few days long) sequences, unsteady state calculations were used, while with the longer (a few weeks to a few months long), the quasi-steady state principle, as described above, was used.

Although most of the mercury is transported to the Gulf in its particulate form, more measurements for dissolved mercury were available; therefore, dissolved mercury was used to calibrate the model.

3.2 Input data for the simulations

3.2.1 Wind
Seasonally averaged wind force and direction above the Gulf was evaluated from the official data for the Beli Križ Measuring Station.

Regarding the frequency of occurrence, there are two peaks of the *bora* wind (NE) in a typical year. These two peaks in February and November coincide almost exactly with the wind force peaks and undoubtedly contribute the most to the mercury transport in the Gulf of Trieste.

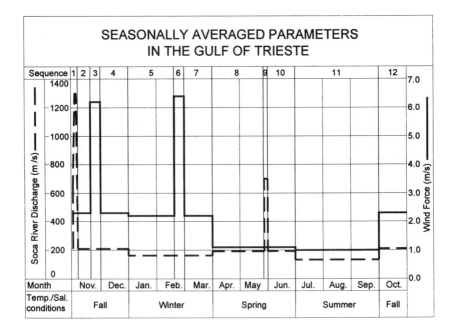

Figure 3: Seasonally averaged parameters in the Gulf of Trieste.

3.2.2 The River Soča

Seasonally averaged discharges, as well as the River Soča flood-peak inserts, are based on measurements in the cross-section at Solkan, near the Slovenian-Italian border (Table 1). Measurements of discharges downstream of Solkan are not available. There are another two important tributaries of the River Soča between Solkan and the river mouth. The River Vipava, which mainly flows through Slovenian territory, and is well elaborated; and the River Torre, which represented the most significant unknown. The total catchment area of the River Soča in Slovenia is 2235 km², while the catchment area in Italy is 1065 km². The mean discharge at the river mouth is, therefore, considered to be equal to 150 % of the sum of the mean discharges of the River Soča at Solkan and the Vipava River at the confluence with the River Soča. The annually averaged discharge of the River Soča at its mouth (168 m³/s) can be compared to the values of other authors, which are between 165 m³/s (Mosetti, [9]) and 172 m³/s (Benini, [10]).

The same relationship between discharges as described above was used to determine discharges during the flood-peak inserts. The intensity of the inserts was evaluated from the mean high discharges. At Solkan, between the years 1926 and 1975, mean high discharges have two peaks, which occur in May and November respectively. The recurrence of a five day long insert with a 715 m³/s discharge is two years; therefore, the length of the spring-insert was set to five days. The autumn-insert is shorter, as the recurrence of a five day long insert with 1370 m³/s discharge is about 50 years. A two day long autumnal insert was adopted. A discharge of 345 m³/s for a day before and a day after were added to the two day long autumn insert.

Table 1: Monthly and seasonally averaged discharges of the River Soča

Month	Discharge				
	River Soča [m³/s]	Vipava River [m³/s]	Σ Soča + Vipava [m³/s]	The River Soča mouth [m³/s]	Seasonally averaged [m³/s]
January	72	22	94	141	
February	70	20	90	135	150
March	94	21	115	173	
April	109	20	129	194	
May	116	16	132	192	190
June	109	13	122	183	
July	69	9	78	117	
August	59	7	66	99	120
September	82	14	96	144	
October	109	20	129	194	
November	144	27	171	257	209
December	94	26	120	180	
Annually averaged	94	18	112	168	168

The water temperature of the River Soča at its mouth is another important factor in the simulation of the stratified conditions. Measurements of water temperature under the last bridge, situated less than one kilometre from the river mouth, were available. Temperature was measured in about two-week intervals between the years 1974 and 1995, most frequently between the years 1978 and 1987. Seasonally averaged water temperatures were statistically evaluated from measurements (Table 2) and used in 3D modelling.

Table 2: Seasonally averaged water temperature in the River Soča

Season	Temperature [°C]
Winter	7.7
Spring	12.9
Summer	16.3
Autumn	9.2

There were several measurements of dissolved mercury in the River Soča available: during the flood-wave in November 1997, the mean low discharge in December, 1998 and the low discharge in October 1997 respectively [Horvat et. al. [1]). In the Gulf of Trieste, measurements were performed in May and

November 1995 (Horvat et. al. [1]). It is evident from the data that the interdependence between the discharge of the River Soča and the concentrations of dissolved mercury is very low. In the River Soča, concentrations vary between the range of 1.6 and 3.5 ng/l, while in the Gulf, near the river mouth (point D6 in Fig. 1), concentrations between 4.5 and 5 ng/l were measured. Higher concentrations in the seawater are due to mercury release from particulate to its dissolved form within the freshwater and saltwater mixing zone. It was not possible to include the process itself in the model; therefore, a mercury concentration of 5 ng/l was taken into account with all simulations in any season.

3.2.3 The Gulf of Trieste
The effect of tide on the circulation in the Gulf was not taken into account with the 3D simulations. Typical tide-induced residual currents in the Gulf were, at least for an order of magnitude smaller than the wind-induced residual currents (Širca & Rajar [4]). Moreover, the error due to tidal forcing exclusion is significantly exceeded by the error caused by false diffusion, which can not be avoided using the existing numerical scheme.

Stratified conditions within the Gulf and their impact on the dispersion of mercury along the water column are also a very important factor, particularly during the warmer half of the year. Temperature and salinity distribution along the entire computational domain had to be evaluated for the four main seasons. Measurements were performed simultaneously in 27 sampling points within the Gulf (Fig. 1), in the middle of each main season (February, May, August, November) in 5-m intervals along the depth. Afterwards, temperature and salinity in all layers were interpolated from the data measured at each sampling point. From these values, the temperature and salinity distribution along the entire computational domain for the four main seasons was calculated using *QuickSurf* software.

The temperature and salinity matrices calculated in that manner need to be 'smoothed' to avoid stability problems with the numerical scheme during computation. Additionally, with long-term simulations, the final distribution of temperature and salinity would be unreal due to the River Soča inflow, density-driven flows and, most importantly, due to advection. To smoothen the temperature and salinity fields, all forcing factors (wind, river inflow and density-driven flow) were taken into account. The computation was performed for a set amount of time (from 8 hours with strong wind and/or high discharge of the River Soča, up to two days with weak wind and lower discharges), until the expected velocities were reached within the entire computational domain. Such matrices were used to simulate the transport of mercury in its dissolved form within the Gulf of Trieste.

In the bottom sediment of the Gulf, mercury concentrations are highly increased and reach about 25-30 μg/g dry weight near the River Soča mouth. Mercury bound to sediment particles is gradually being released into the pore water and, due to molecular diffusion, also proceeds to the surrounding water. Therefore, pore waters are a significant source of dissolved mercury for a long time after the inflow of mercury has begun to reduce or has even ceased. Mercury concentrations in the pore water of the bottom sediment of the Gulf were determined by an in situ benthic chamber experiment at location AA1 (Fig. 1) during 1995 and 1996 (Covelli et. al., [11]).

These measured values cannot be applied to the whole area of the Gulf, as only one location was observed, due to the difficulty and expense of the experiment. The high amount of mercury being released (according to [11] about 470 kg/year) from the bottom sediment is very important and needs to be taken into account with future modelling and simulations.

3.3 Results of the simulation

Input data for the simulations were prepared and the final partitioning of a typical year was carried out using the collected data (Fig. 3). Fully unsteady state simulations (real-time modelling) were applied with the shorter sequences and quasi-steady state modelling with the longer ones. The duration of unsteady state treatment was different with individual longer sequences, as it depends on the discharge of the River Soča, the wind force and temperature / salinity conditions in the Gulf, respectively. Unsteady state simulations in duration from 8 hours with strong wind and/or high discharge of the River Soča, up to a few days with weak wind and lower discharges were applied.

Measurement of total dissolved mercury was performed on 25 June, 1995 in 14 sampling sites at the surface and at the bottom. Figure 4 shows a comparison between measurements and the results of the model simulation.

4 Conclusions

The following conclusions can be made from the results of the simulation and measurements:

At the surface, a very good qualitative agreement of modelling results and measurements was achieved. Relatively good quantitative agreement, always within a factor of two, was a significant improvement in comparison with the 2D modelling. The dispersion of mercury at the surface was somewhat too high, mostly due to false diffusion, which cannot be avoided using the existing numerical scheme.

In spite of an additional increase of dissolved mercury concentrations in the River Soča due to mercury release from particulate to dissolved form, agreement of the measurements with the simulation is worse near the river mouth. Additional research of the of mercury release from particulate matter to its dissolved form in the freshwater / saltwater mixing zone is needed to include the process in the 3D model.

Agreement in the bottom layer is somewhat worse, but still within a factor of two. However, the reliability of the measurements with such low concentrations is limited by several factors, and the accuracy does not exceed the limits of ± 20 %. By taking into account the unreliability of the analytical methods, uncertainty of the input data and the inaccuracy of the modelling, the agreement of the results and measurements can be considered good.

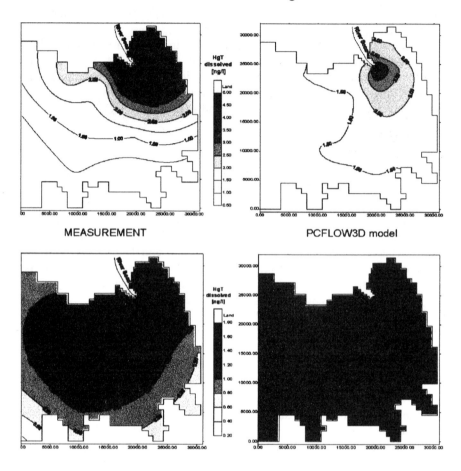

Figure 4: Comparison of measured and simulated concentrations of total dissolved mercury (June, 1995) in the surface layer (above) and the bottom layer (below)

Besides the false diffusion, the results are also influenced by the use of a relatively simple model of turbulence. Despite adapting the eddy diffusivity to stratified conditions, the values of the vertical coefficients are somewhat too high, particularly in weak wind conditions. A two-equation turbulence model (k-ε model) and a numerical scheme of a higher order of accuracy (e.g. *Quickest*) will be included in the model as soon as possible.

The increase of concentrations at the sea-bottom due to benthic fluxes, which can significantly change mercury concentrations within the bottom layer, was not taken into account in the present state of the model.

References

[1] Horvat, M., Covelli, S., Faganeli, J., Logar, M., Mandić, V., Rajar, R., Širca, A., & Žagar, D. Mercury in Contaminated coastal environment; a case study: the Gulf of Trieste. *Science of the Total Environment,* **237/238,** pp. 43-56, 1999.

[2] Horvat, M., Faganeli, J., Planinc, R., Logar, M., Mandić, V., Rajar, R., Širca, A., Žagar, D., & Covelli, S. The impact of mercury mining on the Gulf of Trieste. *Proc. of the 2nd Int. Conf. Coastal Environment,* Cancun, 8-10 Sept. 1998.

[3] Rajar, R., Četina, M., & Širca, A. Hydrodynamic and Water Quality Modelling: Case Studies. *Ecological Modelling,* **101,** pp. 209-228., 1997.

[4] Širca, A., & Rajar, R.: Calibration of a 2D mercury transport and fate model of the Gulf of Trieste. *Proc. of the 4th Int. Conf. Water Pollution 97,* Eds. Rajar, R. and Brebbia, M., Computational Mechanics Publication, Southampton. pp. 503-512. 1997.

[5] Rajar, R., Žagar, D., Širca, A., & Horvat, M. Three-dimensional modelling of mercury cycling in the Gulf of Trieste. *Science of the Total Environment.* Submitted.

[6] Rajar, R., Žagar, D., Širca, A., & Horvat, M. Two- and three-dimensional modelling of mercury transport in the Gulf of Trieste. *Proc. of the 2nd Int. Conf. Coastal Environment,* Cancun, 8-10 Sept. 1998.

[7] Rajar, R., & Četina, M. Hydrodynamic and Water Quality Modelling: An Experience. *Ecological Modelling,* **101,** pp. 195-207, 1997.

[8] van Rijn, L.C. *Principles of Sediment Transport in Rivers, Estuaries and Coastal Sea,* Aqua Publications, Amsterdam, 1993.

[9] Mosetti, F. Sintesi sull'idrologia del Friuli – Venezia Giulia. Quaderni dell'ente tutela pesca – Udine, *Rivista di Limnologia,* **6,** 1983.

[10] Benini, G. *Fiume Isonzo: Atti della Comissione Interministeriale per lo studio della sistemazione idraulica e della difesa del suolo, Vol. II, Parte I,* Roma, Italia, 1974.

[11] Covelli, S., Faganeli, J., Horvat, M., & Brambati, A. (1998) Benthic fluxes of mercury and methylmercury in the Gulf of Trieste. *Estuarine, Coast Shelf Sci,* **48,** pp. 415-428, 1999.

Control and protection of marine environment by Acitrezza town sewage

F. Patania[1], G. Siracusa[2], A. Gagliano[2] & A. Siracusa[2]
Engineering Faculty of Catania University:
[1]*Applied Physic Institute.*
[2]*Physical and Chemical Methodological Department for Engineering*

Abstract

In accordance with Italian legislation about matter, many coastal cities have planned to discharge partly treated sewage into the sea by submarine pipe.
The problem in the case of Acitrezza town sewage became really complex as outfalls of pipes are very close to an attractive wildlife marine reserve called "Cyclops Island", i.e. the mythological land of the giant song by Omerus Odysses.
Owing to the site's peculiarity, it was judged necessary both to investigate the possible dangerous effects arising from sewage and to preserve from irreversible alterations marine environment by monitoring systems.
With regard to the aim of this paper, research was carried out as follows:

- One year of experimental measurements (bathimetry, sea temperature, salinity, velocity vector of marine current, wind direction and velocity, chemical parameters pH, OD, TTS, BOD, COD) have been performed under different meteorological conditions into the investigated area.
- A two-dimensional shallow water finite element model has been used both to solve, by means of finite element methods, vertical-average form of convection-diffusion equations and to forecast the main contaminant concentrations as a function of discharged mass flow, wind velocity vector and biochemical composition of sewage.

From previous procedures it has been possible both to find the critical conditions for contaminant dispersions and the key to control marine conditions to preserve the genuine environment of the wildlife marine reserve.

Introduction

Environmental effects on the sea due to discharge of partly treated sewage have been evaluated by mean of following procedure:
1. survey of chemical composition of sewage
2. survey of physical and chemical characteristic of sea-receiver
3. diffusion of sewage and evaluation of pollutant concentration in the investigated area
4. monitoring system

Investigated area, shown in figure 1 is located in coastal region between Acitrezza and Capo Mulini, presents rectangular of and 8.25 Km^2 surface.

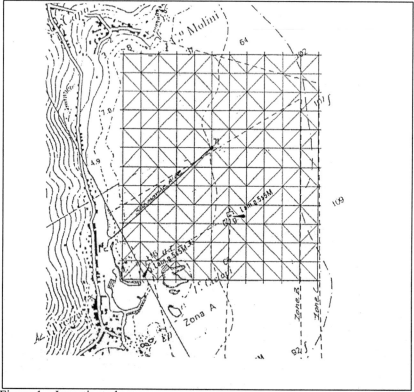

Figure 1 – Investigated area

Sewage composition

Chemical composition of sewage has been defined by mean of experimental measurements carry out by authors during the year 1998.

Table 1 shows both the composition of sewage and the average of pollutant concentration. Results of analysis show that concentrations of pollutant were always lower than law limit [1].

Colour n.p. dil.	Odour	BOD (ppm)	COD (ppm)	Crude materials	Settling Materials (ppm)	Suspended Materials (ppm)	Ammoniacal Nitrogen (ppm)	Nitric Oxide ppm	Nitrous oxide (ppm)	Total Nitrogen
1:20	Bother some	103	2445	absent	2.80	103.8	17.90	2.07	0.32	20.3

Total phosphor (ppm)	Surfactants (ppm)	Test of toxicity percent	Total Coliiformes MPN/100ml	Faecal-Coliiformes MPN/100ml	Faecal-Streptococcus MPN/100ml	Oil and vegetables fats (ppm)	Mineral Oil (ppm)	pH	Arsenic (ppm)	Cadmio (ppm)
6.80	7.45	12.6	45400	20.2	35	1.62	< 0.1	7.70	<0.02	<0.02

Iron (ppm)	Lead (ppm)	Tin (ppm)	Barium (ppm)	Vanadium (ppm)	Boron (ppm)	Chrome (ppm)	Manganese (ppm)	Nickel (ppm)	Copper (ppm)	Zinc (ppm)
0.11	<0.02	<0.02	<0.02	<0.02	1.62	<0.02	0.073	<0.02	<0.02	0.07

Table 1 – composition and average of pollutant concentrations

Physical and chemical characteristic of sea-receiver

The depth of the sea, the marine-currents, the velocity and wind directions and meteorological parameters have been carry out by mean both experimental measurements and bibliography data [1],[2]. Results are shown in Table 2
Physical and chemical characteristics of sea-receiver have been carry out by mean of a campaign of measurements. Measurements have been executed in 16 points located both along perimeters and diagonals of squared area (sides = 700 m) centred respect outfall of submarine pipes as request by Regional Law.

Table 3 shows the average of measurements carried out for points (1, 3, 14, 16) in the vertices of monitored area and for point 9 located in the centre of the area. [1], [3].

Survey day	17/09/1998	21/09/1998	19/10/1998	26/10/1998	5/11/1998
Cloud cover	Cloudy Sky	Scattered Sky	Clear sky	Cloudy Sky	Scattered Sky
Sea – state	Slight	Slight	Smooth	Smooth	Slight
Wind direction	45°	120°	225°	45°	200°
Wind Velocity [m/s]	3,0	1,5	2,0	3,5	7,5

Table 2 – Meteorological and sea conditions.

Sample		1	3	14	16	9
Depth[m]		18.5	45.5			58.5
Visibility[m]		16,0	15.9	16.1	19.3	15.8
Temperature [°C]	Sea floor	23.5	23.14	23.64	23.68	23.46
	Medium	23.7	23.42	23.96	24.04	23.6
	Surface	23.9	23.62	24.16	24.30	24.00
Salinity' [g/Kg]	Sea floor	38,52	38.46	38.53	38.48	38.48
	Medium	38,48	38.52	38.54	38.54	38.50
	Surface	38,52	38.54	38.47	38.55	38.58
.pH	Sea floor	8,36	8.38	8.38	8.4	8.38
	Medium	8,36	8.40	8.36	8.4	8.36
	Surface	8,28	8.38	8.32	8.36	8.38
O.D. [sat. %]	Sea floor	118.6	118.4	118.2	118.8	118.8
	Medium	118.2	118.8	118.4	117.8	118.0
	Surface	119	118.8	118.0	118.2	118.6
Surfactants [ppb]		Absent	Absent	Absent	Absent	Absent
Ammoniacal Nitrogen [ppb]		0,75	0.75	0.75	0.70	0.75
Total Residual Chlorine		Absent	Absent	Absent	Absent	Absent
Sample		1	3	14	16	9
Nitric Oxide [ppb]		0,4	0.36	0.42	0.38	0.38
Nitrous Oxide [ppb]		0,04	0.048	0.04	0.042	0.036
OrthoPhosphates [ppb]		0,086	0.086	0.086	0.088	0.088
Silica [ppb]		7,93	7.95	7.98	7.99	7.95
Oil and vegetal fats [ppb]		0,3	0.28	0.36	0.3	0.32
Total Coli /100 ml		7.4	8	8.4	8	7.8
Faecal Coli / 100 ml		2.6	1.8	1.6	1.8	2
Faecal Streptococcus /100ml		0	0	0	0	0
Chlorophyll α [mg/m^3]		< 1	<1	< 1	<1	<1
Feopigment degr.%		0	0	0	0	0

Table 3 - Physical and chemical characteristics of sea-receiver

Results of measurements allow to observe that initial condition of sea-receiver don't present alarm factors. In fact it presents microbes-charge and feopigment negligible, also nutrients have typical values of not contaminated sea-water.

Estimation of pollutant concentration

AQUASEA computer code have been used in order to calculate time and spatial diffusion of sewage and pollutant concentration. AQUASEA code has been developed by Vatnaskil Consulting Engineers to solve the shallow water flow and transport equations using the Galerkin finite element method.
It consists of two modules: hydrodynamic flow model and transport-dispersion model
The flow model calculates, on the basis of bathymetry, bed resistance coefficients, wind field and boundary conditions, the water level variation and flows
The transport-dispersion model simulates the spreading of a substance into the environment under the influence of the fluid flow and the existing dispersion processes.

Main equations
Equation of continuity [4]

$$\frac{\partial}{\partial x}(uH) + \frac{\partial}{\partial y}(vH) + \frac{\partial \eta}{\partial t} = Q \tag{1.1}$$

Momentum equations [4]

$$\frac{\partial u}{\partial t} + u\frac{\partial u}{\partial x} + v\frac{\partial u}{\partial y} = -g\frac{\partial \eta}{\partial x} + fv - \frac{g}{HC^2}\left(u^2 + v^2\right)^{1/2}u + \frac{k}{H}W_x|W| - \frac{Q}{H}(u - u_0) \tag{1.2}$$

$$\frac{\partial v}{\partial t} + u\frac{\partial v}{\partial x} + v\frac{\partial v}{\partial y} = -g\frac{\partial \eta}{\partial y} + fu - \frac{g}{HC^2}\left(u^2 + v^2\right)^{1/2}v + \frac{k}{H}W_y|W| - \frac{Q}{H}(v - v_0) \tag{1.3}$$

Transport equation [4]

$$\frac{\partial}{\partial x}\left(HD_x\frac{\partial c}{\partial x}\right) + \frac{\partial}{\partial y}\left(HD_y\frac{\partial c}{\partial y}\right) - \frac{\partial}{\partial x}(HcU) = \frac{\partial}{\partial t}(Hc) + S - QC_0 \tag{1.4}$$

where:
u_0, v_0 = velocity of injected water in x and y direction (m/s)
u, v = velocity component in x and y direction (m/s)
C = Chezy bottom friction coefficient ($m^{1/2}$/s)
h = mean water depth (m); η = change in water level (m); H = h+η
Wx, Wy = Wind velocity in x and y direction (m/s); $|W|$ = wind speed (m/s)
Q = injected water (m^3/s)
k = $\rho_a C_D / \rho$ = wind shear stress parameter

f = 2ωsinφ = Coriolis parameters
Dx, Dy = longitudinal and transversal dispersion coefficients (m²/s)
U = velocity within each element taken from the solution of the flow problem
c_o = concentration of the injected water
S = mass flux term (kg/m³).
NUMERICAL MODEL

In AQUASEA code all model equations are approximated using a Galerkin finite element method on triangular elements. Continuous approximations are used for water elevation (h and H) and concentration, temperature or suspended sediments (c), linear within elements, but piecewise constant approximations for the velocities (u and v).

Domain and mesh geometry

The dimensions of domain have been chosen as following:
X direction = 2750 m; Y direction = 3000 m
The grid have constant side with dimension 250 m and the number of cells are 11 in X direction and 12 in Y direction.

Release characteristic
Source of pollution co-ordinates: X=1250 m; Y=1750 m. (node 71)
Amount of constant simulated polluting flow rate: Q = 0.383 m³/s
Submarine cross section = 0.336 m²

Boundary conditions

The domain has been considered with closed boundaries along coastal region and open boundaries in correspondence out to sea where it is possible mass-transport ("flow calculation boundaries"). The following boundary condition have been used:
1) Specified time variation of water level (flow model)
The time variation of the water level has been specified by sinusoidal variation:
h = c + a sin(ω(t + a)),
where c = 0, a = 1, w and a are given constants.
2) Zero concentration gradient, indicating just convective transport of mass through the boundary.

Operative procedure

Several simulations [5] have been performed in order both to expose the most dangerous effects sprang from sewage in the examined area and to verify the code sensitivity caused by variation of input parameters. The contaminant chose for simulation has been lead because it is representative of pollution due to metal compounds. Table n° 4 shows the values of input parameter utilised for each simulation.

Results

Diffusion of pollutant have been calculated in function of the variation of wind velocity and direction, diffusion coefficient and so on. In this way it has been possible to verify that:

a) the increase of wind velocity causes a reduction of contaminant concentration

b) by varying of wind direction (ψ) the more critical condition for marine reserve and coastal region are represented for simulation 1.36 ($\psi = 250°$), that is wind coming from 10° NE.

c) not significant variations have been generated by different flow directions (θ) of sewage discharge

d) by increasing of diffusion coefficients (D_x and D_y) has been originated a growth of the area interested by more big value of contaminant concentration and by simultaneous reduction of contaminant concentration nearby the source. This result permit to observe that the most critical conditions for marine reserve and coastal region have been represented for simulation 1.5.5 ($D_x = D_y = 10$)

e) by decreasing of value of mixing depth the value of contaminant concentration increase

	Paramete r	C [mg/m³]	W [m/s]	φ	θ	D_X [m2/s]	D_Y [m2/s]	ΔH [m]
Sim 1.1	Lead	100	3	0°	0°	1	1	100
Sim 1.2.1	Lead	100	4	0°	0°	1	1	100
Sim 1.2.2	Lead	100	2	0°	0°	1	1	100
Sim 1.3.1	Lead	100	3	30°	0°	1	1	100
Sim 1.3.2	Lead	100	3	45°	0°	1	1	100
Sim 1.3.3	Lead	100	3	60°	0°	1	1	100
Sim 1.3.4	Lead	100	3	90°	0°	1	1	100
Sim 1.3.5	Lead	100	3	100°	0°	1	1	100
Sim 1.3.6	Lead	100	3	250°	0°	1	1	100
Sim 1.3.7	Lead	100	3	270°	0°	1	1	100
Sim 1.3.8	Lead	100	3	315°	0°	1	1	100
Sim 1.4.1	Lead	100	3	0°	45°	1	1	100
Sim 1.4.2	Lead	100	3	0°	90°	1	1	100
Sim 1.4.3	Lead	100	3	0°	135°	1	1	100
Sim 1.5.1	Lead	100	3	0°	0°	2	1	100
Sim 1.5.2	Lead	100	3	0°	0°	1	2	100
Sim 1.5.3	Lead	100	3	0°	0°	2	2	100

Sim 1.5.4	Lead	100	3	0°	0°	3	3	100
Sim 1.5.5	Lead	100	3	0°	0°	10	10	100
Sim 1.6.1	Lead	100	3	0°	0°	1	1	80
Sim 1.6.2	Lead	100	3	0°	0°	1	1	60
Sim 2.1	Lead	100	2	250°	0°	1	1	60
Sim 2.2	BOD	103000	0	250°	0°	1	1	60
Sim 2.3	BOD	80000000	0	250°	0°	1	1	60

Table 4 - Characteristic of simulation

Figures 2, 3, 4, 5, 6, and 7 show diffusion of contaminant for several condition of input parameters.

Fig. 2 - Sim 1.1 (V=3m/s)   Fig. 3 - Sim1.2.1 (V=4 m/s)

Fig. 4 - Sim.1.2.2 (V=2 m/s)   Fig. 5 - Sim.1.3.8 (φ = 315°)

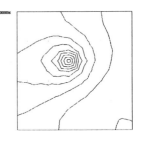

Fig. 6 - Sim 1.5.3 (Dx=Dy=3)   Fig. 7 - Sim 1.5.5 (Dx=Dy=10)

Remote control system

Area interested by sewage discharge will be monitored by mean of continuos surveying of primary parameters that characterizing diffusion of pollutant. Stored information will be elaborated by computer code in order to define corrective intervention to neutralize possible overshooting alarm limit. Figure 8 show the scheme of proposed remote control system.

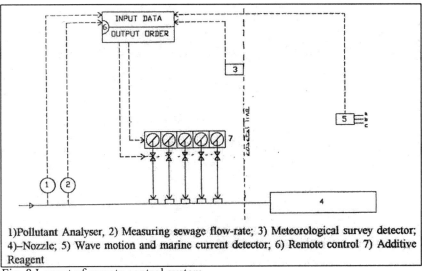

1)Pollutant Analyser, 2) Measuring sewage flow-rate; 3) Meteorological survey detector; 4)–Nozzle; 5) Wave motion and marine current detector; 6) Remote control 7) Additive Reagent

Fig. 8 Layout of remote control system

Conclusion

On the ground of carried out results it has been possible to verify the status of "Cyclopes Island" marine reserve with reference to environmental impact caused by sewage discharge.

Simulations have permitted to define the particular values of wind vector and diffusion coefficients that cause the most unfavourable conditions for sea-water quality into marine reserve, so that in such cases it will be possible to define preventively forecast control techniques.

Results carry out by experimental measurements and numerical simulations have permitted to show following reported:

a) the map of probable diffusion of pollutants inside the investigated area for different meteorological and marine conditions;

b) pollution produced by sewage discharged don't bring significant alteration of marine environment for the totality of examined simulations.

However, it is Authors opinion that it need to built in opportune way remote control system, as like as that one proposed in Fig. 8, to have the possibility to carry out control actions both in the case of meteorological unfavourable conditions and in the case of sewage composition changing in the next future.

References

[1] Cormaci M; Di Geronimo S. *"Studio preliminare sulle caratteristiche del mezzo recettore in corrispondenza di quattro scarichi con condotta sottomarina proposti in progetto per il litorale del comune di Acicastello"*. Catania 1987
[2] Siracusa G, Siracusa A. *" Campagna di indagini marine"* Physical and Chemical Methodological Department for Engineering Catania 1998
[3] Patania F, Siracusa G, Gagliano A. Licitra M. *"Analisi sperimentale, verifica predittiva e tecniche di controllo dell'impatto da condotta sottomarina di scarico sulla componente idrica della riserva marina dei Ciclopi."* Università di Catania – Facoltà di Ingegneria 1999.
[4] Vatnaskil Consulting Engineers " *Aspect of AQUASEA – A software package for two dimensional flow and transport computation"* 1998
[5] Legge Regione Sicilia 15 Maggio 1986 n. 27 *"Disciplina degli scarichi delle pubbliche fognature e degli scarichi degli insediamenti civili che non recapitano in pubbliche fognature"* Palermo 1986.

Sediment transport in the Barra Nova Inlet, Portugal

B.A. O'Connor[1], I.P. Hale[1], S. Pan[1], C.P. Rose[1], J.J. Williams[2],
P.S. Bell[2] & P.D. Thorne[2]
[1]*Department of Civil Engineering, University of Liverpool, United Kingdom*
[2]*Centre for Coastal and Marine Studies, Proudman Oceanographic Laboratory, United Kingdom*

Abstract

The movement of sediment through a newly dredged tidal inlet, the Barra Nova, situated in the Ria Formosa National Park, which is located on the Algarve coast of Portugal, has been studied using Q3D numerical models to simulate hydrodynamic conditions in the vicinity of the inlet. The hydrodynamic model shows that the inlet is ebb-dominant and that wave action is severely depth-limited. Examination of bed sediments show the presence of coarse material of some d_{50} = 600 - 1200µm. Use has been made of Van Rijn's [5] bed load transport approach, which measurements show provides a realistic fit to observations, Williams et al [6], to determine net tidal transport through the inlet as well as the effect of waves. Tidally-induced transport is found to be lower than observed lagoon sediment losses while the presence of wave action at the sides of the inlet reduces the ebb loss of material. However, approximate calculations for a fully ebb-dominant inlet give realistic comparison with lagoon losses based on survey chart analysis and suggests that the hydrodynamics of the inlet has adjusted since its inception towards a more equilibrium state whereby sediment movement through the inlet is more in balance than at earlier times.

1 Introduction

The problems arising from an accelerated rise in sea level due to global warming are particularly important for island communities and those living in low-lying coastal areas. One such site is located in the Ria Formosa National Park in the Algarve region of Portugal, where a series of spits and barrier islands provide protection to a large lagoon area, which itself provides protection to the town of Faro and its international airport, see Figure 1. There is also a significant marine culture industry in the lagoon and the western-most spit, Pria de Faro, contains a large number of holiday homes, tourist restaurants and hotels.

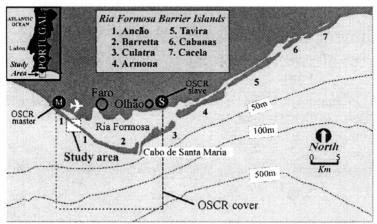

Figure 1: Location of the field site, Algarve, Portugal

In 1997, the western-most natural inlet between the Peninsula do Ancão and the Isle de Barretta became the focus of an international scientific study sponsored by the EU under its Marine Science and Technology Programme (MAST). The three year study, the INDIA Project, aims to obtain a better understanding of interactive coastal processes associated with tidal inlets and is using a combination of computer models and field studies involving state-of-the-art remote sensing equipment, including acoustic and radar equipment, to measure sediment movements, tide, wave and wind-induced currents and wave climate as well as the associated beach and seabed changes, O'Connor et al [1], [2] and Williams et al [3]. The Project contains multi-disciplinary research teams from Australia (James Cook University); France (Bordeaux University); Holland (Amsterdam University); Korea (Kookmin University); Poland (IHE, Gdansk); Portugal (Algarve, Coimbra, IST Universities; LNEC, IH, Hidromod); the UK (BODC, Liverpool, Plymouth, Southampton Universities; CCMS-POL) and the USA (CERC; USC; WHOI) and industrial advisors from the UK Consulting Engineers, Bullen and Partners and Halcrows as well as Valeport Ltd.

Prior to 1997, the Ancão Inlet had been observed to move in an easterly direction downcoast at a rate of some 100m per year under the action of the prevailing south-westerly wave climate. However, in June 1997 the Local Authorities decided to dredge a new inlet, the Barra Nova, at the western end of the Peninsula do Ancão so as to provide easier access to the lagoon and to Faro for local fishing boats, see Figures 2 and 3. The INDIA Project has, therefore, confined its attention to the Barra Nova Inlet, including an intensive field campaign by the Project's field research teams between January and April 1999, including deployment of a large Seacore jack-up barge as a field measurement platform in the inlet itself, see Figure 4.

Figure 2: Barra Nova Inlet, Aug 1997

Figure 3: Dredging Barra Nova Inlet

Topographic and bathymetric surveys of the new inlet following its opening have shown a rapid increase in high water width from some 70m to 240m over the period up to March 1998, although little net down coast movement was found. Results from the intensive field campaign on the 1st February 1999 show a pronounced ebb delta being formed with the axis of the main channel inclined to the general longshore direction of the islands since the opening of the new inlet. Changes between bathymetric charts also suggest a loss of sediment from the lagoon-side vicinity of the inlet.

In order to investigate the possible reasons for a loss of material from the lagoon, a fully-interactive tide and wave computer model system has been applied to the Barra Nova Inlet to simulate flow conditions on the 1st February 1999. Model results have then been used to predict net tidal sediment movements through the inlet. Further details are given below.

Figure 4: Jack-up barge in the inlet

2 Methodology

Using the measured bathymetry for the 1st February 1999 together with offshore tide and wave conditions the Liverpool Q3D tidal coastal model, O'Connor et al [4], was set up to reproduce tide and wave conditions throughout the inlet. The computational domain covered an area of 4 km in the longshore direction by 3.5 km in the cross-shore direction centred at the inlet with a 20 m by 20 m grid size. Tide and incident wave conditions were given at the offshore boundary located at the 18 m water depth contour related to the mean water level. In order to establish a correct water surface slope through the inlet and include the effect of the channels in the lagoon, part of the lagoon was also included in the computational domain. The bathymetry used in the computer model and a typical flow pattern at slack water level of the ebb phase of a 3 m tide are shown in

Figure 5. Time series of the hydrodynamics through the tidal cycles at a number of points across the inlet were extracted from the model for computing sediment transport.

Examination of bed samples taken over the study area show that sediment is generally coarser than some 350μm. Consequently, the predominant mode of sediment transport through the inlet is likely to be as bed load. The tidal velocity model results have been used together with Van Rijn's [5] bed load transport formula, modified for wave action, to calculate the bed load transport both through the tidal cycle and across the width of the inlet. In order to test the accuracy of the bed load equation, use has been made of direct measurements of bed ripple movement at the jack-up barge site within the inlet [3]. Finally, transport conditions have been performed for a series of tide and wave conditions so as to determine the net tidal transport rates through the inlet by relating the flux predictions to the range of tide and wave conditions experienced at the site.

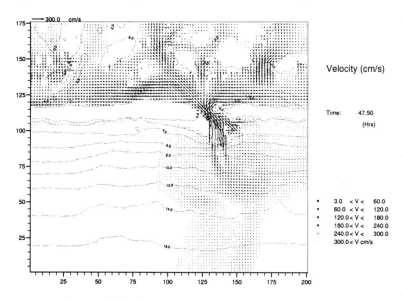

Figure 5: Bathymetry around the Barra Nova Inlet and
flow pattern at slack water of a spring tide

3 Results

3.1 Offshore Wave Conditions

Annual wave records from the nearshore Faro wave buoy, located at the 15m contour, in the vicinity of the inlet were examined to provide an assessment of the incident wave climate. Figure 6 shows the results of this appraisal with two principal incident wave directions being identified; (a) 245 - 275 Deg. (N) and (b) 120 - 140 Deg. (N), of which (a) is more closely defined.

The data corresponding to these two sectors have been analysed separately to assess the range of wave height and wave period conditions that affect the inlet site. Figure 7 show plots of significant wave height (H_s) against wave steepness for each of the wave direction sectors identified above.

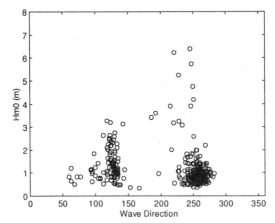

Figure 6: Wave direction scatter plot

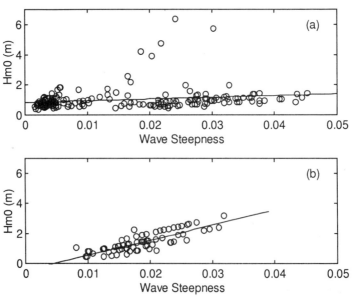

Figure 7: Wave steepness: mean wave height plots

The wave steepness is defined as the wave height divided by the offshore wavelength (L_o) which can be represented as a function of wave period (T):

$$L_o = \frac{gT^2}{2\pi} \qquad (1)$$

A regression line is fitted to the H_s vs (H_s/L_o) plot to enable a wave period to be determined corresponding to any offshore wave height that is selected from either wave direction section.

3.2 Tidal Conditions

An annual tide record for Faro was analysed to provide details of the mean sea level and tidal range. The variation of tidal range between 1.0 m and 3.0 m is shown in Figure 8.

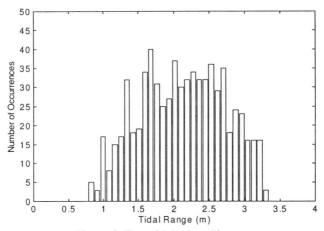

Figure 8: Faro tidal range histogram

3.3 Inlet Conditions

Using Liverpool's Q3D dynamic tidal coastal model the wave and current conditions were determined at the inlet site for a series of different conditions ie. wave height, wave period, wave directions and tidal range. The offshore conditions chosen covered the wide range of possibilities. The modelling provided information of the wave height at a number of locations across the chosen inlet cross section in the throat of the inlet, together with tidal current speed and any wave induced currents. Direction information of the net currents and wave propagation was also output.

The resultant conditions within the inlet showed little variation. Wave conditions were broadly aligned with or against the current through the inlet and the significant wave height was found to generally to be less than 0.3m due to breaking over the ebb delta. Typical maximum depth mean velocities were found to be 2.0 m/s on the flood tide and 2.5 m/s on the ebb tide.

Sediment characteristics across the inlet were determined from sediment grab sample data taken for the whole of the inlet, flood and ebb delta areas. The results for the chosen inlet cross section site are shown in Figure 9, together with the adopted cross section of the inlet bathymetry acquired from hydrographic and topographic surveys.

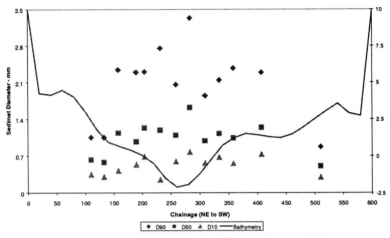

Figure 9: Inlet bathymetry and sediment characteristics

3.4 Sediment Transport Calculations

Predictions of the bed load sediment transport at the inlet site using Van Rijn's bed load transport formula have previously been shown to give reasonable agreement with site observations, Williams[6], being approximately a factor of 1.7 over-estimated, see Figure 10. Van Rijn's formula was adjusted by this factor to simulate sediment transport conditions within the inlet.

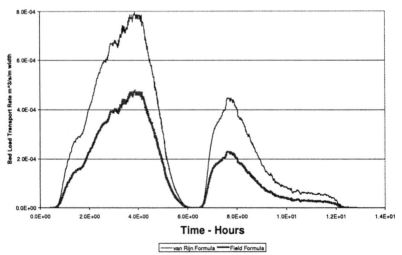

Figure 10: Comparison of bed load transport rate formula

3.5 Tidal Sediment Flux through the Inlet

The tidal sediment flux was determined using the modified Van Rijn sediment transport formula for the range of inlet condition determined as described above. Figure 11 shows the tidal sediment flux through the inlet for a 2m spring tide, negative fluxes indicating seaward direction. The calculated net flux through the inlet for this one tide is −81 m³. By considering the full range of possible tidal condition an net annual flux of −60,320 m³ has been estimated as well as -4,630 m³ for the spring-neap tidal cycle flux. The effect of wave action on this flux rate is to reduce it principally by increasing the net landward movement of sediment at the sides of the inlet as shown in Figure 11.

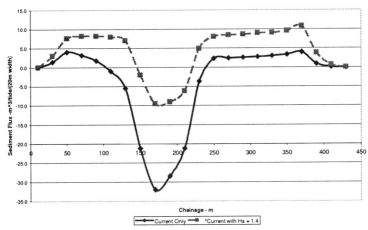

Figure 11: Sediment flux through the inlet for a 3m tide

4 Discussion and Conclusions

Examination of bathymetric charts for the vicinity of the inlet by researchers from the University of the Algarve indicate that there is a likely loss of sediment of some 228,000 - 628,000 m³ over a two year period (O. Ferrierra, personal communication). Clearly, such changes are larger than existing conditions. However, the majority of sediment loss is likely to have occurred in the early stages of the development of the inlet. Using the data of Figure 10 and assuming that the full width of the inlet was ebb dominated gives a net outflow of sediment of some 600 m³/tide or some 235,000 m³/year which is clearly of the same order as the observed values.

The effect of waves is to decrease the ebb predominance of the inlet, see Figure 11, such that over a medium term (say, 5 years from the present) it would be expected that the inlet would eventually achieve a net zero transport so that material transported in by tide and wave action of the flood tides would be balanced by material taken out on the ebb tides. Over longer periods of time, probably of the order of twenty years, the ebb efficiency of the inlet will reduce as it moves down coast and lagoon drainage channels become less hydraulically efficient so that large storms are able to choke the inlet while at the same time beach draw down near Pria de Faro may encourage the establishment of a new inlet.

Acknowledgements

The work was partially supported by the Commission of the European Communities Directorate General for Science and Education, Research and Development under contract number MAS3-CT97-0097. The authors are grateful to Mrs J Price for preparation of the manuscript.

References

[1] O'Connor B. A., Williams J. J., Arens S. M., Davidson M. A., Dias J. M. A., Howa H., Sarmento A. and Voulgaris G. "The INDIA Project", *Third European Marine Science and Technology Conference*, eds. K-G Barthel, H. Barth, M. Bohle-Carbonell, C. Fragakis, E. Lipiatou, P. Martin, G. Ollier, M. Weydert (European Commission, Brussels) pp. 601-609, 1998.
[2] O'Connor B. A., Williams J. J., Dias J. M. A., Collins M., Davidson M. A., Arens S. M., Howa H., Sarmento A. A. J., Seabra-Santos F., Aubrey D., Salles P., Smith J.S., Heron M., Pires H. O., Silva A., Bell P. and Pan S. "Tidal Inlet Monitoring/Modelling Project (INDIA)", *Oceanology Int. 99, Pacific Rim*, Singapore, 27-29 April 1999, pp. 325-335, 1999.
[3] Williams, J. J., Arens, B., Aubrey, D., Bell, P., Bizzaro, A., Collins, M., Davidson, M. A., Dias, J. M. A., Ferreira, O., Heron, M., Howa, H., Hughes, Z., Huntley, D., Jones, M.T., O'Connor, B. A., Pan, S., Sarmento, A., Seabra-Santos, F., Shayler, S., Smith, J.S. and Voulgaris, G. "Inlet Dynamics Initiative: Algarve", Proceedings Coastal Sediments'99, (eds.) N.C. Kraus and W.G. McDougal, ASCE, New York, pp. 612-627, 1999.
[4] O'Connor B. A., Pan S., Nicholson J., MacDonald N. J., and Huntley D. A. "A 2D model of waves and undertow in the surf zone", Coastal Eng. 1998, (ed.) B L Edge, ASCE, New York, pp. 286-296, 1999.
[5] Van Rijn, L. C. *Principles of sediment transport in rivers, estuaries and coastal seas*, Aqua Publications, Amsterdam, The Netherlands, 1993.
[6] Williams J. J., Bell P. S. and Thorne P. D. "Measurement of hydrodynamic conditions, sediment transport and bed morphology in a tidal inlet", to be published at Coastal Environment 2000, WIT, Southampton.

Section 7
Remote Sensing

Environmental management instruments: topics for atmospheric pollution control strategy

M. Lopes, N. Barros & C. Borrego
Department of Environment and Planning, University of Aveiro, Portugal

Abstract

This paper focuses on the analysis of environmental management strategies, covering decision and control, economic and information instruments as well as the fundamental principles of an environmental policy including: the different available tools, their acceptability and applicability, their advantages and disadvantages. Technical measures for control and management are assessed taking into account their implementation, their integration with other measures.

As a study case the analysis of Portuguese atmospheric emissions is presented. Major conclusions are the importance of coastal zone with a contribution of over 80% on total emissions for the most common pollutants. The most relevant emissions sources are power generation, road traffic and industry. Despite of projected economic growth, Portugal still presents a great potential to reduce emissions if an environmental management strategy is defined and implemented. A quantitative approach is needed in order to establish specific measures for the most sensitive areas and to integrate different issues related with scientific, economic and social aspects.

1 Introduction

Environmental management instruments and strategies developed from a passive perspective - that use the load capacity of ecosystems - to a reactive attitude, where the main approach is the use of clean-up technologies. More recently, in the 80s, the integration of the polluter pays principle in the legislation stimulated a proactive perspective and simultaneously new instruments have been developed and applied, particularly those related with market and information.

From the political point of view, the environmental questions have been introduced on the political agenda only quite recently. In fact the fast environmental degradation verified in the last 100-150 years, as well as the increase of public information, has woken people up to fight for a better quality of life and the welfare of the population.

Although human activities throughout time have released pollutants into the atmosphere, undoubtedly the most significant impacts were experienced after the Industrial Revolution and mainly due to fuel combustion. Same of the most important atmospheric problems currently identified are:

- acidification related with nitrous oxide (NO_x) and sulphur dioxide (SO_2) emissions;
- photochemical smog related with emissions of NO_x and volatile organic compounds (VOC), specialy as precursors of secondary ozone production in the troposphere;
- eutrophication related with the increase of ammonia (NH_3) and NO_x in the atmosphere;
- depletion of stratospheric ozone concentration due to the emissions of halons such as clorofluorcarbons (CFC's);
- enhanced greenhouse effect due mainly to carbon dioxide (CO_2), methane (CH_4) and halocarbons (SF6, PFC and HFC) emissions and its impact on climate change.

The recognition of these problems was involved on great controversy and results from the outcry of the scientific community, environmental non-governmental associations and the public in general. Thus, the international political class and, more specifically, the governments of different countries assumed the need for a global strategy to solve these problems. Some examples of those commitments are the conventions, protocols and agreements made in different areas and with focus on various objectives and targets. Those commitments have been transposed to the national level, environmental authorities have been created and environmental policies and strategies developed. More recently, new philosophic approaches have been developed and introduced based in economic and on information related instruments.

This paper analises the different type of instruments used in environmental management and pollution control strategies, as well as the technological development applied to pollution control and prevention.

The Portuguese panorama concerning atmospheric emissions is analysed, identifying the current emissions inventory and air quality management strategy, future emissions scenarios and emission control strategies needed to reach the targets agreed in international commitments.

2 Environmental management instruments

The "dilute and disperse" practice used in pre-industrial society was based only on the assimilative capacity of the natural environment. After the Industrial Revolution, this approach seemed to be adequate for atmospheric emissions and wastewater discharges. Nevertheless, the new approach "concentrate and contain" proved to be more successful for solid wastes (EEA [1]).

From the 1960s onwards, it became obvious that "dilute and disperse" was no longer effective for important point or concentrated sources. The globalisation of environmental problems and the recognition of the planetary effects of pollution, particularly the effects on climate change and on the stratospheric ozone levels, contribute to the development of clean-up technologies, based on end-of-pipe approaches.

At a national level, and resulting from first environmental policies, thousands of regulations applied to air pollution management and control have been produced. Legislation focused on the emissions control and air quality standards and was oriented to the limitations and obligations of the polluters and a strict control by the authorities. This "command and control" strategy had unsatisfactory results due to the need of a heavy bureaucratic system and iniquity in the treatment of different pollutant activities face to their market opportunities.

Thus, a new philosophic approach has been developed and introduced based on economic instruments, which promotes the decrease of wastes production, based on cost-effectiveness options and the development of more eco-efficient and eco-effectiveness products. In fact, these instruments incentive pollution reduction to a balanced level between pollution control costs and environmental costs (externalities) which results sometimes in higher pollution reduction levels than the application only of "command and control" instruments. On the other hand this type of instruments integrates the polluter pays principle quite well and agrees with the concept of sustainable development. In fact, the improvements on production processes or even on the life cycle of products induced by this type of instruments results in saving on raw materials and energy which leads to an increase of economic profits (EPA [2]). The list of economic instruments includes taxes, revenues, incentives and charges.

Due to their environmental performance, the European Union (EU) in the 5th Environmental Action Program, started in 1992 recommended the use of economic instruments. Nevertheless, and despite of the little progress registered since then in member states, in the last 5-6 years the use of environmental taxes has increased, particularly in Scandinavian countries (EEA [3]).

More recently, voluntary information-related instruments have been developed taking into account the increase of public information on environmental issues and, on the other hand, new consumption patterns where environmental performance of products and economic activities plays an important role. As a result of this new approach, polluters need to improve their market image in order to satisfy a more ecologically-conscience consumer. Information instruments for this purpose includes eco-labelling and environmental management standards like ISO 14000 and EMAS.

The co-ordination of these three types of instruments together in the definition of an environmental strategy results to be more effective in both environmental and economic senses than when they are applied in separate, since polluters are stimulated to reach environmental targets in a cost-effective and flexible way.

3 Technology applied to environment management

The evolution of technical solutions for environmental problems agrees with the conceptual evolution from early environmental attitudes over clean-up approaches towards precautionary and prevention principles included in sustainable development concept.

The first technological approach, which results from the reactive concept of environmental management, focused on end-of-pipe systems. Since the 1960s a whole technology and business was developed to install purification units at the end of emission pipes of various production processes in order to reduce the impact of discharges on the environment. But purification units works on the basis of chemical, physical and biological processes which, in the end, generate new sub-products that need to be treated. So this approach results in a continuous cyclical transference of pollution which combined with the increase in population and use of natural resources has became unsustainable. On the other hand, end-of-pipe technologies have revealed to be expensive and sometimes the costs are also unsustainable due to a strong competitiveness market.

Along with the introduction of the polluter pays principle into legislation, some improvements have been registered in the technological field. A more holistic approach of environmental issues has been implemented and new challenges have been posed to the scientific community in order to look at environmental problems in a more integrated way and to refuse the undesirable pollution transference from air to water and from water to soil and so on.

In the last 10 to 15 years, new pro-active ideas have emerged oriented wards pollution prevention and waste minimisation. These new ideas are integrated in the concept of cleaner production which is a preventive, integrated, continuous strategy for modifying products, processes or services to enhance effectiveness, which improves environmental performance and reduces costs. In fact, the major goal of cleaner production is to reduce the quantities of inputs of raw materials and energy and get the same or even greater output (EEA [1]). Cleaner production includes technological measures like life cycle assessment, best available technologies, eco-design and industrial symbiosis.

4 Portuguese atmospheric emissions control and management

In Portugal, the first steps on air quality management began with the institution, in 1966, of the so-called Working Group of Atmospheric Pollution, whose main purpose was the development of a program to fight against atmospheric pollution in most problematic places. In the 1970s, the Services for Environmental Studies were implemented.

From the strategic point of view, the first instrument for air quality management was the creation in 1980 of five Air Management Commissions, with jurisdiction over five mainly urban and/or industrialised-stressed regions (Lisbon, Barreiro/Seixal, Porto, Estarreja and Sines).

Currently, the major authority is the Ministry of Environment that is made up of General Directorate of the Environment, five Regional Delegations one for each of NUT II territory unit (North, Centre, Lisbon and Tagus Valley, Alentejo and Algarve), four thematic Institutes (Nature conservation, Water, Solid wastes, and Environmental promotion) and the Inspection of Environment.

In the legislative field, during the last 15-20 years, several legal diplomas have been produced and EU Directives transposed. First of all was the Basic Law of the Environment that cover all environmental fields including the need for an environmental impact assessment for big new projects and construction. Further legislation and regulation on the atmospheric field has included air quality standards, air emission standards and more recently, in 1999, the implementation of a fiscal incentive for investments made in clean-up technology.

4.1 Analysis of air pollutant emissions inventory

The last national air pollutant emission inventory published in Portugal with more detailed information was CORINAIR 90 and it is the basis of the analysis presented in this paper. Taking into account the major level of spatial downscale of emissions, which is the NUT III territorial units, further developments were made in order to obtain emissions at municipal (NUT IV) and sub-municipal levels (NUT V). This downscaling exercise was based on specific factors related with source activities, such as fuel consumption and population statistical data (Borrego et al. [5]).

Figure 1 presents the spatial distribution of NO_x and SO_2 annual emissions associated with mobile sources and all other sources excluding traffic and large point sources, calculated for the NUT V level. The analysis of both maps shows that major anthropogenic emissions are near the coastline corresponding to preferential distribution of population and economic activities, including large point sources, on the coast. This heterogeneous distribution has been discussed in previous work (Borrego and Lopes [4]) and following world tendencies.

The most relevant sources in the national emission budget of major pollutants (CO_2, SO_2, NO_x) are those related with combustion processes such as public power production and co-generation, district heating, industrial combustion, road traffic and other mobile sources (figure 2). Agriculture, nature and waste treatment and disposal are significant sources of non-methane organic compounds (NMHC), CH_4 and nitrous oxide (N_2O). In the Portuguese national panorama, the 29 large point sources classified according to CORINAIR guidelines (that include paper pulp production plants, refineries, power plants, petrochemical industry, cement plants, nitric acid production plants, sulphur acid production plants and steel production), have a major contribution on CO_2, SO_2 and NO_x emissions with, respectively, 45%, 80% and 30%.

It must be noted that according to the European Environmental Agency (EEA) [6]) Portugal had the lowest CO_2 emissions per capita but the 3rd highest value of CO_2 emissions per unit of GDP in 1994. This fact could be explained by the outdated technology used in a great number of production and combustion processes that are less efficient from the environmental and economic point of view.

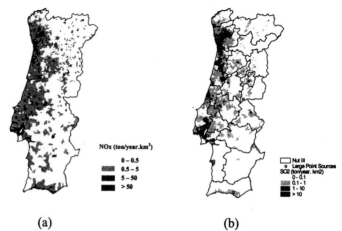

(a)   (b)

Figure 1: Spatial distribution of Portuguese large point sources and area emissions for the year 1990: (a) NO_x emissions from mobile sources; (b) SO_2 from all anthropogenic activities excluding mobile sources, agriculture and large point sources.

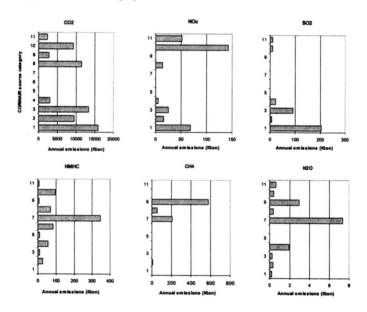

Figure 2: 1990 Annual emissions by pollutant and CORINAIR source category: 1- public production and co-generation; 2 – District heating; 3 – Industrial combustion; 4 – Industrial processes; 5 – Fuel extraction and distribution; 6 – solvent use; 7 – Agriculture; 8 – Nature; 9 – Waste treatment and disposal; 10 – Road transport; 11 – Other mobile sources.

4.2 Portuguese targets related with air pollutant emissions

Portugal is a signatory country of various internationals Conventions and protocols such as for example:

- The 1979 Convention on Long-Range Transboundary Air Pollution;
- The Sofia Convention (1985) for the Protection of Stratospheric Ozone;
- The Framework Convention on Climate Change (1992);
- The Protocol of Montreal (1987) on substances that cause depletion in the ozone layer;
- The Agenda 21 (1992), a political statement concerning environmental development and co-operation.
- The Sofia Protocol (1988) related to the Convention on Long-Range Transboundary Air Pollution;
- The Kyoto Protocol (1997) concerning the emission mitigation of 6 specific greenhouse gases (GHG);
- The Gothenburg Protocol (1999) to the Convention on Long-Range Transboundary Air Pollution to abate acidification, eutrophication and ground-level ozone;

As a member-state of the EU, Portugal has assumed the responsibilities, obligation and guidelines proposed and approved by the European Commission and Parliament, including the application of EU Directives.

At national, regional and local levels, Portuguese government promotes specific measures conducive to the accomplishment of those targets. Despite this attempt at improving the quality of the environment, economic development is still the priority in order to increase the welfare of the population and to achieve the average quality of life levels of the EU. The integration of these two issues is often quite difficult.

From the air quality management point of view, this type of contradiction could be found on the commitments assumed in recent protocols. For example, according to the targets assumed in Kyoto and in the scope of EU "burden sharing" agreement, Portugal undertook, as a national objective, a 27% limit on the increase of GHG and a 40% increase limit concerning CO_2 emissions from the 1990 "baseline" level, until the 2008-2012 period. In the Gothenburg Protocol, Portugal agreed to reach specific national emission ceilings (NEC) for four pollutants (SO_2, NOx, NH_3 and VOC) and more recently the European Commission proposed a Directive on NEC for the same pollutants (COM(99)125), to limit the negative impacts of acidification, eutrophication and tropospheric ozone which is still under discussion (see table 1).

Looking across all those targets, the results appear to be contradictory and difficult to harmonise. In fact, some of the involved pollutants such as CO_2 and NO_x are produced by the same activities like fuel combustion. Furthermore, the projected emission for the year 2010 (figure 3) indicates an increasing tendency on CO_2 emissions for the 3 analysed scenarios (low, medium and high economic development). According to those projections, Portugal is close to reaching or even surpassing the values agreed upon in Kyoto (Borrego et al [7]). This increase in CO_2 emissions due to the development of economic activities indicates an increase on other atmospheric pollutants released by the same sources of CO_2. But the NEC proposal under discussion intends to be more ambitious than the Gothenburg Protocol with a reduction that reaches up to 65 % for the same period.

Table 1. Emission levels for 1990, emission ceilings and percentage of reduction for the pollutants covered by the Gothenburg Protocol and the NEC Directive proposal.

Pollutant	Emission levels in 1990	Emission ceilings for 2010 (kton)		Percentage on emission change	
		Gothenburg Protocol	NEC Proposal	Gothenburg Protocol	NEC Proposal
SO_2	344	170	141	- 51 %	- 59 %
NO_x	303	260	144	- 14 %	- 52 %
NH_3	77	108	67	40 %	- 13 %
VOC	294	202	102	- 31 %	- 65 %

Thus, there is no doubt that an enormous political and strategic exercise must be done in the next few years. An integrated approach between NEC and Kyoto is needed in order to develop a strategic plan for mitigation of atmospheric emissions. On the other hand this strategic plan must be focused not on the pollutant itself but on the pollution-related sources which are fuel combustion, in particular on power plants and industry, and road traffic.

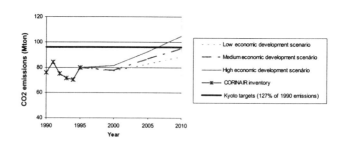

Figure 3: Projected scenarios of CO_2 emissions (Borrego et al. [7]).

4.3 Technical and management measures to mitigate atmospheric emissions

According to international commitments and EU orientations Portugal must control the increase or even reduce their atmospheric emissions. A more detailed analysis of most important economic activities, such as industry, shows that the majority of industrial processes are outdated from the technological point of view. Along with the technological improvement, investments in end-of-pipe technology must be made in the upcoming years, in order to reach short-term targets of environmental air quality. According to the 6th Interim Report related with the cost-effective control of acidification and ground-level ozone (IIASA [8]), Portugal still has great potential to reduce NEC pollutants in a cost-effective perspective. Some strategic options include the implementation of EU Directives and Programmes, technological improvement using best available technologies, implementation of end-of-pipe treatment systems, fuel substitution and more efficient use of energy.

The implementation of the cleaner production concept and the application of some related techniques such as life cycle assessment and eco-design could produce good results and promote a more sustainable development.

From a source-oriented point of view, voluntary agreements between government and economical groups should be implemented with specific economic activities to promote sectorial targets, specific goals and mitigation measures to be adopted according to the most realistic time schedule. Governments also play an important role in what concerns the implementation of existing regulation, promotion of an efficient inspection system and development of financial and fiscal incentives for specific investments on technological improvements.

Economic instruments should have an important role in an integrated strategic plan that conciliates climatic and NEC targets. Although some economic instruments have been implemented in Portugal in the last years they are more oriented to the energy products than to environmental issues, and the income taxes has been canalised to the national budget. Thus, their impact on the public and polluters in particular has been low. The application of new pollution oriented taxes such as SO_2 or CO_2 emission taxes has a great potential specially if implemented in conjunction with a revenue taxes system, where the income is directly administrated by a commission and applied to technological improvement or to more energetic efficient activities.

The abatement measures for NEC pollutants would have necessarily beneficial effects on CO_2 and other GHG emissions. On the other hand, taking into account that the next step on the climate change international policy agrees with recent developments in the adoption and implementation of market mechanisms foreseen in the Kyoto Protocol, this indirect mitigation effect creates an extra emission quota that could be used in future negotiations for an emissions trade system.

5 Conclusions

In what concerns atmospheric emissions, Portugal presents a very complex panorama. The analysis of emissions inventory shows that emissions and related sources are concentrated in coastal areas according to the preferential distribution of population and economic activities. Sources of major concern are fuel combustion, particularly associated with power plants and industrial combustion, road transport and industrial processes. The projected scenarios show that air emission tends to increase quickly in the next decade due to the need for the economic development of the country in order to reach EU average levels.

Portugal also assumed specific commitments in international conventions and protocols that are conducive to a limitation on the increase (Kyoto protocol) or even reduction (Gothenburg Protocol) of some specific pollutants. Furthermore the new EU NEC Directive currently under discussion, proposes a more restricted target for the same pollutants. To achieve those goals some key issues are: technological update using BAT, implementation of environmental taxes, a more efficient use of energy and a fuel switch to less pollutant fuels. The co-ordination of all those targets will be an enormous political exercise that must be put in practice as soon as possible. Nevertheless the development of a strategic integrated plan oriented for the sustainable development of economic activities and supported by emission reduction measures, could be an add-value for the negotiation on an international emission trade system.

Acknowledgements

The authors would like to express their gratitude to CZCM – Centre of Coastal Zone and Sea of the University of Aveiro for the Post-Doctorate grant to N. Barros, the PRAXIS XXI PhD grant to M. Lopes and DGA – General Directorate of Environment for the support to the Directives work.

References

[1] EEA – European Environmental Agency. Cleaner Production: A guide to information sources. European Environmental Agency, 1997.
[2] EPA – United States Environmental Protection Agency. *A Guidebook of Financial Tools: paying for sustainable environmental systems,* Environmental Financial Program, 1999.
[3] EEA. *Environmental Taxes: implementation and environmental effectiveness,* European Environmental Agency, Copenhagen, Denmark, 1996.
[4] Borrego, C. and Lopes, M. Climate change and coastal zone: the importance of atmospheric pollutant transport. *Environmental Coastal Regions,* ed. C.A.Brebbia, Computational Mechanics Publications/WITPress, Southampton, pp. 265-275, 1998.
[5] Borrego, C.; Barros, N.; Lopes, M.; Conceição, M.; Valinhas, M. J.; Tchepel, O.; Coutinho, M. and Lemos, S. Emission inventory for simulation and validation of mesoscale models. *Proc. of the EUROTRAC-2 Symposium 98,* ed. P.M. Borrell and P. Borrel, WITPress, Southampton, pp. 8-10, 1998.
[6] EEA. Europe's Environment: The second assessment report. Elsevier Science Ltd., Oxford, UK, 1998.
[7] Borrego, C.; Miranda, A. I.; Conceição, M.; Carvalho, A. C.; Lopes, M. and Tchepel O. Climate change and air quality management: emissions and consequences. *Proc. of the 6th Conference on Environmental Science and Technology,* Vol. A. ed. T. Lekkas, pp. 438-446, 1999.
[8] IIASA - International Institute for Applied Systems Analysis. Cost-effective control of acidification and ground-level ozone, Part B: Emission control scenarios, 6th Interim Report to the European Commission, Austria, 1998.

Analysis of spatio-temporal fluctuations of East China Sea fishery resources using GIS

F. Su, C. Zhou, Q. Shao, Y. Du & S. Wang
*State Key Lab. of Resource & Environmental Information System,
Institute of Geography, Chinese Academy of Sciences, China*

Abstract

Catch data from the East China Sea from 1987 through 1997, and other environmental factors, were overlain using Geographic Information System (GIS) technology. Maps of the spatio-temporal density of the demersal and pelagic fishery resources were produced. After analyzing the density distribution, we analyzed relationships between spatio-temporal distributions and environmental factors. The Kuroshio Current plays a major role in determining the fishery density distribution, especially for the demersal fishery. The South Yellow Sea Water Mass and the expansion of the Changjiang Diluted Water (ECDW) are two other main environmental factors. The demersal fishery resource is mainly distributed between the 200-m isobath and the central route of Kuroshio Current. Density shifts of the demersal and pelagic fisheries were analyzed by calculation of centers of gravity. Trend analyses showed that the fishery resources generally decreased in the study area during the 11-year period.

1 Introduction

In our research, we used GIS to analyze the spatio-temporal fluctuation of fishery resources in the East China Sea. The East China Sea is a marginal sea which lies between the Asian continent and the Pacific Ocean. Its current system is composed of littoral currents and the Kuroshio Current system. Therefore, water in this region is composed of littoral diluted water and water from the far sea. The littoral diluted water has low salinity, low transparency, and great inter-annual variability in temperature. The temperature and salinity of the far sea water are high, so that gradients of temperature and salinity in the East China Sea are steep. Biomass in this region is low compared with other regions in the world, only 3.92 tonnes per square kilometer. Approximately 11 percent of production in this region is pelagic fishes and 40 percent is demersal fishes [1]. This region is the most important fishing area for China, generating 40 percent of China's total fishery production [2].

Weizhong Chen analyzed the change of fishery resources in the East China Sea based on the biographic characteristics of sample fish, total catch production, and production of the main species[3]. Weihai Liu and Benyi Zhan calculated the correlation between CPUE and catch effort, and analyzed the utilization of fishery resources using two surplus production models, the Schaefer Model and the Fox Model. [2]However, none of these analyses contains a spatial component.

GIS is still not widely used in fisheries management, as the typical stock assessment models do not include spatial information. There are several areas in which the use of GIS in management of marine fisheries is emerging [4-5]. One area is for Pacific salmon in the USA. Olivier et al. [6] analyzed the relationship between fishery resource fluctuations and environment factors, including fishing effort. Several studies have analyzed spatial and temporal variation of groundfish assemblages off the east coast of North America [7-13]. Also, GIS is now being used in for modeling fish habitat in U.S. estuaries [14].

2 Data

Sample data for 1987-1997 were obtained from four of the biggest fishery companies in China. The catch of these four companies totals 12 percent of the total catch from the region by all Chinese companies. The East China Sea was divided into regular grid with 10' x 10' each cell (Figure 1). Each grid cell has one catch record per day for each gear type. Trawl catches were used as a measure of demersal fish density; catches in purse seines were used as a measure of the density of pelagic fishes. There was a 54 percent probability that two companies were fishing in the same cell on the same day, with an over 80% overlap if the low-production records are deleted. The results are only slightly changed if the records of any one company are deleted, which demonstrates the robustness of the sample data set.

3 Analysis

3.1 Fishery resource distributions and environment in the East China Sea

3.1.1 Demersal fish distribution
Demersal fishes are caught primarily on the shelf (Figure 1). There are three areas where the density of demersal fishes is high: Changjiangkou Zhoushan, Kuroshio and Dasha. The Kuroshio area is divided into four subareas in Figure 1. Densities in the majority of cells in the Changjiangkou Zhoushan and Kuroshio areas is 0.5-2 tonnes per tow, with some over 2 tonnes Per tow. In the Kuroshio area, high density occurs strictly between the central line of the Kuroshio Current

and the 200-m isobath, resembling a belt along the Kuroshio. The Changjiangkou Zhoushan area lies between the 50-m and 100-m isobaths. In half of the cells in the Dasha area, the density of the demersal fishery is in the 0.5-2 tonne range. The last two areas resemble two slices of a pie .

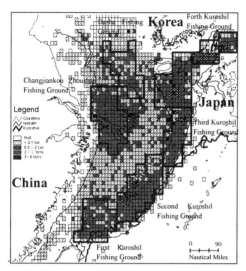

Figure 1: Density distribution of demersal fishes

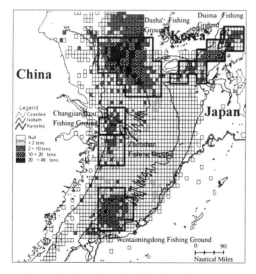

Figure 2: Density distribution of pelagic fishes

3.1.2 Pelagic fish distribution

The pelagic fishes occur mainly near the coast (Figure 2). There are four areas where the density of pelagic fishes is high: Dasha, Changjiangkou Zhoushan, Wentaimindong, and Duima. The density of pelagic fishes per cell in the Dasha area ranges from 2-50 tonnes per purse seine. One-third of these cells has densities between 2 and 10 tonnes per purse seine, the middle third has densities of 10-20 tonnes per purse seine, while the highest third has densities over 20 tonnes per purse seine. Densities in the other three areas are between 2 and 10 tonnes per purse seine.

3.1.3 Relationships between fish density and the environment

The fishing ground for demersal fishes in the Kuroshio area occurs in the updraft current, which lies to the left of the Kuroshio Current. The bottom drops steeply east of the 200-m isobath, which enhances the intensity of the ascending current. These factors have important influences over the demersal fishing grounds. The fishing grounds in Changjiangkou Zhoushan area lie in the temperature-front zones between the Changjiang Diluted Water (ECDW), the Yellow Sea Cold Water Mass, and the Taiwan Warm Current. There are two temperature fronts in the upper waters of the Dasha area. One, on the west, is formed by the Yellow Sea Warm Current and the Yellow Sea Matrix Water. The other, to the east, is formed by the littoral water of Korea and the Duima Warm Water Mass[15-16].

The Kuroshio Current plays a major role in determining the density of the bottom-fishery resource. High biomass areas typically lie to the left of major currents in the Northern Hemisphere [17]. The catch data indicates that fishery productivity is high where high primary productivity occurs. Xuechuan Wong, Fenqi Li and Beiwei Lu's [18-19] have shown that chlorophyll-a concentrations are high in the middle of the East China Sea's shelf, and distributed discontinuously in a belt along the 200-m isobath as the same as the ditribution of high fish density areas(Figure 1).

3.2 Changes in fishing ground location

Changes in the location of fishing grounds over the 1987-1997 period were evaluated by calculation of the annual "center of gravity" for each fishing ground [20]. The following formula was used to calculate the center of gravity.

$$xc = \frac{1}{M} xu(x,y)dxdy, \; yc = \frac{1}{M} yu(x,y)dxdy$$

D is the area being analyzed, (xc, yc) are the coordinates of the center of gravity, M is the total biomass, and μ (x, y) is the density at the point of (x, y). For use with the sample data, we re-arranged the above two formulas as follows.

$$xc = \frac{\sum_d x\rho(x,y)}{\sum_d \rho(x,y)}, \; yc = \frac{\sum_d y\rho(x,y)}{\sum_d \rho(x,y)}$$

In these formulas, d is the area of the fishing ground and ρ (x, y) is the density of fish in the cell (x, y).

3.2.1. Demersal resources

For the demersal resource, centers of gravity were calculated for the four fishing grounds shown in Figure 1. The position shifts of each fishing ground are shown in Figure 3. The center of the Dasha fishing ground shifted to the northwest almost continuously during the study period. The net distance shifted was 56 nautical miles to the west and 28 nautical miles to the north. The center of Changjiangku Zhoushan fishing ground shifted primarily to the south. The net distance moved was 22 nautical miles. The center of the first

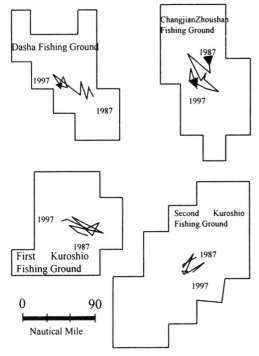

Figure3:Locations of the centers of gravity for the East China Sea demersal fishing grounds

Kuoshio fishing ground drifted erratically, with a net change of 24 nautical miles to the west and 15 nautical miles to the north. The center of the second Kuoshio fishing ground also drifted erratically. The net distance shifted was 29 nautical miles to the south and 10 nautical miles to the east.

3.2.2 Pelagic resources

For the pelagic resource, centers of gravity were calculated for the four fishing grounds shown in Figure 2. The position shifts of each fishing ground are shown in Figure 4. The center of the Dasha Fishing Ground shifted to the west continuously from 1987 to 1997, with a net change of 31 nautical miles. Over the entire 1987-1997 period, the center of the Changjiangkou fishing ground shifted to the south slightly from 1987 to 1997, but the inter-annual shifts were considerable, about 20 nautical miles. The center of the Zhoushan fishing ground shifted about 30 nautical miles inter-annually, but without a discernable pattern. The center of the Wentaimingdong fishing ground shifted to the northeast, with a net position change of 30 nautical miles to the east and 15 nautical miles to the north.

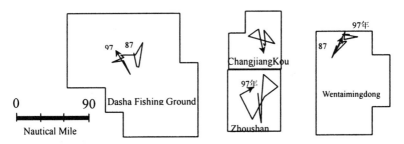

Figure 4: Locations of the centers of gravity for the East China Sea pelagic fishing grounds.

3.3 Spatio-temporal analysis of the fishery-stock fluctuations

Temporal trends in density over the study period were analyzed by grid cell using least-squares linear regression. The data were organized as in Table 1. The regression equation was as follows:

$$Y = ax+b,$$

where Y is the density in a given cell and year, x is the number of years from 1987 (e.g., the value of x for 1991 is 1991-1987 = 4), and a and b are the coefficients. The value of "a" is the slope of the line, and indicates the temporal trend in the cell. A positive slope would indicate an increasing trend in catch, while a negative slope would indicate a decreasing trend. The absolute value of the slope indicates the magnitude of the trend.

Table 1.   Density of demersal fishes by grid cell and year.

Year Cell No	87	88	89	90	91	92	93	94	95	96	97
15727	34	64	45	58	50	14	30	24	7	9	27
15728	44	81	56	32	73	24	17	26	6	31	26
16426	51	70	36	62	103	20	58	19	19	3	18
16427	106	46	51	46	157	40	28	27	108	42	64
...											...

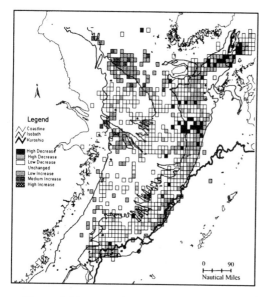

Figure5 Change of density of demersal fishes

The results of the trend analyses were divided in seven categories: high decrease , medium decrease, low decrease , unchanged, low increase , medium increase, and high increase (Figure 5). Most of the cells show decreases for the demersal fishery, especially in the third and fourth Kuroshio fishing grounds. In the southeast of the second Kuroshio fishing ground, there are some cells in the low increase category, located southeast of the Kuroshio Current. In the Dasha fishing ground,

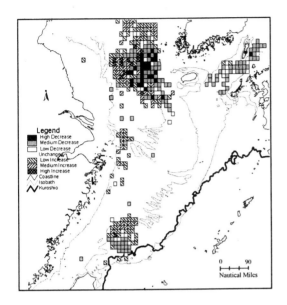

Figure6 Change of density of pelagic fishes

most of the cells are also in the low increase category, and some are in the medium increase category. In the Changjiangkou Zhousha fishing ground, some cells are in the low increase category. In the first Kuroshio fishing ground, there are some cells along the Taiwan Warm Current in which density has increased slightly. In the whole East China Sea, there are some cells in the high decrease category, but no cells in the high increase category. These patterns are similar to the changes in location discussed in Section 3.2.1.

Density of the pelagic fisheries decreased in two-thirds of the cells in the East China Sea during the study period (Figure 6). In the Dasha fishing ground, there are some cells in the high decrease category, and the number cells in the low decrease and low increase categories is the same. In the Duima fishing ground, most of the cells are in the medium decrease category. In the Wentaimingdong fishing ground, the density in half of the cells are in the low increase category , while that in the other cells are in the medium category. These patterns are similar to the changes in location discussed in Section 3.2.2.

4 Conclusion

There are three areas in which the density of demersal fishes is high. The Kuroshio area resembles a belt between the 200-m isobath and the center of the Kuroshio Current. The other two areas lie in the Dasha and Changjiankou Zhoushan areas, respectively. Fish distributions in the East China Sea are influenced by the Kuroshio Current, the Yellow Sea Water Mass, and the expansion of the Changjiang Diluted Water (ECDW). Over the 1987-1997 study period, the Changjiangkou Zhoushan fishing ground shifted to the south slightly, the Dasha and the first Kuroshio fishing grounds shifted to northwest, and the second Kuroshio fishing ground shifted to the southeast. The trend analysis showed that the demersal fish resources in the East China Sea have decreased. There are four areas in which the density of pelagic fishes is high: Dasha, Changjiangkou Zhoushan, Wentaimindong, and Duima. Overall, the Dasha fishing ground shifted to the west; the Changjiangkou fishing ground shifted to the south slightly, but with inter-annual variability; the Zhoushan fishing ground did not have a substantial net change in location, but shifted long distances inter-annually; and the Wentaimingdong fishing ground shifted to the northeast. The trend analysis showed that the pelagic fish resources in the East China Sea have decreased slightly.

Acknowledgements

We wish to thank Shixian Han, Weizhou Chen and Vincent Lyne for discussions about this work. We are grateful to Dr. Stephen K. Brown who edited carefully the paper and provided some relevant references. An anonymous reviewer also contributed to the presentation. This work was supported in part by Program 863 of China (project no. 818-07-02).

References

[1] Chuanyin Zhao etc (eds). *Marine Fishery Resources of China*, Zhejiang Science & Technology Publication House, 1988.
[2] Weihai Liu & Bingyi Zhan, The dynamic analysis on the fishery stocks in the East China Sea, *Journal of Shanghai Fisheries University*, 8 (1), pp. 19-24, 1999.
[3] Weizhong Chen, Changsong Li & Fen Hu,A review of the fisheries resource status in the East China Sea, *Journal of Fishery Sciences of China*,4(3), pp. 39-43,1997.
[4] Rubec, P.J. and J. O'Hop (eds.) *GIS Applications for fisheries and coastal resources management.* Gulf States Marine Fisheries Commission, Ocean Springs, Mississippi, USA. Publication No. 43, pp. 202, 1996.
[5] Nishida, T. *First international symposium on GIS in fishery sciences, Seattle, Washington, USA (abstracts).* National Research Institute of Far Seas Fisheries, Shimizu, Shizuoka, Japan, pp. 149, 1999.
[6] Olivier Maury & Didier Gascuel, ,SHADYS,a GIS based numerical model of fisheries. Example application: the study of a marine protected area, *Aquatic Living Resource*,12(2), pp. 77-88, 1999.
[7] Mahon, R., S.K. Brown, K.C.T. Zwanenburg. D.B. Atkinson, K.R. Buja, L. Claflin, G.D. Howell, M.E. Monaco, R.N. O'Boyle, and M. Sinclair. Assemblages and biogeography of demersal fishes of the east coast of North America, *Canadian Journal of Fisheries and Aquatic Sciences,* 55:1704-1738. 1998.
[8] Gabriel, W.L. Persistence of demersal fish assemblages between Cape Hatteras and Nova Scotia, Northwest Atlantic. *Journal of the Northwest*

Atlantic Fisheries Society, 14:29-46, 1992.

[9] Overholtz, W.J. and A.V. Tyler. Long-term responses of the demersal fish assemblages of Georges Bank. *Fishery Bulletin* 83(4): 507-520. 1985.

[10] Murawski, S.A. and J.T. Finn. Biological bases for mixed-species fisheries: Species co-distribution in relation to environmental and biotic variables. *Canadian Journal of Fisheries and Aquatic Sciences* 45: 1720-1735. 1988.

[11] Gomes, M.C., R.L. Haedrich, and J.C. Rice. Biogeography of groundfish assemblages on the Grand Bank. *Journal of the Northwest Atlantic Fisheries Society* 14: 13-27, 1992.

[12] Gomes, M.C., R.L. Haedrich, and M.G. Villagarcia. Spatial and temporal changes in the groundfish assemblages on the north-east Newfoundland/Labrador Shelf, north-west Atlantic, 1978-1991. *Fisheries Oceanography* 4(2): 85-101,1995.

[13] Langton, R.W., P.J. Auster, and D.C. Schneider. A spatial and temporal perspective on research and management of groundfish in the Northwest Atlantic. *Reviews in Fisheries Science* 3(3): 201-229,1995.

[14] Brown, S.K., K.R. Buja, S.H. Jury, M.E. Monaco, and A. Banner. Habitat suitability index models for eight fish and invertebrate species in Casco and Sheepscot Bays, Maine. *North American Journal of Fisheries Management* 20:408-435, 2000.

[15] Kentang Le, Source of the Yellow Sea Warm Current, *ACTA OCEANOLOGICA SINICA*, 14(2), pp. 9-19, 1992.

[16] Yuxiang Tang, Distributional features and seasonal variations of temperature fronts in the East China Sea, *OCEANOLOGIA ET LIMNOLOGIA SINICA*, 27(4) , pp. 436-444, 1996.

[17] Qinshanhenxiong, *Demersal fishery resource,* Zhejiang Publish House, pp. 26, 1980.

[18] Beiwei Lu, Rong Wang, Distribution characteristics of surface layer chlorophyll-a concentration on different spatial scales in the East China Sea in Spring, *OCEANOLOGIA ET LIMNOLOGIA SINICA*, 27(5), PP 487-492, 1996.

[19] Chunseng Wang & Dehu He, Distribution characteristics of the biomass of zooplankton in the southeast of the East China Sea, *ACTA OCEANOLOGICA SINICA*, 18(3), pp. 66-77, 1996.

[20] Yiming(eds). *Enchiridion for mathematics.* Science Publishing House. pp. 354, 1996

Section 8
Pollution Studies

Measurement of hydrodynamic conditions, sediment transport and bed morphology in a tidal inlet

J.J. Williams, P.S. Bell & P.D. Thorne

Centre for Coastal and Marine Sciences, Proudman Oceanographic Laboratory, Bidston Observatory, Bidston Hill, Prenton CH64 7RA, UK

Abstract

In many tidal inlets strong tidal currents and/or large waves generally prevent the deployment of instrumentation using benthic frames or boats and consequently measurements of hydrodynamic and sediment parameters necessary for calibrating numerical models remain scarce. Recently a solution to this problem has been found using a multi-sensor instrument deployed from a small 'jack-up' barge. Observations demonstrating successfully the acquisition of a range of hydrodynamic and sedimentological data are presented.

1 Introduction

As part of the International Project *INDIA* (Inlet Dynamics Initiative: Algarve), fieldwork has been undertaken in the *Barra Nova* tidal inlet located in southern Portugal (Figure 1), during January to March, 1999. The inlet is approximately 150 m wide with an average depth of 4 m. The spring tidal range of 4 m gives rise to depth-mean currents $O(2 \text{ m/s})$. D_{50} for the inlet sediments is 1.2 mm.

The three principal objectives of the field study were: a) to measure in detail hydrodynamic and sediment parameters pertaining to sediment transport processes; b) to quantify the sediment budget and migration rate of the channel; and c) to provide data sets required to calibrate, validate and verify a suite of generic numerical models of the site. In order to help develop strategies for future coastal zone management, observed and predicted results have been combined with data from other inlet studies to examine generically processes in and adjacent to tidal inlets.

The acquisition of data at this site presented a number of difficulties: a) the depth mean current speed exceeds 2 m/s during spring tides; b) the sites is exposed to waves exceeding 3.0 m during storms; and c) the channel bed, comprising deep, highly mobile coarse sand is subject to very rapid morphological changes. In addition, an environmental assessment indicated that during storm conditions, the inlet had a potential to migrate along the coast several tens of metres over a period of one or two days, O'Connor et al., [1].

Figure 1. (a) Location of the field site; and (b) aerial view of the *Barra Nova*

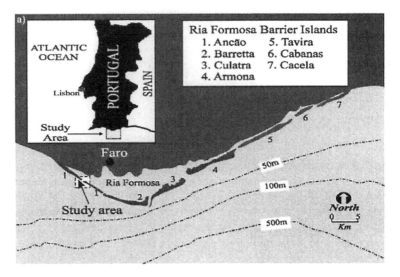

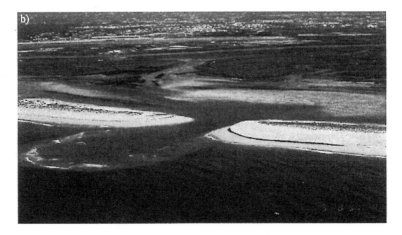

The present paper describes the use of a small jack-up barge to deploy a multi-sensor instrument at locations in the *Barra Nova*. Selected data pertaining to flow turbulence, suspended sediments and bed morphology at locations in the inlet are then presented and discussed. Since this paper is aimed primarily at conveying information about the new deployment technique, the intention here is not to present or to discuss in detail results from the present series of experiments. In the sections below only selected data illustrating PIP capabilities are presented.

2 Measurements

2.1 The 'jack-up' barge

The Seacore Skate II jack-up barge used in the Barra Nova study (Figure 2) consisted of four pontoon sections that were linked together to form a working platform of size 14.65 m x 9.62 m. A drilling tower fixed to the deck facilitated the lowering and raising of the equipment described below into the water beneath the barge through the moon pool in the deck. Generators provided

Figure 2. The jack-up barge used in the *Barra Nova* study

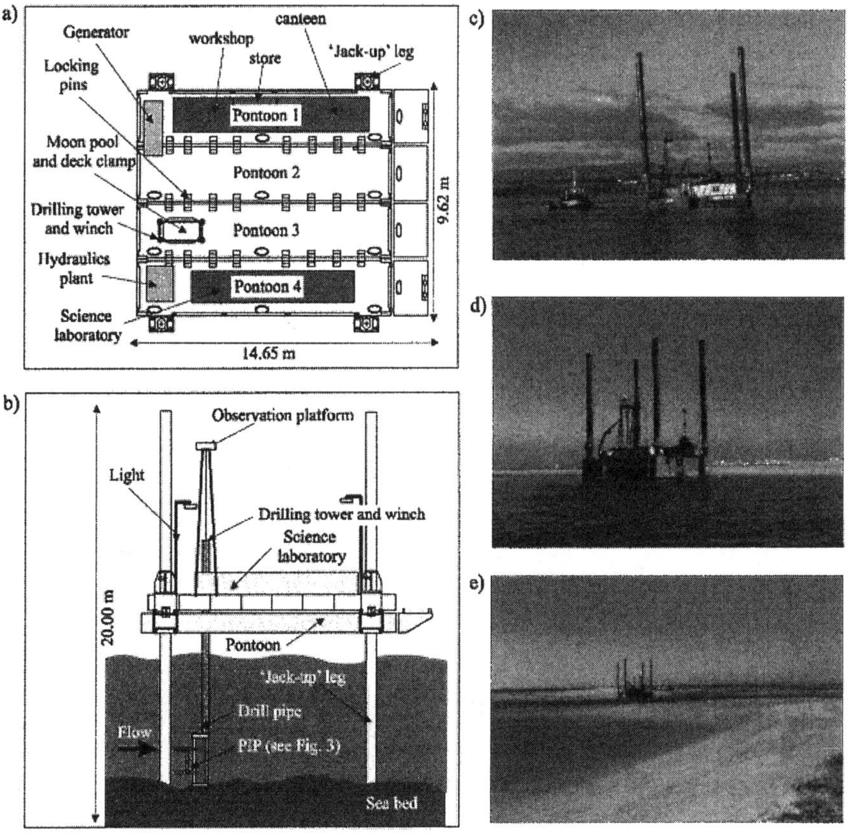

electrical and hydraulic power and cabins were installed to house a workshop, a canteen and scientific equipment. Hydraulic legs of length 20 m were positioned at each corner of the platform and could be raised or lowered independently of the others. Once raised above the water surface, the platform was ably to withstand the strong tidal currents and wave action present in the *Barra Nova*.

2.2 The CCMS-POL Instrument Package (PIP)

In a number of past studies, the present authors and others have used data gathered by the autonomous, multi-sensor instrument STABLE (Humphery & Moores, [2]) to investigate and model wave – current – sediment interactions (e.g. Williams et al., [3]). Recently, these technologies have been refined and brought together as the POL Instrument Package (PIP). The PIP, shown in Figure 3, consisted of: a) an adjustable deployment frame; and b) a suite of sensors and sampling devices.

Figure 3. The POL Instrument Package (PIP)

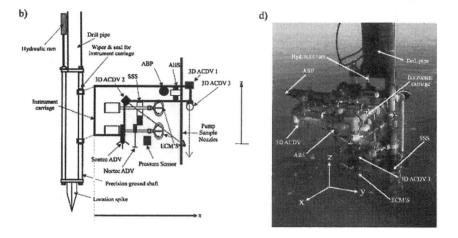

The PIP deployment frame comprised two parallel precision ground stainless steel shafts fixed vertically using ridged steel plates with a movable instrument carriage fabricated using lengths of scaffold pole and elbow joints connected to the shafts. A hydraulic ram was used to raise and lower the carriage thereby allowing the instrumentation to be positioned at any desired height above the bed in the range $0 \text{ m} < z < 1.0 \text{ m}$. Specially designed seals and wipers were used to ensure sand did not enter the points of contact between the carriage and the shafts. The upper plate of the PIP was attached to a 15 m length of standard drill pipe sections (diameter 35 cm) and then connected to the winch on the barge drilling tower by mean of a steel cable. In this way, the PIP could be lowered to the bed of the inlet through the moon pool and raised onto the deck when required. A spike approximately 1 m long was fixed beneath the lower plate to locate the PIP rigidly into the bed. During deployment, as the locating spike penetrated the bed under the combined weight of the PIP and drill pipe sections, contact between the lower surface of this plate and the sea bed was confirmed using a sensor. Above the PIP, the drill pipe was held in position by the combined forces of the tidal flow against the PIP and support pipe and a hydraulic deck clamp around the supporting drill pipe as it emerged from the moon pool.

A suite of sensors to measure hydrodynamic, sedimentological and morphological parameters were fixed to the instrument carriage. These consisted of the following instruments to measure hydrodynamic conditions: two pairs of Valeport Series 800 electromagnetic current meters, ECM's, and a Digiquartz pressure sensor, sampling at 8 Hz; a Sontek acoustic Doppler Velocimeter, ADV, sampling at 25 Hz; a Nortek ADV, sampling at 25 Hz; and the CCMS-POL 3D Acoustic Coherent Doppler Velocity Profiler, 3DACDV, sampling with a vertical resolution of 4.6 cm up to approximately 80 cm above the sea bed at 16 Hz. A triple frequency (1.0 MHz, 2.0 MHz and 4.0 MHz) acoustic backscatter system, ABS, sampling at 4 Hz was deployed to measure instantaneous vertical suspended sediment concentration profiles (C-Profiles). Pump-sampling equipment on the PIP was used to measure time-averaged and horizontally spatially averaged C-Profiles and to obtain samples of suspended sediment for the provision of information on suspended sediment grain size, shape and density. For any given experiment, the collection of pump-samples from all sampling locations above the bed took approximately 15 minutes to complete.

Beneath the PIP acoustic images of the bed morphology in a circular area of radius approximately 2.5 m were obtained using a sector-scanning sonar device, SSS. In addition, a scanning acoustic ripple profiler, ARP was used to measure the morphology of the bed along a single line of length approximately 3 m centred beneath the PIP and running along the x axis. Cables from the instruments were run into a clean laboratory cabin where data were logged using a suit of PC's. Pump-sample tubes were connected to the pumping and sampling system on the deck of the barge. All samples were inspected *in situ* and then bagged for more detailed analysis in the laboratory.

3 Results and discussion

Data presented and discussed below were obtained during the flood tide on 28 February 1999 between approximately 10h00 GMT and 13h30 GMT (hereafter referred to as Day 59). During this time, depth-mean tidal flows exceeded 2 m/s and resulted in the transport of appreciable quantities of sediment as bedload for a period of approximately three hours.

3.1 Hydrodynamics

The 3DACDV was deployed in the field for the first time during these experiments. Intercomparison between U, and V flow components measured over a period of 100 s at $z \approx 4$ cm by the ADV and by 3DACDV are shown in Figure 4a. Agreement between U and V time-series is good ($R_c = 0.94$) demonstrating the accuracy of the 3DACDV. Differences between U or V time series are considered real and attributable primarily to the physical separation between the two measurement locations. Figure 4b shows time-averaged vertical current profiles from the 3DACDV. The profiles exhibit a linear increase in current speed, S, with $\log_e(z)$ up to $z \approx 40$ cm. In contrast, profiles shown using open squares and diamonds have two distinct gradients: the first shallower gradient in the range 2 cm $< z <$ 10 cm; and a second steeper curve in the range 10 cm $< z <$ 50 cm. In all cases, the gradient of the profile decreases at $z > 50$ cm owing to complex near-field effects discussed by Lohrmann et. al [4]. Vertical profiles of the time-averaged bed shear stress derived from turbulent kinetic energy, TKE, profiles are shown in Figure 4c. The profiles are consistent with theoretical expectations for conditions at the field site.

Figure 4a Comparison between the 3DACDV and the Sontec ADV

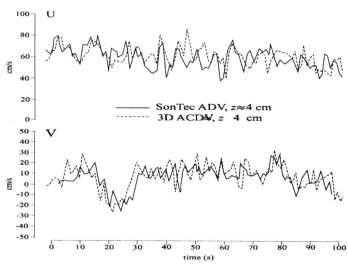

Figure 4b & c Results from the 3DACDV

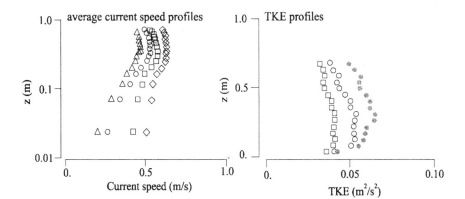

3.2 Bedforms and bedload

The ARP measured the migration of ripples beneath the PIP for a period of approximately 4.5 hours on Day 59 (Figure 5). To assist assessment of scale in Figure 5, the position of the PIP framework at an arbitrary point in time is also shown schematically in this figure. Figure 5 shows an initial reversal in ripple asymmetry shortly after the start of the run in response to a change in flow direction. Thereafter, rates of ripple migration increase to reach a maximum rate $O(5$ m per hour) during the peak flow. The presence of such ripples in the coarse inlet sediments is not predicted by theory.

Also evident in Figure 5 is scour associated with flow around the PIP support pile and the apparent deposition of sediment downstream of the framework attributable to interactions between flow and the PIP. Despite the interruption to sediment transport processes downstream of the PIP, bedload transport is shown to proceed essentially unhindered at all positions beneath the PIP. Clearly the rates of ripple migration shown in Figure 5 account for the transport of appreciable quantities of bed sediments.

If it is assumed that all grains in transport roll over the stoss slope and down the lee face of ripples before coming to rest, it is possible to calculate the volumetric bedload transport rate, Q_b, expressed in units of m^2/s, by considering the ripple migration speed, S_{Rip}, so that $Q_b = k_{Rip} . h_r . S_{Rip}$. Here k_{Rip} is an empirical constant related to ripple shape and generally assumed to be 0.32, Soulsby, [5]. Here S_{Rip} is estimated using cross-correlation analyses between successive ARP profiles measured from the PIP so that $S_{Rip} = X_{cor}/dt$, where X_{cor} is the spatial lag

Figure 5 Ripple migration beneath the PIP measured by the ARP

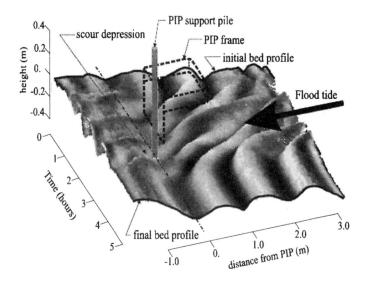

corresponding to the maximum cross-correlation coefficient for successive ARP profiles and dt is the time interval between ARP profiles (63 seconds).

Using regression analysis, a simple expression to predict rate of bedload transport at the present study site has been derived using the present observations of ripple migration rates. The most statistically significant correlation was found between Q_b and U^4 giving the simple expression $Q_b = 1.75U^4 \times 10^{-5}$ m²/s, where U is the depth-mean current speed. Q_b values predicted by this simple expression agree favourably with predictions from Van Rijn [6]. Agreement between observed and predicted Q_b values indicates that the PIP has little influence upon the processes driving bedload transport in the inlet.

3.3 Suspended sediments

Given the magnitude of the tidal flow through the inlet, one might expect the suspended load to be correspondingly high. However, in Figure 6a, showing typical examples C-profiles derived from laboratory analysis of pump samples obtained at approximately 10h00, 11h00, 12h00 and 13h30 on Day 59, suspended sediment concentrations are shown to be only $O(0.1$ g/l). A measured C-Profile has been compared with C-Profiles predicted by the well-known

Power-Law profile and a semi-empirical expression derived by Williams et. al [3], W&R, in Figure 6b. Profiles *1* (Power-Law) and *2* (W&R) are calculated using a reference concentration given by Zyserman & Fredsøe [7] and profile *3* (W&R) is fitted to the data using a single concentration measurement.

Figure 6 (a) Measured time-averaged vertical suspended sediment profiles; and (b) modelled C-Profiles.

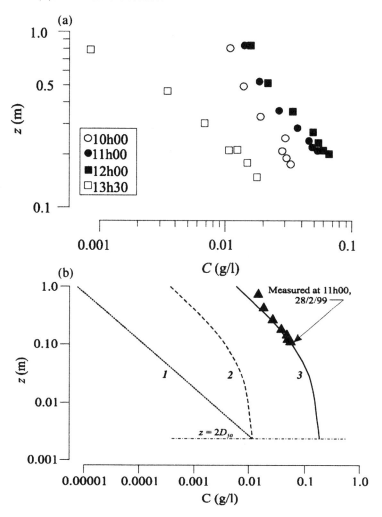

All terms in the relevant equations used to derive the profiles shown in Figure 6 have been derived from PIP measurements including bed shear stress, *in situ* settling velocity and ripple dimensions. Clearly, the power-law is not valid in the present situation and reference concentrations given by the Zyserman & Fredsøe expression are too low. The fitting if the Williams et al. expression to a single measured concentration value gives the best predicted results.

4 Conclusions

Using a small jack-up barge, a new technique for the deployment of instruments in a tidal inlet has been developed successfully. Data pertaining to hydrodynamic conditions and sediment transport at locations in an inlet have been obtained successfully using state-of-the-art instrumentation and have been used to study processes and to calibrate and validate numerical models. For reasons of practicality, safety and cost, future deployments of the present instrument package from a similar jack-up barge in water depths up to 20 m is recommended.

Acknowledgements

The work was supported jointly by the European Communities by the MAST 3 project '*Inlet Dynamics Initiative: Algarve*' (*INDIA*), contract MAS3-CT97-0106, and by the UK *NERC*. The authors would like to acknowledge the Director of the Ria Formosa Natural Park and Port Authority at Faro for granting special permissions to work in the *Barra Nova*.

References

[1] O'Connor B. A., Williams J. J., Dias J. M. A., Collins M. B., Davidson M. A., Arens S. M., Howa H., Sarmento A. J., Seabra-Santos F., Aubrey D., Salles P., Smith J. S., Heron M., Pires H. O., Silva A., Bell P. S. AND Pan S. (1998) Tidal Inlet Monitoring/Modelling Project (INDIA). *Proceedings Oceanology International,* Singapore, 14pp.

[2] Humphery J.D. & Moores S.P. (1994) STABLE II - An improved benthic lander for the study of turbulent wave-current-bed interactions and associated sediment transport. *Electronic Engineering in Oceanography,* IEE Conference Publication No. **394**, 170-174.

[3] Williams J. J., Rose C. P., Thorne P. D., Humphery J. D., Hardcastle P. J., O'Connor B. A., Moores S. P., Cooke J. A. & Wilson D. J. (1999) Field Observations and Predictions of Bed Shear Stresses and Vertical Suspended Sediment Concentration Profiles in Wave-Current Conditions. *Continental Shelf Research,* **19**(4), 507-536.

[4] Lohrmann A., Hacket B. & Roed L. P. (1990) High resolution measurements of turbulence, velocity and stress using a pulse to pulse coherent source. *Journal of Atmospheric and Ocean Technology,* **7**, 19-37.

[5] Soulsby R. L. (1997) *Dynamics of marine sands: a manual for practical applications.* Thomas Telford Publications, 249pp.

[6] Van Rijn L. C. (1989) *Handbook of Sediment Transport by Currents and Waves.* Delft Hydraulics Report H461.

[7] Zyserman J. A. & Fredsøe J. (1994) Data analysis of bed concentration of sediment. *Journal of Hydraulic Engineering,* **120** (9), 1021-1042.

A numerical modelling study of the impact on the sediment regime of a proposed Marina at Haifa

T.J. Chesher[1] & E. Hatarsi[2]
[1]HR Wallingford Ltd, Wallingford, Oxon, UK
[2]Haifa Economic Corporation Ltd, MTM Scientific Industries Center Haifa Ltd, Haifa 31905, Israel

Abstract

Haifa Economic Corporation is planning to construct a new marina at the resort of Haifa, Israel. As part of a series of studies during this planning stage HR Wallingford carried out studies to investigate the potential impact that the marina would have on the sediment regime of the adjacent areas.

An impact assessment study was devised which involved simulating the existing wave-driven sediment transport regime over a large area, using the HR coastal area modelling framework, PISCES. The combination of high resolution and large model area made simulation of all possible wave conditions and the associated sediment transport impractical. By analysing the wave, flow and sand transport equations of PISCES this large number of wave conditions was reduced to a more manageable set of representative conditions. By re-running PISCES including the proposed marina an accurate and reliable estimate of the impact was made.

Following this initial impact assessment, PISCES was used to assess the long-term impact of the marina and to investigate the potential for sand which may accumulate to the west of the marina to be transported under storm wave action into the nearby Shikmona Marine Nature Reserve.

1   Introduction

The Haifa Marina resort project was initiated by the Haifa Economic Corporation as the most important of one of several projects aimed at attracting tourism to the area. The site will cover a total area of some 1,000,000m^2, comprising a marina with 850 berths and 6km of frontage, a new 650m long sandy beach, a 90,000m^2 artificial lake, hotels and apartments, restaurants and other tourist facilities.

The proposed project site occupies the headland at the end of a long mountain range that extends north from Tel-Aviv and finishes at Mount Carmel (Figure 1). The currents and sediment transport patterns around this headland and within Haifa Bay are complex and spatially irregular, arising from the combined effects of wind and wave action, as well as the influence of larger-scale circulations. This aspect, together with the requirement to consider the potential impact on all areas of the coast that could be affected meant that the study required the application of a numerical model capable of simulating the above processes in two horizontal dimensions. Application of the HR Wallingford coastal area modelling framework, PISCES, enabled a comprehensive numerical modelling procedure to be undertaken to ensure an accurate prediction of the potential impact.

As with all models of this nature, the specific application requires the detailed analysis of the site and assessment of the appropriate input parameters including detailed topography and wave climate.

2   Site Description

The town of Haifa is situated some 80km north of Tel Aviv at 35°58'E,32°50'N, and marks a major deviation from the otherwise relatively gently curving coastline of Southern Mediterranean Israel. To the south of Haifa the beaches are characterised by a limited supply of the Nile-derived sediments, and the seabed slopes gently offshore to a depth of about 20m in 1500m. The headland at Haifa is marked with a number of reefs where waves break in moderate swell. As expected, the tide range is very small, ranging from 0.6m (spring) to 0.3m (neap).

The wave climate of the Southeastern Mediterranean is characterised by three wave seasons (Goldsmith & Sofer[1]), with largest wave heights in the December to March period, an intermediate climate in summer (June - September), and low wave heights in April/May and October/November. This wave climate, coupled with the gradually curving coastline from a north-facing coast in the south to a west-facing coast at Haifa, gives rise to a sediment regime which is characterised by strong net easterly transport of the Nile deposits in the south, modifying to net northerly transport that becomes gradually weaker as one moves northwards up the coast. At Haifa the sharp curve of the coastline gives rise to strong easterly potential transport of any available sediment (see for example Carmel et al[2]).

Significantly, the location of the marina is characterised by a rocky domain interspersed with pockets of sand. Winter storm conditions often give rise to removal of this cover of sand leaving only the base rock, which is subsequently re-covered during the calmer summer months.

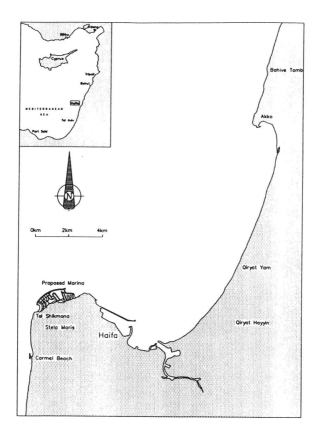

Figure 1 Location map

3 Data Collation

3.1 Bathymetry

A new bathymetric database covering an area some 15km by 9km was constructed from a database of soundings made in 1995, supplemented with data from the latest Admiralty Chart 1585 for Haifa Bay. All bathymetry was specified relative to

Israeli land survey datum (ILSD).

3.2 Wave data

Although measured wave data from a buoy offshore of Haifa was available, it only covered the period since December 1993 and was therefore not representative of the average annual wave climate. Consequently, data from the UK Meteorological Office European Wave Model was commissioned. This data is based on wind climate data since October 1986.

This wave climate, summarised in the form of wave height exceedence as a function of direction is given in Table 1. This clearly shows that the waves are largest from the westerly sector and sectors adjacent. Almost 70% of the waves are smaller than 1m, and storm waves in excess of 6m occur occasionally from the west. Typical *mean* period is of the order 4-6s.

```
Data in parts per hundred thousand
Hs is the significant wave height in metres
P(H>H1) is the probability of Hs exceeding H1
```

```
                    Annual

                    ------

        Total number of hours =      81096
        Based on UKMO predictions for     October    1986 - December    1995

   H1 To H2    P(H>H1)                     Wave direction in degrees North
   --------    ------                      -------------------------------
```

H1	H2	P(H>H1)	SSW	SW	WSW	W	WNW	NW	NNW	N	NNE	NE	ENE	E	ESE	SE	SSE	S
0.00	0.50	0.98190	382	753	1617	4597	7820	4120	2676	2000	1275	747	613	575	513	363	323	342
0.50	1.00	0.69474	499	1019	2245	9374	15850	6397	3755	2815	1512	940	732	720	1072	826	581	416
1.00	1.50	0.20721	196	527	1439	3673	3550	887	540	582	263	189	127	118	279	306	141	102
1.50	2.00	0.07803	47	210	520	1477	773	207	182	131	147	36	21	35	75	88	20	6
2.00	2.50	0.03829	18	138	472	695	296	49	59	12	38	31	5	4	7	10	0	0
2.50	3.00	0.01993	10	112	216	403	100	16	6	1	16	15	5	4	4	5	0	0
3.00	3.50	0.01080	4	97	164	239	46	5	4	0	0	6	1	0	0	0	0	0
3.50	4.00	0.00514	0	44	65	128	26	4	7	0	1	4	0	0	0	0	0	0
4.00	4.50	0.00234	0	0	49	58	12	9	0	0	0	1	0	0	0	0	0	0
4.50	5.00	0.00105	0	2	18	57	4	7	0	0	0	0	0	0	0	0	0	0
5.00	5.50	0.00016	0	0	0	0	1	0	0	0	0	0	0	0	0	0	0	0
5.50	6.00	0.00015	0	0	11	0	0	0	0	0	0	0	0	0	0	0	0	0
6.00	6.50	0.00004	0	0	0	4	0	0	0	0	0	0	0	0	0	0	0	0

```
Parts per thousand         12   29   68  207  285  117   72   55   33   20   15   15   19   16   11    9
for each direction
```

Table 1 UKMO average wave climate for HAIFA, 1986-1995

3.3 Sediment data

The sediment size data was deduced from the information supplied by the client for the previous physical model study, where the median grain size, D_{50} was 0.25mm. An important aspect of the sediment regime at this site is the limited supply of sand covering the otherwise rocky foreshore, which was confirmed by jetting experiments carried out in April 1993. Clearly this was an important requiring careful interpretation of the sediment transport predictions.

4   PISCES model study

4.1   Model description and layout

PISCES is a fully-interactive coastal area modelling framework, capable of simulating the various processes of wave propagation, current distribution, the resulting sediment flux, and the associated response of the coastline and bathymetry in complex coastal areas.

The PISCES model covered an area spanning 20km of the coastline and extending 5km offshore from the headland at Haifa, centred on the Port of Haifa. This region was covered with a regular model grid of size 50m in order to resolve the detailed flow patterns expected around the headland and marina structure, whilst allowing a large area to be modelled.

All wet areas were prescribed with an abundant supply of 0.25mm size sand. As described in Section 3.3, this is not always the case, and particularly following stormy conditions there is limited supply of sand over the rocky substrata. However, since this covering of sand varies temporally, it is not possible to determine the correct sand supply over all model areas, and hence it is assumed in the model that there is always sand available to be transported.

4.2   Selection of representative wave conditions

The generated wave data was used to define six representative wave conditions, which were be used to test the impact of the proposed marina.

Choice of appropriate wave height is made by appealing to the specific wave-driven current algorithm and sand transport algorithm used in PISCES in order to determine a suitable weighting factor. By these means the representative wave height H_{rep} was calculated according to:

$$H_{rep} = \left[\frac{\sum \left(fH_s^{1.95} \right)}{\sum f} \right]^{\frac{1}{1.95}}$$

where f is the specific frequency of occurrence of each wave and H_s is the significant wave height.

Each directional band therefore has a calculated representative wave height associated with it. Most of the significant waves occur in a wave mean period band 4-6 seconds. Choice of the specific wave period for each sector was made according to the relative frequency of each period, and the associated representative wave height. Having carried out this analysis for all wave directions it was clear that the dominant wave directions were from the west and adjacent areas. Using this analysis the following wave conditions were chosen:

Offshore Direction	Wave Height Hs (m)	Wave Period Tm (s)	Frequency	Frequency (waves > 1m only)
WSW	1.36	4.5	0.067	0.029
W	1.33	4.5	0.156	0.030
	1.76	5.5	0.051	0.036
WNW	1.30	4.5	0.243	0.028
	1.64	5.5	0.041	0.019
NW	1.30	4.5	0.116	0.012

The summed frequency of all six waves is 0.674; 32% of the time waves come from directions outside these sectors, but these waves are always relatively small. Clearly, the frequency of the waves larger than 1m from each sector are small, and larger waves than those chosen do occur, but the analysis described above yields the conditions most representative for sediment transport. Larger waves would transport more sediment, but the frequency of such conditions is much lower than those above.

4.3 Impact assessment

Standard application of PISCES comprises specification of the offshore wave conditions (height, period and direction) along the model offshore boundary. The FDWAVE model within the PISCES framework carries out a propagation analysis of the waves from this offshore boundary over the model domain, including the important physical effects of wave shoaling, refraction and breaking. Where the waves break a current-driving force is calculated. These wave forces are passed to the current module that solves the shallow-water equations of conservation of mass and momentum to obtain the wave-driven current field for the specified wave condition. Finally, the wave and current fields are used to define the sediment transport patterns for these conditions. These sediment fluxes are processed in terms of associated areas of potential deposition and erosion. By running the model for pre- and post-construction scenarios the effects of the proposed marina development can be readily demonstrated.

The results indicated the following broad scale effects of the marina:
- No significant effect on the coast to the south of the marina
- Deposition to the west of the marina in the vicinity of the proposed Golden Beach
- Reduction in the supply of sediment to the east of the marina
- An increase in wave activity and hence potential transport to the east of

the marina
* Little effect beyond the headland to the east of the marina
* No discernable effect beyond the tip of the Haifa port breakwater

The study identified a number of impacts that the marina could potentially make on the adjacent coastline which require interpretation in the light of site-specific information. Slight deposition is predicted to the west of the marina, and potential erosion to the east. As described above, the impact assumes an abundant supply of sediment, and the jetting information confirms that at many times during the year there is a limited supply of sediment available for transport. Hence, these results represent the worst case scenario.

4.4   Assessment of potential long term impact

Concerns of the Ministry of Environment were based upon the potential long term impact of the development on the Shikmona Marine Nature Reserve to the south.
The deposition footprint is shown in Figure 2. This footprint of the accumulated sediment is delineated by the areas of potential erosion to the west and south of the footprint and by the current close to the breakwater head, and changed little for all wave conditions modelled. The potential long-term impact was investigated by considering the supply of sediment to the footprint and running further tests with PISCES with a modified bathymetry. It was concluded that this bar reaches dynamic stability at a crest level of about −2m ILSD which could be obtained within 5 years. The footprint of the bar is outside the boundaries of the Marine Nature Reserve.

4.5   Effect of storm events in the long term

Further concerns were expressed regarding the potential for sand to be transported under storm wave action from the footprint into the rocky areas of the Marine Nature Reserve. To investigate this PISCES was re-run for pre- and post-construction scenarios to simulate storm waves of limited duration with 5m, 6m and 7m wave height. These storms with return periods of up to 1 in 10 years, move sediments from their natural position into the rocky parts of the Marine Nature Reserve forming potential deposits of up to 0.2m. The impact of the marina on this naturally occurring process in the 20 years following construction was shown to be negligible.

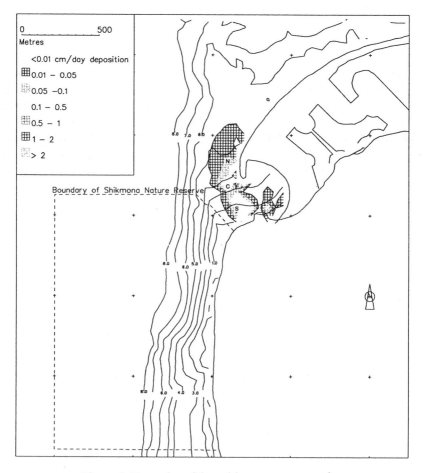

Figure 2 Footprint of deposition post-construction

5   Conclusions

The PISCES coastal area model was used to determine the wave-driven sediment transport regime in the Haifa region under existing conditions and following construction of a proposed marina.

This wave climate was analysed in terms of the sediment transporting capacity, and six wave conditions were selected in order to test the impact of the proposed scheme, whilst being largely representative of the annual sediment regime.

The impact assessment of the marina assumed an abundant supply of sediment and hence represents a worst-case scenario. As a result deposition is predicted to the west of the marina, and potential erosion to the east.

The potential long-term impact was investigated by considering the supply of sediment to the deposition footprint and running further tests with PISCES with a modified bathymetry. It was concluded that this bar reaches dynamic stability at a crest level of about −2m ILSD which could be obtained within 5 years.

To investigate the effect of storms on this long-term impact PISCES was re-run for pre- and post-construction scenarios to simulate storm waves of limited duration with 5m, 6m and 7m wave height. The impact of the marina on this naturally occurring process in the 20 years following construction was shown to be negligible.

6 References

[1]   Goldsmith, V. & Sofer, S.   *Wave climatology of the southeastern Mediterranean.* Israel Journal of Earth Sciences, Vol 32, 1983.

[2]   Carmel, Z., Inman, D.L., & A. Golik.   *Transport of Nile sand along the southeastern Mediterranean coast.* Proc 19th Coastal Engr. Conf., Amer. Soc. Civil Engr., 1984.

Predictive modelling of hydrodynamics and marine water quality: three applications along the South African coastline

S.A. Luger & R.C. van Ballegooyen
Environmentek, CSIR, P O Box 320, Stellenbosch, 7599, South Africa

Abstract

The predictive capability of hydrodynamic and water quality models is illustrated by three applications of the Delft3D modelling suite from WL|Delft Hydraulics along the South African coastline. The first application involves dredging-induced turbidity in Saldanha Bay on the west coast. Modelling of the hydrodynamics and the dispersion and deposition of the turbidity plumes was performed to assess the impact of the plumes on the sensitive ecology in the bay. The second application investigates the feasibility of using seawater for industrial cooling at Saldanha Bay. Since this study involves a thermal discharge into a highly stratified environment, the Delft3D model was applied in three-dimensional baroclinic mode including air-sea heat fluxes. The model outputs of absolute temperature, increase in temperature due the thermal discharge and dilution enabled the ecological impacts to be assessed and the optimum discharge location to be identified. The final application is the design of a marine outfall discharging a buoyant effluent near Durban on the east coast. Since the initial dilution process exhibited significant temporal and spatial variability, a dynamic coupling of near- and far-field hydrodynamic models was developed to accurately simulate the impact of the outfall. The results indicated that, under stratified conditions, a local optimum in the required pipeline length with respect to the impact of the effluent at the shore exists - the so-called "pipeline paradox".

1 Introduction

As in other countries, the coastal zone in South Africa is coming under increasing pressure due to rapid growth in domestic and industrial development along the coastline. This is placing conflicting requirements on the water quality in these areas, requiring an increased understanding of the underlying processes and a predictive capability to optimise solutions to these problems. Hydrodynamic and water quality models are well suited to this task and in this paper we discuss three applications of the Delft3D modelling suite from WL|Delft Hydraulics along the South African coastline.

The first application involves modelling dredging-induced turbidity plumes in Saldanha Bay on the west coast of South Africa. The second is a study to determine the feasibility of a thermal discharge into Saldanha Bay, while the third involves the design of a marine outfall on the east coast. In this paper we highlight some of the interesting or unique aspects in each of these studies.

2 Description of the models

The Delft3D modelling suite from WL|Delft Hydraulics in the Netherlands was used in the three applications described in this paper. The two modules applied in these applications were the Delft3D-FLOW hydrodynamic model and the Delft3D-WAQ water quality model.

2.1 Delft3D-FLOW hydrodynamic model

The Delft3D-FLOW model (WL|Delft Hydraulics [1]) solves the unsteady shallow water equations in three dimensions. The system of equations consists of the horizontal momentum equations, the continuity equation and the transport equations. The model includes formulations which take into account the following processes: tidal forcing, wind shear stress, air-sea heat fluxes, the effect of the earth's rotation (Coriolis force), free surface gradients (barotropic effects), horizontal gradients in pressure due to changes in water density (baroclinic effects), water with variable density due to temperature or salinity differences (equation of state), bed shear stress at the seabed, drying and flooding on tidal flats, discharge and withdrawal of mass (discharges and intakes), transport of heat, salt and other conservative constituents (advection-diffusion equation) and turbulence induced mass and momentum fluxes (k-ϵ turbulence closure model). An irregularly-spaced orthogonal curvilinear grid is used in the horizontal direction and a σ-coordinate grid is used in the vertical. The time integration method is the Alternating Direct Implicit type.

2.2 Delft3D-WAQ water quality model

The Delft3D-WAQ model (WL|Delft Hydraulics [2]) solves the advection-diffusion equation in three dimensions. Over one hundred substances may be modelled, including sediment particles, heavy metals, phytoplankton, macro algae, nutrients, organic waste, oxygen and bacteria. Processes that may be modelled include settling-deposition-resuspension of particles, adsorption-desorption of trace metals to these particles, primary production, mineralization, oxygen demand and bacterial die-off. These processes are implemented as additional source or sink terms in the advection-diffusion equation. The model obtains the hydrodynamic database through an offline coupling to Delft3D-FLOW.

3 Dredging-related turbidity in Saldanha Bay

3.1 Background

Saldanha Bay is located on the west coast of South Africa approximately 100 km north of Cape Town. It is an important tourist and industrial resource, supporting activities such as mariculture, fishing, cargo handling and large industries such as the steel industry, all of which place conflicting requirements on the water quality in the bay.

A proposed upgrading of the oil terminal presently operated in Saldanha Bay will require dredging up to 2.5 million m^3 of material from the existing entrance channel. Although dredged material is to be disposed in a confined disposal area it is anticipated that "leakage" associated with the dredging operation will place a quantity of fine sediment into suspension. Transported by the ambient flow regime, the suspended material can potentially move throughout the bay in the form of turbid plumes and so impair water quality and the surrounding habitat, in particular the adjacent mariculture areas and the ecologically-sensitive Langebaan Lagoon (an international Ramsar site). In order to assess these impacts, computer modelling of the hydrodynamics and the dispersion and deposition of the turbid plumes was performed.

3.2 Hydrodynamic modelling

The Delft3D-FLOW hydrodynamic model was used to simulate the three-dimensional flow regime in the semi-enclosed Saldanha Bay system. The processes included in the model were tidal forcing, wind forcing, Coriolis effects, baroclinic flows due to thermal stratification and the drying and flooding of tidal flats in the lagoon. An orthogonal curvilinear grid with cell sizes ranging from 200 m in the areas of interest to 1 000 m near the model boundary was used in the horizontal direction, with eight σ-coordinate layers used in the vertical.

The model was calibrated based on current and water temperature data measured at four locations in the bay for a 12 day period. Measured wind, water level and thermistor string data were used at the model boundaries. The following coefficients were found to give the best correlation to the measured data: Chezy coefficient for bottom friction = 65 $m^{0.5}/s$, horizontal eddy viscosity = 1 m^2/s, horizontal eddy diffusivity = 0.5 m^2/s, wind coefficient = $6.3 \times 10^{-4} + 6.6 \times 10^{-5} \times U_{10}$ (based on Smith and Blanke [3]) where U_{10} is the wind speed 10 m above the water surface. Figure 1 indicates a good comparison between the measured data and model currents. The water temperature is modelled less accurately since in this case the air-sea interaction module was not used and the thermal stratification is driven

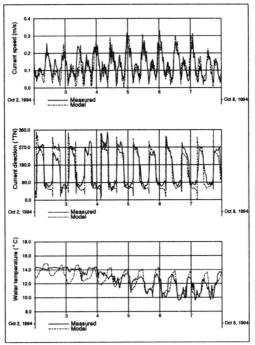

Figure 1: Comparison between measured data
and model results in Saldanha Bay.

by the imposed open boundary conditions only. A significantly improved calibration was possible when running the model in three-dimensional baroclinic mode compared to three-dimensional barotropic or two-dimensional modes. Figure 2 depicts the simulated three-dimensional current structure. Predicted current magnitudes are generally below 0.5 m/s, except in the constriction at the entrance to Langebaan Lagoon where currents may exceed 1.0 m/s at spring tide.

3.3 Turbid plume modelling

The Delft3D-WAQ water quality model was used to simulate the advection, dispersion, settling and deposition of the turbid plumes arising from the dredging operations. The model solves the advection-diffusion equation in three dimensions including settling of particles and deposition or erosion based on specified critical shear stresses. The bed shear stress is computed as the sum of the stress due to currents and waves. A wave refraction-diffraction model was used to simulate the wave heights throughout the model domain under the dominant wave conditions.

The turbidity loading rate due to "leakage" of fine sediment at the dredger was based on the measured characteristics of the sediment, the type of dredger and the proposed dredging plan. A best estimate loading of 9 kg/s for the silt/clay material (< 63 microns) and an extreme loading of 70 kg/s due to a failure of the pipeline

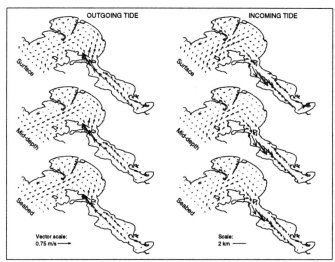

Figure 2: Modelled currents in Saldanha Bay at neap tide with a 10 m/s SW wind.

from the dredger were determined (refer to Luger [4]). The silt/clay material was subdivided by mass into three fractions each with a characteristic settling velocity and critical shear stress for deposition. Settling velocities of 0.05 mm/s, 0.5 mm/s and 2.0 mm/s (based on measurements) and corresponding critical deposition shear stresses of 0.1 Pa, 0.2 Pa and 0.3 Pa were used in the simulations.

The following outputs were obtained from the model simulations: contour plots of maximum turbidity, time series of turbidity levels and exposure times at the ecologically-sensitive sites in the bay as well as contour plots of the deposition thickness throughout the bay. Figure 3 shows the predicted turbidity plume due to the extreme loading case of 70 kg/s. The turbidity at the ecologically sensitive sites was predicted to be below 25 mg/l, which is within the range of the natural background turbidity. Wave-generated bottom shear stresses were found to have a significant influence on the results by inhibiting deposition of the finer mud fractions in the exposed areas of the bay, resulting in a pervasive spreading of these particles into the lagoon and also out to sea. The maximum deposition thickness in the lagoon was, however, predicted to be less than 2 mm over the dredging duration of 4 months.

3.4 Conclusions

When running the hydrodynamic model in three-dimensional baroclinic mode, a good correlation between observed and modelled currents was achieved. The wave-generated bottom shear stresses were found to have a significant influence on the turbid plume behaviour by inhibiting deposition of the finer mud fractions in the exposed areas of the bay.

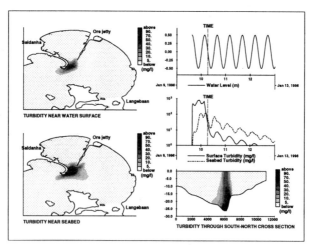

Figure 3: Predicted turbidity plume 3 hours after a simulated failure of the dredge line has been repaired.

4 Thermal discharge into Saldanha Bay

4.1 Background

This study investigated the feasibility of using seawater for industrial cooling as an alternative to ground water or river water. The thermal effluent would be discharged into Saldanha Bay at a temperature 27°C above intake temperature at a flow rate of 5 m^3/s. The potential impacts of this effluent include: increased temperatures and the associated impact on the ecology, biocidal action of possible residual chlorine in the effluent and modification of the food supplies to mussels due to displacement of natural waters by the effluent stream (which will contain negligible particulate organic matter). The objectives of the study were thus to predict the extent of these impacts and determine a discharge position which minimises these impacts.

4.2 Set-up of the hydrodynamic model

Except for a few months during winter, the Saldanha Bay system displays strong thermal stratification with temperature differences up to 8°C measured across the thermocline. This stratification is maintained by atmospheric heat fluxes into the surface waters and the inflow of cold bottom waters from upwelling on the adjacent open shelf. The water temperature in the system also displays a high temporal variability due to wind-induced vertical mixing and the advection of cold water from the shelf and warm water from the lagoon. This temperature structure has a significant influence on the hydrodynamics in the system and thus on the advection-dispersion of the thermal plume. In addition, the impact of the thermal plume on the ecology will depend on the relative temperature change compared to ambient variability. It was thus important to simulate the ambient stratification realistically.

The simulations were performed using the three-dimensional baroclinic Delft3D-FLOW model and the following processes were included: tidal forcing, wind forcing, Coriolis effects, drying and flooding of tidal flats, as well as air-sea heat fluxes due to insolation, back radiation and turbulent heat fluxes across the air-sea interface. The scenarios modelled included representative spring, summer, autumn and winter conditions as well as periods of exceptionally calm conditions which were anticipated to constitute a worst case scenario. Each scenario comprised a 28 day period of simulated bay hydrodynamics forced by measured winds, predicted tidal levels at the mouth of the bay and seasonal atmospheric inputs of insolation, relative humidity, air temperature and thermal stratification at the mouth of the bay. Five different discharge locations were investigated including both surface and bottom discharges.

4.3 Results

To determine the increase in temperature due to the thermal discharge, the simulated temperatures with and without the discharge were compared at each time step in each computational segment. These temperature increases were presented as exceedance contours at the 50%, 80% 95% and 99% levels for each of the discharge options. Figure 4 shows the 99% exceedance contours near the seabed for the five different discharge locations. Also of importance in determining thermal impacts are the occurrence of high short-term increases in temperature as well as the time of exposure to various water temperatures. Time series of the temperatures at a number of ecologically sensitive monitoring sites were therefore produced.

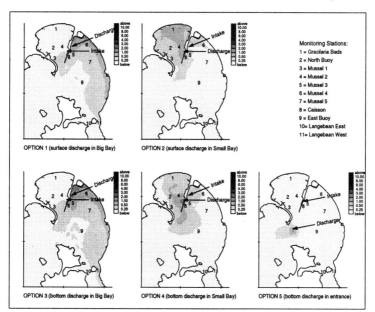

Figure 4: Predicted increase in water temperature near the seabed for five possible locations for a thermal discharge.

The stratification present in the bay was found to play a significant role in both the near- and far-field behaviour of the thermal plume. Strong stratification in the vicinity of the discharge together will strong flows tended to trap the thermal plume in the subsurface layers for bottom discharges and restrict the plume to the surface layers in the case of surface discharges. In the far-field, temperature reductions occurred by heat loss to the atmosphere and mixing of the thermal plume with the cold bottom waters of the bay, the latter being the most effective process. The surface discharges therefore lead to larger impacts than the bottom discharges.

4.4 Conclusions

In this study, involving a thermal discharge into a highly stratified environment, a three-dimensional baroclinic model including air-sea heat fluxes was required to take all the relevant processes into account. The model outputs of absolute temperature, increase in temperature and dilution enabled the ecological impacts to be assessed and an optimum discharge location to be identified.

5 Design of a marine outfall near Durban

5.1 Background

Durban is located on the east coast of South Africa. The coastline in this region is relatively straight with an average seabed slope of 1:140. A study has been performed to design a marine outfall discharging buoyant wood pulp effluent. The objective was to determine the outfall length and diffuser configuration which minimised the visibility of the effluent at the adjacent beaches.

5.2 Modelling approach

In far-field hydrodynamic models based on the three-dimensional shallow water equations and the hydrostatic assumption, the immediate effect of buoyancy on vertical acceleration is in general not considered and density differences are taken into account only in the horizontal pressure gradients and in the vertical exchange coefficients. The grid size is also large (typically 50 to 100 m) compared to the spatial scales involved in the initial dilution processes (diffuser port diameters in the range 0.1 to 0.5 m). These models are thus generally not capable of accurately modelling the initial dilution process, which may lead to inaccuracies in the far-field predictions. For this reason, a coupling of near- and a far-field models was developed for this study.

The far-field model (Delft3D-FLOW) was run in three-dimensional baroclinic mode and included the following process: wind forcing, Coriolis effects, baroclinic flows due to thermal stratification and remote forcing due to the Agulhas current. The latter forcing was included as a velocity boundary condition using time series of measured currents. The far-field model was used to predict time-series of current and density profiles at the diffuser position. These data are used to compute the near-field rise height and initial dilution for each diffuser port at each time step using a diffuser hydraulics model coupled to the UOUTPLM near-field model developed by the US EPA [5]. These results are used to compute into which elements in the three-dimensional far-field grid the effluent should be released.

5.3 Results

An example of the input and output from the near-field model is shown in Figure 5. Note how the effluent surfaces when the stratification is reduced by mixing and the associated increase in initial dilution due to the increased length of buoyant rise. The far-field model results were analysed to determine the three-dimensional zone impacted by the effluent for various percentages of time (Figure 6). The model simulations were performed for outfall lengths between 3 and 9 km. Comparing the predicted impact of the effluent at the shore, a local optimum of approximately 5.5 km is found (Figure 7). This unanticipated result is caused by the increasing strength and persistence of the thermal stratification with distance offshore, which tends to trap the plume within 15 m of the seabed and reduces the initial dilution for the longer pipelines to an extent which cannot be compensated for by the increased secondary dilution associated with the longer travel time to shore.

5.4 Conclusions

The initial dilution process exhibited significant temporal and spatial variability which required a dynamic coupling of near- and far-field hydrodynamic models to accurately simulate the impact of a marine outfall. The results indicated a local

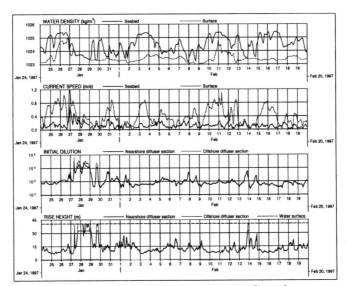

Figure 5: Time series of the input and output from the near-field model

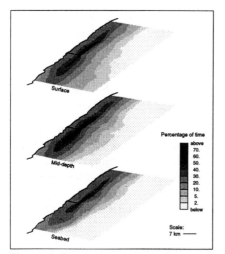

Figure 6: Occurrence zones where effluent will have an impact for 6.5 km pipeline

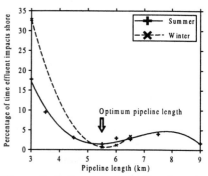

Figure 7: Influence of pipeline length on percentage of time the effluent will have an impact on the shore

optimum in the required pipeline length with respect to the impact of the effluent at the shore existed - the so-called "pipeline paradox".

6 Conclusions

Hydrodynamic and water quality models are able to provide a predictive capability which can assist in optimising solutions to water quality problems. While these models are capable of adequately addressing the *"what-if"* questions with regard to the physical and bio-geochemical impacts of pollutants, further research is required in addressing the *"so-what"* questions regarding the corresponding ecosystem responses.

References

1. WL|Delft Hydraulics, *Delft3D-FLOW User Manual release 3.00*, Delft, The Netherlands, 1997.
2. WL|Delft Hydraulics, *Delft3D-WAQ User Manual release 3*, Delft, The Netherlands, 1998.
3. Smith, A. and Blanke, E. Variation of the sea surface drag coefficient with wind speed. *Quart. J. R. Met. Society*, No. 101, pp. 665-673, 1975.
4. Luger, S.A., Schoonees, J.S., Mocke, G.P. & Smit, F. Predicting and Evaluating Turbidity Caused by Dredging in the Environmentally Sensitive Saldanha Bay, *Proc ICCE 98*, Copenhagen, 1998.
5. EPA. *Initial mixing characteristics of municipal ocean discharges*, Environmental Research Laboratory, Report EPA/SW/MT-86/012(a+b), 1985.

3D model system for hydrodynamics, eutrophication and nutrient transport

E. Kock Rasmussen, I. Sehested Hansen, A.C. Erichsen,
D. Muhlenstein & J. Dørge
DHI – Water and Environment, Denmark

Abstract

The existing MIKE 3 model system consisting of modules for hydrodynamics and eutrophication has been improved to include a description of exchangeable pools of nitrogen (N) and phosphorus (P) in the sediment. It is now possible to make a complete mass balance for N and P in a model area and to make calculations of the transport of nutrients across sections in the model. Besides the new feature the model system is also able to simulate water levels, current fields, concentrations of nutrients, chlorophyll, dissolved oxygen, biomass of eelgrass and benthic macroalgae.

The model system has been calibrated (1998) and validated (1997) on a temperate shallow Danish fjord, (Odense Fjord, Funen). The fjord, which covers 64 km^2 and has an average depth of 2.5 m, is connected to the sea through a narrow entrance.

The hydrodynamic module was able to reproduce the water levels and the salinity satisfactorily at three stations in the fjord for both years.

At two stations weekly data of water chemical parameters, such as total P & N and inorganic nutrients, were available for the calibration and validation of the eutrophication module. Based on model results a nutrient budget was set up for 1997 and 1998. Due to low precipitation the 1997 load of N and P was about 60 % lower than in 1998. The yearly retention of N for the two years was 35 % and 15 % and was caused by the burial of N in the sediment and denitrification. The yearly retention of P was negative,–121% and –28 %, for the two years due to the release of P from the sediment. This is consistent with the implementation of the nutrient treatment of domestic sewage in the early 1990's.

1 Introduction

With the introduction of faster PC's it has become possible to make 3-dimensional (3D) simulations of water levels, currents, water quality and the transport of nutrients. In many systems, like dimetic lakes and marine areas with haloclines or termoclines, a (3D) resolution of the transport processes is needed to be able to make a sensible prediction of, for example, the water quality.

In connection with an application for an extension of the emission of cooling water from a coal-fired power plant, a 3D model system was set up on Odense Fjord (Funen, Denmark) with the purpose of assessing the effect on the environment.

The model system used was DHI's 3-dimensional model, MIKE 3, consisting of a hydrodynamic (HD) module simulating water level and currents, a advection/dispersion (AD) module for the calculation of the salinity and a eutrophicaton module for the calculation of water quality.

In order to simulate the important ecological parameters in the fjord and the transport of nutrients in and out of the area, the eutrophication module has been extended with a description of the benthic macro vegetation and a description of nitrogen and phosphorus pools in the sediment.

2 Materials and methods

2.1 The model system

The hydrodynamic module and the advection dispersion modules used in the model system are described in Abbott, McCowan & Warren [1], Vested et al. [2] & DHI [3].

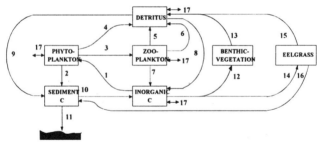

1. Production, Phytoplankton	10. Mineralization of Detritus
2. Sedimentation, Phytoplankton	11. Accumulation in Sediment
3. Grazing	12. Production, Benthic/Vegetation
4. Extinction, Phytoplankton	13. Extinction, Benthic/Vegetation
5. Excretion, Zooplankton	14. Production, Eelgrass
6. Extinction, Zooplankton	15. Sloughing, Eelgrass
7. Respiration, Zooplankton	16. Sedimentation of Eelgrass
8. Mineralisation of Suspended Detritus	17. Exchange with surrounding Waters
9. Sedimentation of Detritus	

Figure 1: Processes and state variables in the ecological model exemplified by the carbon cycle.

The eutrophication model used in this application consists of the standard eutrofication model from DHI for the pelagic system, DHI [4], with the addition of rooted vegetation (Zostera marina & Ruppia) and annual macroalgae (mainly Ulva sp.), Bach [5], Bach & Rasmussen [6]. In figure 1 the state variables and processes for carbon (C), nitrogen (N) and phosphorus (P) are presented for the pelagic system including benthic vegetation. The state variables in the standard module include: phytoplankton C, N & P, zooplankton C, detritus C, N & P, inorganic N & P, oxygen and chlorophyll. The benthic vegetation includes macroalgae C, N & P and for rooted vegetation: the number of shoots pr m^2 and the biomass (C, N & P) of one shoot.

The above eutrophication module, including benthic vegetation but without sediment N and P, has been used in combination with 2-D and 3-D hydrodynamic setups assessing the ecological effects of sediment spill due to dredging, reclamation and a change in nutrient load, Bach, Rasmussen & Foster [7] and Bach, Jensen & Lyngby [8].

The N cycle in the sediment is presented in figure 2. Three state variables are included in describing the conversion of N: sediment organic N (SON, g N/m^2), NH_4-N in pore water (SNH, g N/m^3) and NO_3-N in the pore water (SNO$_3$, g N/m^3). The simulated SNO$_3$ is the concentration in the uppermost few millimetres (KDO$_2$) of the sediment, where oxygen from the water is able to penetrate the sediment. In this layer NO_3-N will reach its maximum value due to the nitrification of NH_4-N. Below this layer NO_3 is denitrified to N_2 and the NO_3 concentration will therefore be 0 a few centimetres from the sediment surface (KDOX).

The O_2 penetration into the sediment is estimated using an empirical relation generated using data from Danish coastal marine sediments.

The penetration of NO_3 into the sediment (KDOX) is estimated using Ficks First Law for the diffusion of NO_3 into the sediment.

The flux of NO_3-N and NH_4-N between the water and the pore water in the sediment is calculated as the product of a diffusion constant and the concentration difference between water and pore water.

The model description of P in the sediment is presented in figure 3. The model includes three state variables: sediment organic P (SOP, g P/m^2), PO_4-P in pore water (SIP, g P/m^3) and P chemosorbed to oxidised ion and manganese (FESP g P/m^2). P is released into the pore water as PO_4-P due to the mineralisation of SOP. PO_4-P can be adsorbed to metals, mainly Fe & Mn, provided these metals are in their oxidised form. If the metals are reduced again the PO_4-P is released into the pore water. In the model the metals are assumed to be in their oxidised form if NO_3 is present in the pore water. In other words, Fe & Mn are assumed to be able to adsorb PO_4-P in a layer down to a sediment depth of KDOX.

As for NH_4-N and NO_3-N, the PO_4-P flux across the sediment surface is calculated as the product between a diffusion constant and the concentration difference of PO_4-P in the pore water and the water.

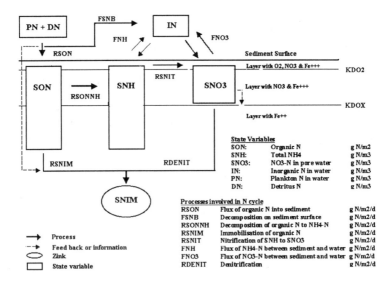

Figure 2: Diagram of state variables and processes in the sediment N model.

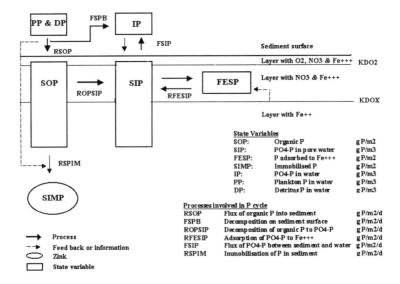

Figure 3: Diagram of state variables and processes in the sediment P model.

2.2 Study site

Odense Fjord is a shallow lagoon with a mean depth of 2.5 m covering an area of 64 km^2, see figure 4 and 11. The fjord is connected to Kattegat through a narrow gap, through which water passes in and out driven mainly by tidal and wind-induced changes in the water level outside the fjord. Occasionally, the water is exchanged through a lock exchange of water due to changes in the salinity outside the fjord. The fjord consists of an inner part, Seden Strand, which receives the discharge from five streams, of which Odense Stream and Stavids Stream are the biggest.

The depth in Seden Strand is between 0.5 to 1 m except for a dredged fairway. An island separates Seden Strand from the outer part of the fjord.

The County of Funen has monitoring stations for water chemistry at two locations: one at Seden Strand and one at the outer part of the fjord. Once a year the fjord is monitored for benthic vegetation cover and biomass is collected at selected locations. Continuous records of water levels have been available from a station situated at the entrance and at two stations in the inner part of the fjord. Likewise, continuous salinity has been monitored at several stations in the fjord. Nutrient loading and monitoring data presented in this article have been provided by the County of Funen, County of Funen [9].

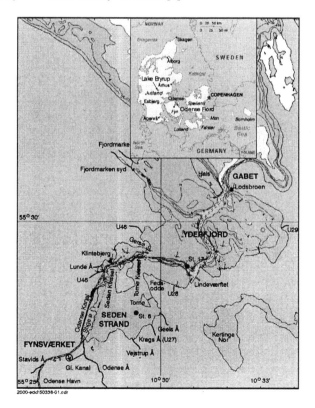

Figure 4: Odense Fjord, Funen Denmark (55 28 N, 10 28 E).

3 Results

The different modules in the MIKE 3 model system have been calibrated and validated separately against measured data.

The hydrodynamic model is calibrated against water level data from two stations and for salinity data from three stations, see examples in figures 5 and 6. Both the simulated salinity and the water levels match the measured data. A similar agreement between measured and simulated water levels and salinity has been achieved at the other stations and at other periods. Based on this information it may be concluded that the hydrodynamic model, as well as the advection dispersion model, is well calibrated and the model is ready to be used for the next step: water quality modelling.

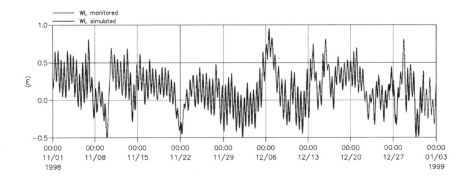

Figure 5: Measured and simulated water levels at Klintebjerg, Nov. - Dec. 1998.

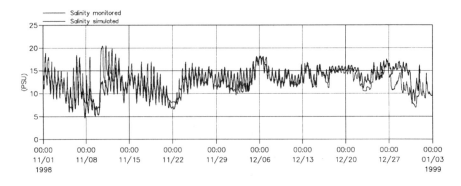

Figure 6: Measured and simulated salinities at Klintebjerg, Nov. - Dec. 1998.

The modelled and measured total N & P and inorganic N &P from station 8 in Seden Strand are presented in figures 7 and 8. There is a good agreement between measured and simulated values for all four parameters. The N pattern is partly due to a decrease in the N load and partly due to the fact that the denitrification of the inorganic N (NO_3-N) is high in the summer period. The opposite trend is seen for PO_4-P with the highest concentrations during the summer despite the fact that the load from the land is at its lowest. The summer maximum of PO₄ has been explained by an internal load of P.

Simulated pore water concentrations of NO_3-N in the uppermost sediment layer with oxygen are presented in figure 9 together with the thickness of the oxidised layer (KDOX). The oxidised layer is dependent on the concentration of NO_3 in the water as well as on the nitrification and denitrification processes in the sediment. During the summer, when the NO_3 concentration in the water is low and denitrification in the sediment is at its highest, the oxidised layer is smallest, whereas it is highest during winter. The simulated seasonal variation of PO_4-P chemosorbed to oxidised metals in the sediment (FESP) has a seasonal variation, which is the opposite of the variation of the oxidised layer. During the summer the oxidised metals in the sediment are reduced and the PO_4-P is released into the pore water, increasing the flux of PO4-P from the sediment to the water, see figure 10.

The model results confirm the idea of the internal load of P being responsible for the summer maximum of PO_4-P in the water.

In a shallow fjord the biomass of the benthic vegetation may be important for the seasonal variation of dissolved nutrients as well as the overall N and P budget. This model includes both rooted vegetation (Zostera marina & Ruppia sp.) and macroalgae (mainly Ulva). Simulated biomasses of macroalgae and rooted vegetation are presented in figure 11.

The biomass of the macroalgae is highest at the innermost part of the fjord, where the N and P loads from the city of Odense, and the rivers, enter the fjord. The few measured biomasses of macroalgae match the simulated biomasses.

Until the beginning of the 1990's, before chemical treatment of wastewater from the city of Odense, massive blooms of Ulva occurred at Seden Strand. After the reduction of the N and P loads the Ulva blooms were restricted and Ruppia vegetation increased. The differences in N and P loads between 1997 and 1998 resulted in a similar pattern, which was also reproduced with the model.

The simulated biomass of rooted vegetation shows the highest biomasses in the outermost part of the fjord, where the nutrient load is lowest. The simulated biomass of the macroalgae and the rooted vegetation had the same spatial distribution as map coverage of eelgrass and macroalgae generated from air photos (not shown), County of Fuen [10]. When the biomass fits the information available on the macro vegetation it can be assumed that the model is simulating the biomass of nutrients in the vegetation correctly.

4 Nutrient budget

Using the model results, it is possible to calculate nutrient budgets for the entire fjord or for sections of the fjord. In table 1 the nutrient budgets for total N and P for 1997 and 1998 are presented. In 1997 the N and P loads were only 37 % and 43 % of the loads in 1998. The significant lower loads in 1997 are due to lower precipitation resulting in lower non-point loads of N and P from the intensively farmed catchments. The budget for phosphorus shows a negative retention for both years. A breakdown of the P budgets (not shown) shows that this negative retention is caused by P released from the sediment. This finding is consistent with the implementation of N and P treatments of domestic sewage at the beginning of the 1990's, reducing the average P load with about 80 %. The N budgets show a net retention of 15-35 % of the load mainly due to the burial of organic N in the sediment and denitrification.

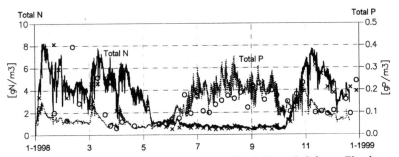

Figure 7: Simulated and measured total N & P, st. 8 Odense Fjord

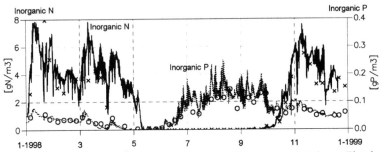

Figure 8: Simulated and measured inorganic N & P, st. 8 Odense Fjord

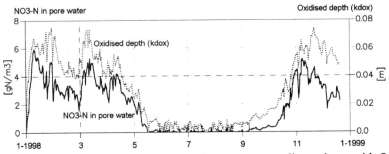

Figure 9: Simulated concentration of NO_3 in uppermost sediment layer with O_2 together with penetration depth of NO_3 in sediment (KDOX).

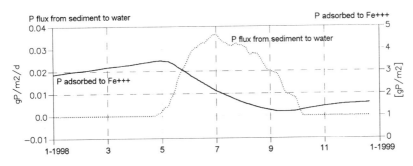

Figure 10: Simulated pools of PO$_4$-P chemosorbed to oxidised metals in sediment together with simulated fluxes of PO$_4$-P from sediment to water.

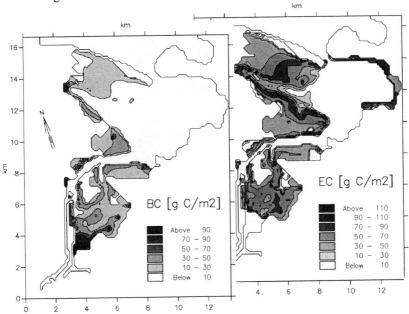

Figure 11: Simulated biomasses of macroalgae (left figure) and rooted vegetation (right figure) in Odense Fjord, mid August 1998.

Table 1: Nitrogen and phosphorus balance for Odense Fjord 1997-98

Year	Nutrient	Loading Tonnes/year	Retention Tonnes/year	Export Tonnes/year	Retention % pr. year
1997	N	1264	446	818	35
	P	30.4	-36.7	67.1	-121
1998	N	3408	502	2906	15
	P	70.0	-19.6	89.6	-28

5 Conclusion

In addition to traditional water quality parameters it is also possible to set up nutrient budgets for a user-defined area using a combined model system including modules for hydrodynamic, advection dispersion and eutrophication extended with a description of the N and P cycles in the sediment.

In the case of Odense Fjord, the model revealed a significant internal loading of P from the sediment after implementation of N and P treatments of domestic sewage years ago and a retention of nitrogen in the range of 15-35 % of the load from land.

6 Acknowledgement

Fynsværket and the County of Funen are acknowledged for their funding and data delivery which have been the basis for the Odense Fjord case study.

References

[1] Abbott, M.B., McCowan A., Warren I.R. Numerical modelling of free-surface flows that are two-dimensional in plan. Transport models for inland and coastal waters. Academic press. pp. 222-283.

[2] Vested, H.J., Justesen, P. and Ekebjærg, L.C. Advection-dispersion modelling in three dimensions. Appl. Math. Modelling, Vol. 16, October 1992

[3] DHI, MIKE 3 Reference manual, A Three-dimensional Hydrodynamic Model, A Short Description. 1998.

[4] DHI, MIKE 3 Eutrophication Module, User guide and reference manual, release 2.7. DHI - Water and Environment. 1998.

[5] Bach H. K., Eciological modelling describing the seasonal variation on growth and distribution of eelgrass (Zostera marina L.). I. Model theory. Ecol. Mod. 65(1) PP 31-51, 1992

[6] Bach H.K., Rasmussen E.K., Riber H., The application of an ecological model to assess the impact of sediment spill on benthic vegetation. In Proceedings of the 12th Baltic Marine Biologists Symposium 1991, ISBN 87-85215-25-2

[7] Bach H. K., Rasmussen E.K., Foster T., Eutrophication modelling of a tidally influenced mangrove area in Bali subject to major dredging and reclamation activities. in Brebbia C.A. (ed.) 1998: Environmental Coastal Regions. PP. 251-251. WIT press, Boston.

[8] Bach H.K., Jensen K, Lyngby J.E., Management of Marine Constructions Works Using Ecological Modelling. Estuarine, Coastal and Shelf Science, 44 (Supplement A), .pp. 3-14, 1997

[9] County of Fuen, Denmark. Monitoring data delivered from ongoing national monitoring programme.

[10] Bio/Consult as, Distribution of vegetation and common mussel, October 1998, Odense Fjord, revised report, 1999.

Laboratory simulation of ocean surface circulation in the Canary Islands area: applications to pollutant transport prediction

M. Pacheco, G. Rodríguez, L. García-Weil, A. Tejera
& E. Pérez-Martell
*Department of Physics, University of Las Palmas de Gran Canaria,
35017 Las Palmas, Spain*

Abstract

The adverse effects of spilling pollutants in water have focused attention on the behaviour of contaminants at sea, especially at coastal and shelf seas due to their key role for a wide range of human activities and interest. In particular, when oil is spilled in the marine environment, a primary concern is where the oil will go. Oil spill models suitable for use in oil spill response and contingency planning, providing rapid predictions of the movement of spilled oil, require an adequate knowledge about the ocean surface circulation in the area around the oil spill launch site to estimate the probabilities of contacting portions of a coastline. Unfortunately, field measurement programs are so expensive and time consuming that usually the existing information is scant and sparse. Numerical and physical models represent an alternative to alleviate this drawback.

This paper presents preliminary results of ocean circulation in the Canary Islands area obtained through laboratory simulations. Experiments were carried out in the SINTEF Coriolis rotating basin, using a simplified topography that reflects the most relevant morphological aspects of the study area in the upper 400 m of the ocean. The tank was filled with water of homogeneous density, in such a way that the density stratification effects were removed. Therefore, the observed flow perturbations might be due to the gradients of fluid depths, or vorticity changes, the blocking effects of the islands on the incident flow and the topographic irregularities of the coastline. Ink and surface drifters were released to visualise relevant structures and to estimate the surface flow by tracking the drifter paths under different flow conditions.

1 Introduction

Marine pollution has two main sources, the land and the sea. Land-based sources include pipelines and storm-water run-off, while marine sources include shipping accidental, deliberate and operational discharge and activities such as offshore oil-drilling operations.

Large oil tanker spills accidents (eg. the Urquiola, May 1976, in La Coruña, Spain; the Amoco Cadiz, March 1978, in France, and more recently the ERIKA, December 1999, in France) have demonstrated the particular susceptibility of the coastal environment to oil spills. These events evoke visions of despoiled coasts, oiled birds, and economic loss from ruined fisheries. Thus, for example, the ERIKA oil spill soaked in heavy crude more than 400 km of coast line and the costs related to the spill are estimated to exceed one billion French Francs (USD 160 million). In addition to pollute the coastal environment and damage marine life, the movement of oil slicks can cause the shutdown of power and desalination plants, cutting off supplies of fresh water and electrical power, with the consequent socioeconomic effects, particularly in small oceanic islands.

Due to their geographical location, the Canary Islands and the Saharan coasts are considerably vulnerable to oil spills because of the significant volume of oil transported through their surrounding waters from the Middle East to Europe around the South African shipping lane. This is one of the most important routes for crude transport in the world, and is known as *"Cabo de Buena Esperanza"* route. Thus, for example, on 19 December 1989 the Khark 5 tanker spilled some 20 million gallons of oil to the Atlantic Ocean, 185 km from Moroccan coast. So, it results evident that these regions should have a number of oil-spill contingency plans and personnel trained to forecast the movement of oil slicks, to ensure an efficient actuation in the event of a spill.

The development of adequate oil-spill contingency plans, to ensure that when oil spills do occur they can be dealt with effectively, requires information about oceanographic and meteorological regional conditions. Measurements of temperature and salinity distributions, wind and wave climatology, and bathymetry help to determine how spills will be transported and dispersed and the conditions that holding teams will encounter as they attempt to contain the oil. Knowledge of winds and ocean currents play a key role in the prediction of likely spill trajectories. However, field measurements of these parameters are seldom available.

Numerical models have been developed by several authors to simulate the time-space evolution of oil spills in the marine environment. These models can be classified into two main groups: At one hand, those considering the oil slick movement a primarily deterministic phenomenon governed, ultimately, by the transport equation derived from the general continuity and momentum conservation laws. On the other hand, those taking into account the random character of oil slicks movement at sea and, consequently, focussing the problem from a statistical point of view. Each group of models present advantages and disadvantages in relation to each other, but a common drawback is that most of the proposed models depend on some empirical relationship or simplification to circumvent theoretical and practical difficulties preventing to arrive at a successful resolution of the problem. Independently of the relative goodness of the different kind of mathematical models, it is clear that no agreement on a general theory to fully explain the oil spill movement exists.

A recent approach that could alleviate some of the existing problems and contribute to significant improvements in predicting oil spill movements is the use of laboratory circulation models, such as suggested by McClimans [7]. According to McClimans [5] a laboratory model is an advanced experiment where the boundary and forcing conditions are as close to natural conditions as the governing conditions allow. Several laboratory models have been used for dispersion studies [3].

In regions where the topography steers the flow, there is often a good validation to field measurements. In these regions, the requirement for geographical resolution of a numerical model is also very stringent. In spite of their good topographic resolution capabilities, differences up to 50% between laboratory and field dispersion measurements have been reported [6]. The lower dispersion values observed in laboratory have been attributed to the wind effects that are seldom included in this kind of model, due to technical difficulties.

Nevertheless, large-scale motions can be simulated accurately in a rotating basin. This is because of such flows, where the force balance is nearly geostrophic, are governed by the gravitational Froude and rotational Rossby similitudes, which are compatible [5]. The viscous Ekman layer near the bottom is only a few millimetres thick in most of these laboratory models. The most important effect of friction is therefore the wind forcing of surface currents, technically difficult to simulate in the laboratory, as mentioned above, and is therefore relegated to numerical models, which play an important role in this context.

Laboratory model approach is used in this study to simulate circulation in the upper layer of the ocean in the Canary Islands region. Results from the present study should improve our understanding of the surface dynamics in the study area and, consequently, our ability to predict the movement of potential oil or other pollutant discharges at sea, even though the stratification effects and the wind shear stress are not considered.

2 Background

The Canary Archipelago is placed at about 28 degrees North in latitude in the North Atlantic Ocean, very close (approx. 200 Km) to the north-west coast of Africa. Along this eastern boundary of the Atlantic Ocean flows the Canary Current, which is associated with the Northwest African upwelling. Furthermore, the Canary Archipelago acts as a barrier of about 500 km for the Canary Current. As a consequence of the African coast proximity and the obstruction of the ocean current flow by the islands, this region is characterised by a large mesoscale activity on the backside of the islands [8].

Surface manifestations of various mesoscale features, such as island wakes, large cyclonic and anticyclonic eddies downstream of the islands, and cool water filaments extending offshore from the upwelling region, sometimes reaching the islands, have been revealed by means of satellite observations [10, 9]. While clearly observed in satellite images and inferred from some in situ measurements (temperature and salinity fields), the physical origin and the necessary conditions for the generation of these mesoscale features is not well understood.

The physical processes underlying the dynamics of the formation and the stability of these features have been partially examined by using numerical models, in situ measurements and remote sensing data. However, the present knowledge of these physical processes does not result enough to give an appropriate explanation of the observed patterns, giving place to discrepancies among different hypothesis.

3 Laboratory model

The laboratory model was developed in the Coriolis rotating basin at SINTEF NHL. It has a size of 5 m diameter and 0.5 m height and was equipped with a video camera rotating firmly with the tank, viewing from above. Experiments were performed with a simple topography representing the most relevant morphological aspects of the study area in the upper 400 m of the ocean.

The tank was filled with water of homogeneous density, removing in this way the effects of density stratification and baroclinic motions. Therefore, observed flow perturbations might be due to the gradients of fluid depths, or vorticity changes, the blocking effects of the islands on the incident flow and the topographic irregularities of the coastline.

The model was run for three different flow conditions (1.5 Sv, 3 Sv and 6 Sv). Sets of 20-30 surface drifters were released, regularly spaced and aligned to the source orientation, at various instants and distances upstream Canary Islands, to track the flow along the African coast and across the islands, under different flow conditions. Ink of different colours was spilled at different places, inside the source, close to the African coast, around the islands, etc., to visualise relevant structures, also under different flow conditions. Tracking individual particles allows the identification of surface oceanographic features too. However, structures such as, eddies and filaments are easier observed by following the dye evolution or the ink plume envelope. Particles used to follow surface currents were black colour while used dyes had different colours (red, blue, green and black) to facilitate visual recognition from different dye spills.

It is worth of mention that particles and ink spills can be considered as imaginary oil spills. So they can be tracked and obtain information about the likely oil slick paths in terms of the point where it was spilled and the flow conditions.

The model scaling was determined by practical considerations of laboratory conditions. The horizontal length scale (ratio) was given as L_r = 1500 km/5 m = 300000 and a convenient vertical length scale (ratio) was H_r = 500 m/50 cm = 1000, leading to a vertical model distortion of 300. Time scale is $T_r = L_r (H_r \, \varepsilon_r)^{-1/2}$

≈ 9500, where ε_r is the ratio of natural to model density gradients, in this case ε_r = 1, and a day is simulated in 9.1 s.

The model bathymetry, the solid boundaries, and the source and sink are sketched in figure 1., where the circle represents the Coriolis basin wall. The shaded rectangle is the camera field of view, and the two zigzag lines stand for the coastline and the 200m isobath, which are overlapped with the true coastline and isobath (wrinkled lines).

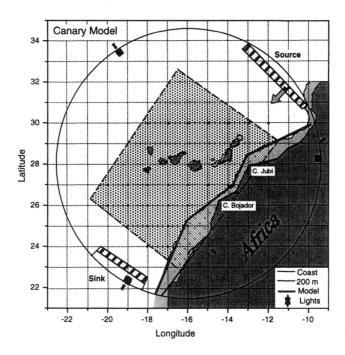

Figure 1: Canary Islands model region.

Note that between these two topographic levels the bathymetry displays, in a simplified but realistic way, two broadening zones, northward of Cape Jubi and southward of Cape Bojador. Below the 200m isobath the shelf falls vertically to the model bottom (400m), which is also a realistic simplification.

4 Results

About 100 particles, from the several hundreds released during the experiment, were chosen to be examined in the image sequences obtained by digitising the video camera film. Each sequence of video frames was manually analysed to establish the paths of individual particles and therefrom determine the velocities.

Velocities are estimated measuring the displacement of particles during the prescribed time between two consecutive video frames.

The selection of particles was conditioned by the interest of examining the possible paths followed by the oil patch spilt from a hypothetical accident of a tanker navigating along the *"Cabo de Buena Esperanza"* route, and occurred northward Canary Islands. This idealised oil spill could be one similar to that, mentioned in section 1, of the Khark 5 tanker, which in 1989 spilled 80,000 tonnes of oil, 185 km from Moroccan coast, just in the zone where is placed the source in the model, marked with an arrow in Fig.1.

Each particle has been considered as an idealised oil patch. Particle paths have been classified in two groups, those travelling close to the African coast and those passing close or through the islands. Furthermore, these two groups of particle paths also allow the detection of two classes of mesoscale structures. At one hand, the structures associated to the Coastal Transition Zone (CTZ) of the Canary Current, produced by the interaction of this current with the Northwest African coastal and shelf topography. On the other hand, those observed backside of the Canary Archipelago, and generated by the islands obstruction of the Canary current (cyclonic and anticyclonic eddies).

4.1 Circulation in Northwest African Coastal Transition Zone

Flow patterns in the CTZ show similar characteristics under different simulated flow intensity. Particles released next to the African coast were transported to the south by the currents showing a strong meandering in their trajectories, such as can be observed in the particle paths depicted in Figs. 2a and 2b, indicating the presence of energetic mesoscale features in this area. Note that, for clarity, only a small number of particle paths have been represented in these figures. Another generally observed remarkable feature is the increase of the velocity of the particles when they move downstream of Cape Yubi.

The trajectories of some particles display one or more loops when trapped by cyclonic and anticyclonic eddies. Dye injected next to Cape Yubi, jointly with cyclonic loops described by particles, shows that cyclonic eddies are found next to the coastal shelf between Cape Yubi and Cape Bojador, and occasionally they are advected offshore principally when they go south of Cape Bojador. Moreover, dye and anticyclonic particle loops reveal that normally anticyclonic eddies are placed rather offshore than cyclonic ones at south of Cape Yubi and they are frequently viewed south of Cape Bojador.

Finally, in several occasions injected dye and particle paths deflect offshore, more frequently when they arrive south of Cape Bojador where continental shelf gets wider, as observed in Fig. 3.

In relation to our idealised oil patch, it is interesting to note that particles of this group flow generally along the African coast, but sometimes they are pulled to the African coast, while in some occasions are transported towards the island coasts, mainly toward those more oriental.

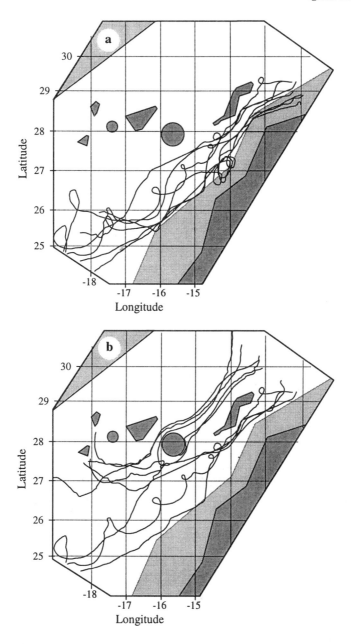

Figure 2: Laboratory model drifter paths; (a) particles moving close to the Northwest African CTZ; (b) particles through the Canary Islands.

4.2 Circulation around Canary Islands archipelago

Particles released off the CTZ also progress southward, reaching the Canary Archipelago. In this case, particles move more rapidly when flow intensity increase, and under these conditions some of them seems to be accelerated through island channels. Generally, particles trajectories continue in a south-westward direction when they go south of the islands, but when they pass through the Gran Canaria – Fuerteventura channel, they can be deflected also to the African coast. Also, it is noticed that particles advancing close to the islands show a higher meandering when crossing the island channels displaying occasionally one loop, indicating a higher mesoscale activity just behind the islands.

This mesoscale activity is also evidenced by injecting dye quite closed to them. Dye injection at both sides of Gran Canaria shows clearly the generation and shedding of cyclonic and anticyclonic eddies sequentially leeward the island. Cyclonic eddies are formed in the western part of Gran Canaria while anticyclonic eddies surge in the eastern one, both of them propagating to the south as vortex streets. Additionally, it is frequently observed that these mesoscale features generated by islands flow perturbation interact strongly with those coming from the CTZ, giving place to a quite complex flow field in the rear side of the islands, such as observed in Fig. 2b and Fig.3.

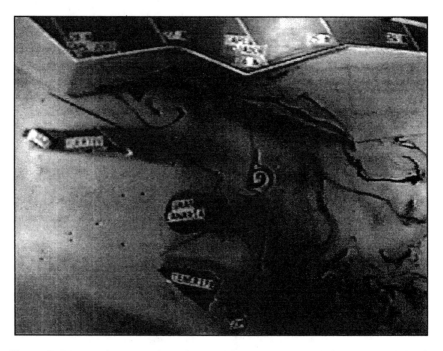

Figure 3: Mesoscale structures, generated by coastal irregularities and by islands blocking effect, marked by dye.

It is worth to note that in this case, a large number of particles (oil patchs) reach the island coasts and often impact coastlines, particularly those of the Gran Canaria, the most populated of the islands.

5 Conclusions

Laboratory model simulations reveal that great part of the observed mesoscale variability in the Canary islands area could be explained by the effect of coastal and bottom topography over the Canary Current flow. Respect to the observed features in the CTZ, laboratory simulations shows that flow around Cape Yubi could generate oceanic eddies downstream by separation of the current from the coast. Gyre formation at a curved coast in a rotating system has been observed in a number of simplified laboratory and computer studies [2, 4]. This supposes an alternative mechanism to that proposed by Barton *et al.* [1]. They suggest the origin of cyclonic eddies is probably vortex stretching of the flow exiting the shallower channel between Fuerteventura and Africa (this effect is not incorporated in the laboratory model).

It has been observed that Canary current is able to transport hypothetical oil patches spit north to the islands over considerable distance and impact their coasts, or deflect it toward African coasts. In addition, the variability of the currents due to mesoscale activity in the Canary Islands area could lead to a large shear dispersion of potential effluents of pollutants in the region. Both, mesoscale eddies and filaments, often seen in satellite images, enhance spreading of pollutants. Furthermore, filaments can transport pollutants from the African coasts to the islands. Then even if wind is attributed an important role in the spreading of surface discharges as oil spills, the variability of the background currents must be modelled correctly to obtain the true spreading [7]. Consequently, laboratory model simulations seems to be a useful tool for the prediction of pollutants transport that should be used jointly with numerical models and adequate databases to improve our ability to predict the movement of potential oil or other pollutant discharges at sea.

Acknowledgments

The experiments were developed at the Coriolis Laboratory of the Civil and Environmental Engineering Department, SINTEF, Trondheim, Norway, within the Trondheim Marine Systems Large Scale Facility, sponsored by the European Community.

Thanks are due to Prof. Egil Sakshaug (LSF manager) and Alexandra Neyts (LSF co-ordinator) for the opportunity offered to perform the experiments and for the hospitality during our staying in Trondheim.

The hospitality, kindness, and the essential theoretical and technical assistance of Prof. Dr. Thomas McClimans during the development of the experiment are deeply acknowledged.

References

[1] Barton, E.D. et al., The transition zone of the Canary Current upwelling region. *Progress in Oceanography*, **41**, pp. 455-504, 1998.

[2] Boyer, D.L. & Tao, L. On the motion of linearly stratified rotating fluids past capes. *Journal of Fluid Mechanics*, **180**, pp. 429-440, 1987.

[3] Fischer, H.B., List. E.J., Koh, R.C.Y., Imberger, J. & Brooks, N.H. Mixing in inland and coastal waters, Academic Press, New York, 1979.

[4] Klinger, B.A. Baroclinic eddy generation at a sharp corner in a rotating system. *Journal of Geophysical Research*, **99**, pp. 12515-12531, 1994.

[5] McClimans, T.A. The role of laboratory experiments and models in the study of sea straits. In: Pratt, L. J. (Ed.), *The Physical Oceanography of Sea Straits*, Kluwer, Dordrecht, pp. 373-388, 1990.

[6] McClimans, T.A. & Gjerp, S.A., Numerical study of distortion in a Froude model. Proceedings 16th Int. Conf. On Coastal Engineering, Vol. III, ASCE, New York, pp. 2887-2904, 1979.

[7] McClimans, T.A. & Johannessen, B.O. On the use of laboratory ocean circulation models to simulate mesoscale (10-100 km) spreading. *Environmental Modelling & Software*, **13**, pp. 443-453, 1998.

[8] Mittelstaedt, E. The ocean boundary along the northwest African coast: Circulation and oceanographic properties at the sea surface. *Progress in Oceanography*, **26**, pp. 307-355, 1991.

[9] Pacheco, M. & Hernández-Guerra, A. Seasonal variability of recurrent phytoplankton pigment patterns in the Canary Islands area. *International Journal of Remote Sensing*, **7**, pp. 1405-1418, 1999.

[10] Van Camp, L., Nykjaer, L., Mittelstaedt, E. & Schlittenhardt, P. Upwelling and boundary circulation off northwest Africa as depicted by infrared and visible satellite observations. *Progress in Oceanography*, **26**, pp. 357-402, 1991.

Section 9
Hydrodynamics and Transport Modelling

Mathematical and numerical modelling of eutrophication process in Masan Bay, Korea

J.L. Lee[1], J.K. Chu[2] & B.-G. Hwang[3]
[1]Department of Civil and Environmental Engineering,
Sungkyunkwan University, Korea
[2]KORDI, Korea
[3]Department of Environmental Engineering,
Sangmyung University, Korea

Abstract

A mathematical model has been developed as an idealized formulation to understand the eutrophication status in Masan Bay. The model predictions were in general agreement with field data in simulating the global growth and decay of phytoplankton. The WASP5 model was used for investigation of phytoplakton-nutrient dynamics as well as the comparison with both mathematical and numerical models developed here.

The water quality simulation with fine resolution has been performed by a coupled hydrodynamic, transport, and eutrophication model. The hydrodynamic component simulates the main tidally-driven advective flows throughout the Masan Bay and the transport component of the model simulates the movement of dissoved materials in response to the tidally-driven advective and dispersive processes coupled with freshwater from 12 streams flowing into Masan Bay. The transport was determined by a new hybrid method [1] which takes forward-tracking for advection and finite difference approximation for diffusion. All submodels including eutrophication model operate on the same grid. Results are compared with field data obtained in 1995 and they show an encouraging degree of similarity.

1 Introduction

During recent two decades, red tide and oxygen deficiency, which are major undesirable phenomena related to the eutrophication, have been continuously observed in Masan Bay and the quantity is growing more and more as it goes. Masan Bay is a large bottleneck-shaped bay with a narrow opening to the sea through the Modo and Gadeogdo channels and generally polluted by untreated municipal sewage and discharges from a heavy-industry complex. Therefore, it has brought out a lot of environmental problems as one of the most polluted water bodies in Korea.

The basin boundary of bay system is shown in Figure 1 and the system is divided into a inner bay part north of Somodo and a outer bay part south of Somodo. The bay has been known to have the maximum tidal range of about 2.5m. Because the bay bends extremely and is semi-enclosed, any significant intrusion of waves and oceanic currents has not been observed. The average depth of Masan Bay basin is about 15m. The inner bay is shallower than the outer. The bay can be considered as a large enough estuary with a constantly changing environmental situation where rapid biological processes take place. The bay catchment covers an area of approximately 232km^2, most of which has been impacted by human activities in some form. 12 small streams (Nam, Changwon, Naedong, Palyong, Yangdug, Sanho, Samho, Hoiwon, Goobang, Chugsan, Jangun, and Changwon) flow into the bay with nutrients and one place of Waste Water Treatment Plant (WWTP) in Dugdong disposes of its waste in the middle of bay.

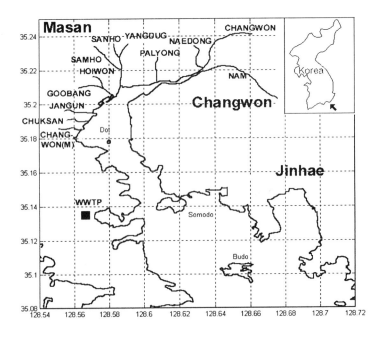

Figure 1: Location of Masan Bay and basin boundary of bay system.

The watershed is divided into three administrative districts constructed by the cities of Masan, Changwon, and Jinhae. Population of Masan City shows a tendency to decrease gradually and in contrast population of Changwon and Jinhae to increase gradually. Therefore, population of the watershed remains nearly constant in total.

Climate is sub-tropical with a mean annual rainfall of 1150mm, most of which occurs in the summer months (March to September). Air temperatures range from -7.9 to 34.9℃ over the season. Wind records show the mean wind speed of 2.1-2.3m/s and a strong seasonal pattern in direction. Over the summer months April to August winds are mostly from SW and NE, the pattern then changing to NW during winter, and the daily average of sunshine is 6.9 hours.

For the effective management of Masan Bay, the following studies were carried out:
1. Analysis of monitored water quality data;
2. Mathematical eutrophication modeling;
3. Application of WASP5 model;
4. Development and calibration of FDM model.

2 Mathematical modelling

The mathematical modeling through zero-dimensional analysis was carried out to understand the eutrophication status in Masan Bay. Although there have been many developments that make it possible to simulate the fate and transport of pollutants in great detail and with more realistic processes, mathematical models are necessary to determine allowable loadings for each plant as a part of their permit or to analyze the primary pollutant effluents causing the environmental problem.

The simple eutrophication kinetics simulates the growth and death of phytoplankton, with its effects on the nutrient cycles and DO balance. Growth can be limited by the availability of inorganic nitrogen, inorganic phosphorus, and light. Light limitation is described by the Di Toro formulation and complete mixing is assumed in the assigned control volume.

The nutrient loads of Masan Bay are divided into one wastewater treatment plant and twelve streams. The wastewater treatment plant (WWTP) discharging into the Masan Bay catchment is administered and licensed by Masan City. Because the WWTP situated in Dugdong is primary level in wastewater treatment, most of nutrients are been discharging into the bay. Particularly, a nitrogen flows to the bay always in large quantities. The nutrients from twelve streams are influent with various distributions in the bay because inflow from streams is in proportion to rainfall. The total loading rate of each nutrient discharging from streams and WWTP is shown in Table 1. The data were reffered by Masan City [2]. Total volume, mean depth, cross-section area, and mixing length of Masan Bay used to zero dimensional analysis are 936,580,000m^3, 13.5m, 88,860m^2, 8000m, respectively. Table 2 presents kinetic coefficients determined through mathematical model study. Temporal variation of phytoplankton is compared in Figure 2, showing the peak in late summer.

Table 1. The loading value of nutrients (unit: kg/day).

	Mar.	Jun.	Jul.	Aug.	Sep.	Oct.
NH	10972.71	14545.58	12250.69	11286.99	15701.11	12370.03
NO	3472.75	4044.72	4540.27	102.30	2885.66	5687.81
OPO	445.86	619.99	751.86	375.33	721.66	680.72
CBOD	24571.05	44034.84	45953.45	89359.97	50745.86	56116.70
DO	2055.51	2401.40	3560.28	4774.99	3096.72	3002.26
ON	9918.08	14447.82	13000.00	12289.84	6227.58	5611.29
OP	225.80	1107.03	1108.54	1168.21	567.44	699.16
Q	528462	687350	1136448	1088000	731668	641361

Table 2. Kinetic coefficients used in simple eutrophication(day^{-1} at 20°C).

Constant	Code	Value
Nonpredatory phytoplankton death rare	XK1D	0.02
Deoxygenation rate constant	XKD	0.01
Reaeration rate constant	THD	1.01
Mineralization rate constant	XK71	0.01
Nitrification rate constant	XK12	0.04
Average phytoplankton growth rate constant	XK1C	2.00
Average phytoplankton respiration rate constant	XK1R	0.125
Denitrification rate	XK20	0.05
temperature coefficient	TH20	1.08
Mineralization rate of dissolved org. N	XK71	0.01
temperature coefficient	TH71	1.045
Mineralization rate of dissolved org. P	XK83	0.03
temperature coefficient	TH83	1.08

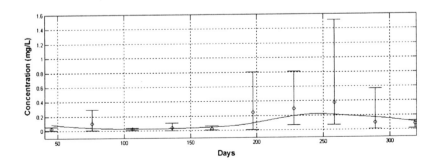

Figure 2: Temporal variation of phytoplankton by zero-dimensional analysis.

3 Application to WASP5 model

WASP5 model was run on the segment system dividing the Bay into nine segments. Three segments of them cover over Dot island, three segments between Dot island and Somodo, and the rest between Somodo and Budo. Segment No. 10 is taken account into the total bottom layer for deposit materials. Figure 3 shows the 9-segment configuration for the study area from Nam, Changwon, and Naedong streams to Budo. Loading by streams and WWTP discharge in segment 1, 2, and 8.

Inflows into segment 1 are discharged from Nam, Changwon, Naedong, Palyong, Yangdug, Samho, and Sanho streams, meanwhile Hoiwon, Goobang, and Jangun streams flow into segment 2. Discharge by WWTP flows into segment 8. As shown in Figure 4, the WASP5 model predictions are in general agreement with phytoplankton field data. The model was also used to investigate phytoplankton-nutrient dynamics. The calculated influence of load reduction on the phytoplankton concentration is presented in Figure 5. The tendency is toward a reduction of biomass with reduction in the nutrient loading. In the inner bay, the 10% reduction shows about a half

reduction of phytoplankton.

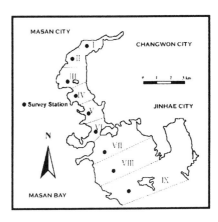

Figure 3: Segment system of Masan Bay.

Algal blooms in Masan Bay are caused by the considerable external load of nutrients. Therefore, the only way to decrease eutrophication is to reduce the nutrient load. Because the removal of nutrients is expensive, it is important to know where, when and by how much we have to put in efforts to get satisfactory results.

4 Water quality prediction by FDM

Although current and wave stresses are the major physical factors in water quality studies, their effects have been given little consideration. Furthermore, most of previous models were restricted to long term predictions with time steps of a day or longer. In the present study because the temporal and spatial plankton distribution inside so big an area is very variable, a depth-integrated two-dimensional FDM model is developed to simulate short-term and fine-resolved spatial variations of eutrophication process and follows three phases; hydrodynamics, transport and algal dynamics. All submodels operate on the same grid.

The hydrodynamic component simulated the main tidally-driven advective flows throughout the bay. The shallow water equation was solved by a fractional step method in conjunction with the characteristic method for an advection step and the approximate factorization technique for diffusion and propagation steps leading to the stable and fast computation.

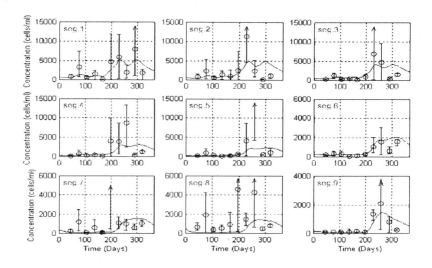

Figure 4: Comparison between model results and field data for phytoplankton.

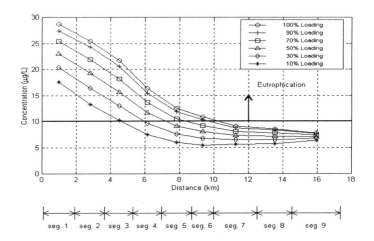

Figure 5: Influence of load reduction on chlorophyll-a.

The transport component of the model uses a subset of the hydrodynamic model configuration. It simulates the movement of dissoved materials in response to the tidally-driven advective and dispersive processes coupled with freshwater from 12 streams flowing into Masan Bay. The pollutant transport was determined by a new hybrid method [1] which is superior in both result accuracy and time-saving ability. It is based on the forward-tracking particle method for advection. However, unlike the random-walk Lagrangian approach, it solves the diffusion process on the fixed Eulerian grids, which requires neither an interpolating algorithm nor large number of particles.

In eutrophication model, two forms of phosphorus (inorganic and organic), three forms of nitrogen (oxides of nitrogen, ammonium nitrogen and organic nitrogen), dissolved oxygen kinetics and CBOD as well as phytoplankton were simulated. The same kinetic coefficients as determined by zero-dimensional analysis (Table 2) were used. Dynamic transfer of dissolved and particulate material between water column and benthic compartments was incorporated in a semi-empirical manner.

The FDM model covers the study area as shown in Figure 6. The study area is discretized by 149×125 rectangular grid mesh of 500×500m resolution and the time interval is given 30minutes.

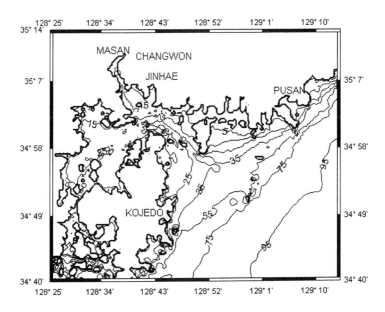

Figure 6: Bottom topography around Jinhae Bay.

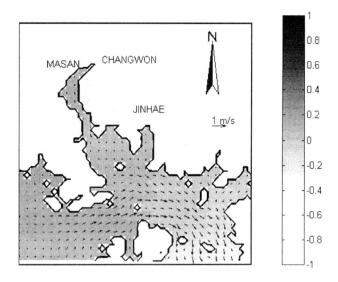

Figure 7: Distribution of highest ebb currents in July 1995.

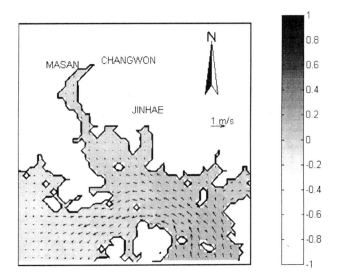

Figure 8: Distribution of highest flood currents in July 1995.

The simulations were carried out for 1995 where the monitored data are available. Figures 7 and 8 show the flow patterns with highest values of ebb and flood currents simulated in July 1995, respectively. Results of phytoplankton by FDM model are shown in Figure 9. Because most of loadings are discharged into the inner bay and matters are transported to the outer bay, the regional distributions of concentration show a tendency that it is high at inner bay and low at outer bay. Comparison between observed data, WASP and FDM model results for the concentration of phytoplankton is shown in Figure 10. The FDM model provides somewhat higher values than those by WASP5 model.

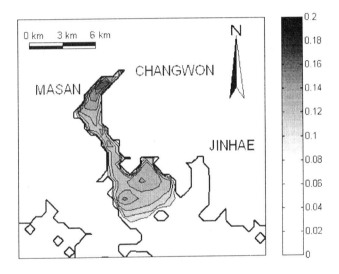

Figure 9: Spatial distribution of phytoplankton concentration by FDM model in July 1995 (unit in).

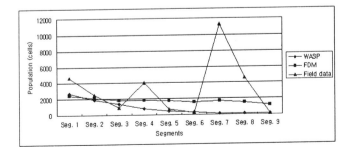

Figure 10: Comparison between observed values, WASP5 and FDM model results for concentration of phytoplankton.

5 Conclusions

The zero-dimensional analysis was carried out through a simple eutrophication process to understand the eutrophication status in Masan Bay.The model predictions were in general agreement with field data in simulating the long-term growth and decay of phytoplankton. The WASP5 model was also employed to ascertain the more detailed eutrophication status in space and to investigate phytoplankton-nutrient dynamics.

Because the temporal and spatial plankton distribution in estuary is very variable, a depth-integrated two-dimensional FDM model is developed to simulate short-term and fine-resolved spatial plankton distribution. The model follows three phases; hydrodynamics, transport and algal dynamics. All submodels operate on the same grid.

In water quality model, two forms of phosphorus (inorganic and organic), three forms of nitrogen (oxides of nitrogen, ammonium nitrogen and organic nitrogen), dissolved oxygen kinetics and CBOD as well as phytoplankton were simulated. Phytoplankton growth is subject to nutrients and light limitation. The pollution sources were given unsteady and their mass loadings were estimated based upon monitoring records. The same kinetic coefficients as determined by mathematical model were used in FDM model. The comparison of FDM results with monitoring data shows that the method could represent the algal growth dynamics and water quality processes reasonably.

In the future the model should be supported by higher resolution data to investigate the meso-scale prediction. In addition, coastal circulation of the bay under the combined action of wind and tide should also be investigated because the influence of wind is of great importance on the ecological model.

Acknowledgement

This work was supported by the Korean Research Foundation (1997-012-E00001).

References

[1] Lee, J.L. A high-accuracy approach for modeling flow-dominated transport. *Environmental Coastal Regions*, ed. C.A. Brebbia, WIT Press: Boston and Southampton, pp. 277-286, 1998

[2] Masan City. *Overall monitoring of ocean environments in Masan Bay.* Kyungnam Univ., Environment Problem Institute, 1996.

Investigation of current pattern and dispersion of the organic pollutants in urban-reservoirs

H. Karahan & N.O. Baykan
Department of Civil Engineering, Pamukkale University, Denizli- Turkey

Abstract

In this study, the current pattern and dispersion of the organic pollutants in the "Tahtalı Dam" under construction is investigated. The Dam is located in the Aegean Region of Turkey, near to the City of Izmir, the third largest metropolitan area with over 3 million population. A 2D mathematical model is used containing the equations of momentum, continuity and dispersion and considering shallow water and well mixing features, based on the long-term available data such as bathimetry, annual average intake, load of organic pollutants, dominant wind directions per season and average wind velocities.

1 Introduction

Generally in Turkey, as a consequence of the effect of resettling to big cities, the infrastructure-system on those are nearly to collapse in a very short period of time. This event has caused to review the re-estimation of population as a main criterion in the infrastuctural design. Under these new circumstances, it is born a necessity to plan and takes into operation those services earlier than predicted. National migration has changed the facade of the big cities considerably. The enlargement of the settlement areas toward country-sides, unauthorized restructuring, delay of infrastructure, increasing of the industrial activities and consequently pollution of the environment, has brought also the pollution of natural ground and surface water resources. The rapid increase of the population and shortage on the water supply, unconscious over-extraction of groundwater and as a result of this, the energy consumption of deeper wells has increased and the groundwater elevation has depleted incredibly (e.g. 10-30 m lowered during the last 7-8 years in İzmir). All these non-positive points nesses have brought earlier utilization of surface waters. Unplanned and uncontrolled construction scheduling of the squatter's shacks in the country-area, the non-existence of the sewerage system of those settlements and the small scale industrial facilities has started to pollute the surface water also. Therefore, early by the planning stage, the new criteria have to be foreseen for a more realistic design by considering the weakness on the control of the dynamic structure of the pollution loads, the creditability conditions of the construction.

2 Aim

With the help of the continuity and momentum equations, which explains the movements of water in lakes and reservoirs, the current pattern and water exchanges are to be solved for reservoirs which are complex and geometrically dynamic in space and time. In addition to hydrodynamic model, by solving of the relevant dispersion equation, the exchange of currents; transferring, mixing, dilution and dispersion of the pollution loads in the reservoir are to be determined.

The dispersion of the pollution in the reservoirs which supply mainly domestic and industrial water, can bring a new horizon by selecting of the location, shape and type of the intake structure taking into consideration of the available regional parameters at the beginning, as well as in the operation stage. With this aim, 2D model for the current pattern and dispersion of the pollution has been developed by the manner of the averaging along the depth using finite difference method. Meteorological, physical and chemical parameters in the model were available from the measured and observed data within the territory.

Tahtalı Dam is located 40 km south of İzmir, 5 km west of the small town Gümüldür. It is going to supply domestic and industrial water up to the year of 2015 (DSİ [1]). It has an annual yield capacity of 128 hm^3 per year (4.06 m^3/s) and regulation ratio of 0.80 (DSİ [2]). The construction of the dam and auxiliary structures have been recenly completed. The purification plant and the transmission lines are still under construction (Fig. 1). A further and detailed information's can be found in Reference's ([2 and 4]).

3 Methodology

Topographical data is available in 1/5000 maps. The maximum elevation of the reservoir is 60.50 m. Considering this elevation, the periphery is identified and the depths of mid points in the selected grid have been read. Fig. 2 shows the bathymetric maps of the large portion of the reservoir, which influences directly the current pattern and dispersion of the pollution. Wind data was available in the meteorological station of "Güzelyalı", close to region and having a continuos observation data of 26 years. By analyzing of the wind data, it has been determined that the exchanges of wind directions occur mainly in two periods, namely May-Oct. and Nov.-Apr. (Table 1).

Table 1: Dominant wind directions and velocities in the Tahtalı Reservoir.

Period	Dominant Wind Direction	Percentage of Representation	Ave. Velocity (m/s)
May-Oct.	W	20 %	3.4
Nov.-Apr.	SE	24 %	4.4

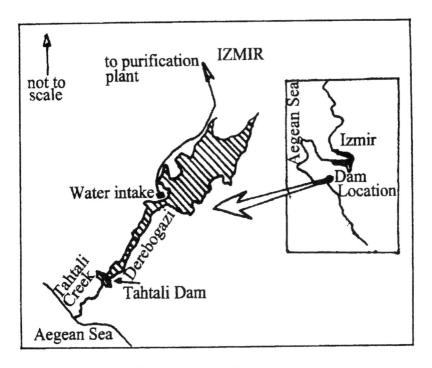

Fig. 1: Location of Tahtalı Dam.

During the period between 1982-1989, some parameter values (i.e. Nitrate, Turbidity, TDS, Clorur, Sulphate, Ph, Hardness, e.t.c.) had been measured by DSI (State Hydraulic Works) and since 1993, IZSU (Izmir Water and Sewerage General Directorate) continues to measure all of the relevant parameters increasing the number of parameters to 20, and sampling locations from 1 to 4. Unfortunately, because of asynchronies' measurements between pollution and discharge, the available pollution-loads have not been used in this study.

The developed model has been run only for the organic load given in the literature, which is 60 g/N/d. For the identification of pollution sources, two different points have been considered, namely the conjunction of the Menemen and Gölcükler Bridges (NE) and Bulgurca Settlement (SE).

The present population of the basin is close to 38.000. A population value has been estimated of 100.000 in the year of 2015 by using an increase ratio of

5%. It is assumed also that 70 % of the pollution shall come from NE and remaining from SE.

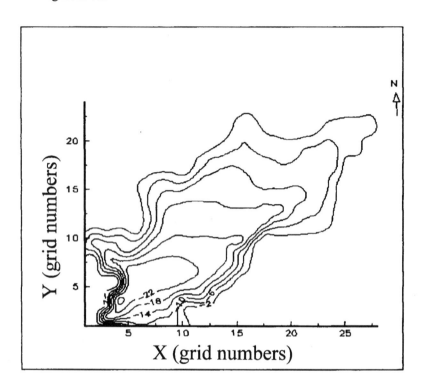

Fig. 2: Bathimetric map of the large portion of Tahtalı Reservoir.

The 2D mathematical model has been developed by considering bathimetry, water depths, dominant wind directions and velocities, pollution's loads, average intake. The model also covers the features of shallow water and well mixing and consists of the following equations of momentum, continuity and dispersion (Falconer [3] and Leendertse [9]):

Equation of momentum in x direction,

$$\frac{\partial U}{\partial t} + U\frac{\partial U}{\partial x} + V\frac{\partial U}{\partial y} - fV + g\frac{\partial \eta}{\partial x} + \frac{gU}{c^2 H}(U^2 + V^2)^{1/2} - A_h(\frac{\partial^2 U}{\partial x^2} + \frac{\partial^2 U}{\partial y^2}) - F_x = 0 \quad (1)$$

Equation of momentum in y direction,

$$\frac{\partial V}{\partial t} + U\frac{\partial V}{\partial x} + V\frac{\partial V}{\partial y} + fU + g\frac{\partial \eta}{\partial y} + \frac{gV}{c^2 H}(U^2 + V^2)^{1/2} - A_h(\frac{\partial^2 V}{\partial x^2} + \frac{\partial^2 V}{\partial y^2}) - F_y = 0 \quad (2)$$

Equation of continuity,

$$\frac{\partial \eta}{\partial t} + \frac{\partial(UH)}{\partial x} + \frac{\partial(VH)}{\partial y} = 0 \qquad (3)$$

Equation of dispersion,

$$\frac{\partial(H\bar{P})}{\partial t} + \frac{\partial(HU\bar{P})}{\partial x} + \frac{\partial(HV\bar{P})}{\partial y} - \frac{\partial\left[HD_x\left(\frac{\partial\bar{P}}{\partial x}\right)\right]}{\partial x} - \frac{\partial\left[HD_y\left(\frac{\partial\bar{P}}{\partial y}\right)\right]}{\partial y}$$

$$+[K]H\,P + H\,S = 0 \qquad (4)$$

In these equations are:
U,V: depth-averaged velocity components in x and y directions, respectively
η: water-surface elevation above or below mean depth,
d: mean depth
H: total depth of water column (d+η)
F: Coriolis parameter
G: gravitational acceleration
A_h: cinematic viscosity
C: Chezy coefficient
F_x, F_y: external forces components in x, y directions
D_x, D_y: diffusion coefficient in x and y directions
K: reaction matrix
P: pollutant concentration
S: source or sink term
t: time

The equations (1) to (4) are solved implicitly by using of finite difference method (Juang [5], Karahan [6,7,8]).

The reservoir volume of Tahtalı Dam at maximum elevation (60.50 m) is approximately 307 hm^3 (Fig.1). The lake can be separated into two portions, lower one is narrower and deeper (down, the Dereboğazı location) than the upper one which is wider and shallower (Fig. 1).

By the first attempt of application, these two portions have been taken into consideration without any separation and used a grid interval by 100*100 m. In Fig. 3, the entire current pattern is given for this type of grid. The pattern has been obtained by using critical wind direction NE which is found by analyzing of total observance set, The average velocity has been calculated as 4 m/sec and average intake has been considered as 4.06 m^3/sec.

The run-time of the model is 24 hours with 45 second intervals. As seen in Fig. 3, the existence of the pass (Dereboğazı) does not effect the current pattern considerably. After seeing this event, the grid dimensions are reduced to 250*250 m. From the Fig. 3, it has been noticed that the currents do accelerate in the upper and shallow portion, while they slow down toward shores. On the other hand, they are negligible in the middle of the lake, along the pass and surroundings of the intake structure (Fig. 3).

By considering 250x250 m grid dimensions, periodical dominant wind directions and velocities (Table 1) and 4.06 m^3/sec average intake, the current patterns and dispersions of organic pollution's are given for the winter season (Dec.-Jan.-Feb.) in Fig. 4 and summer season (Jun.-Jul.-Aug.) in Fig. 5.

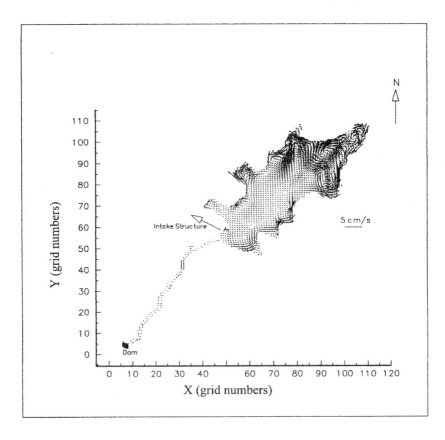

Fig. 3: Current Pattern in Tahtalı Reservoir (grid sizes 100*100 m).

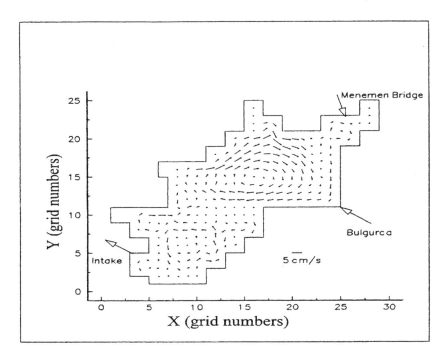

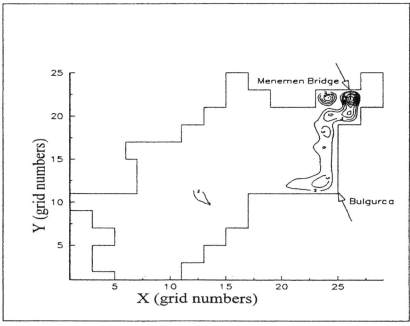

Figure 4: Current Pattern (a), Pollution's Dispersion (b) in Tahtalı Reservoir (Wind Direction, SE; Velocity, 4.4 m/s; Grid sizes, 250 x 250 m)

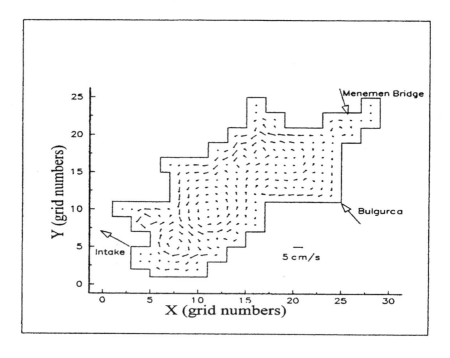

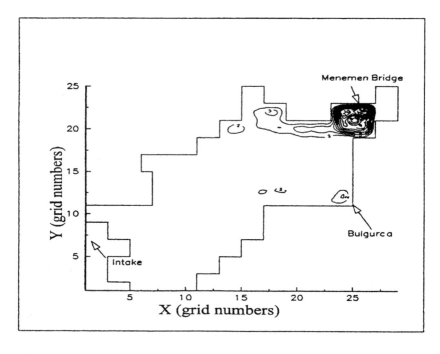

Figure 5: Current Pattern (a), Pollution's Dispersion (b) in Tahtalı Reservoir (Wind Direction, W; Velocity, 3.4 m/s; Grid sizes, 250 x 250 m)

4 Conclusions and proposals

As seen in Fig. 3, under the critical wind direction (NE) for the pass, the exchange of the current pattern is nearly zero.

As seen from Figures 4 and 5, the relation between dominant winds and dispersion of the pollution are perceptible. Particularly, during the summer months, the dominant wind direction (W) empress the dispersions from Menemen Bridge and Bulgurca Settlement towards intake structure.

The parameters that identify the water quality in the reservoir, are not synchronous and to take measurement during dry summer-months is impossible. Therefore, the real pollution loads have not been determined, instead of it, the above mentioned two pollutant sources have been fed by a certain organic load, which is 60 g/N as given in the literature, and their dispersions have been identified.

The tear-off excavation has to be absolutely preceded. Otherwise, this will bring a great organic pollution especially at the first stage of operation. During the interval from the beginning the filling up to the operation time, some important pollutant-parameters have to be measured in the certain points of the reservoir. Considering those measured parameters, if necessary, the capacity of the purification plant or operation policy has to be reviewed and the calibration of the model has to be realized.

In order to estimate water quality in the reservoir and take necessary precautions in the basin, the calibrated model has to be run under various meteorological conditions, pollution's loads and intake ratios. So, the results obtained from the model and from the real measurements can be mutually evaluated and setup an interactive information system.

References

[1] DSİ-State Hydraulic Works-, *İzmir Water Supply Project, Tahtalı System* (in Turkish: İzmir İçmesuyu Projesi Tahtalı Sistemi), Volume 5, General Information, GEBB-TW-WRC-SETAN-KBM, 82 p., 1991.

[2] DSİ-State Hydraulic Works-, *Report for Environmental Impact Evaluation of İzmir Tahtalı Water Supply System* (in Turkish: İzmir Tahtalı Sistemi Çevresel Etki Değerlendirmesi Raporu), 32 p., 1991

[3] FALCONER, R. A., Numerical Modeling of Tidal Circulation in Harbours, *J. of the Waterway, Port, Coastal and Ocean Division*, A.S.C.E., Feb., pp 31-48, 1980

[4] İZSU-İzmir Water Supply and Sewerage General Directorate-, Presentation on Izmir as Medwan Pilot, *Medwan Water Resources Management Program, Institute Mediterranean de l'Eau* (IME), 21 p., 1994 (Prep.: N. O. Baykan and S. Gök).

[5] JUANG, J.T., The Mutuality of Current and Pollution in Yin-Yang Bay (Taiwan), *the First International Conference on the Mediterranean Coastal Environment*, (Ed.: Erdal Özhan), Volume 2, MEDCOAST 1135-1151, Antalya/Türkiye, 1993.

[6] KARAHAN, H., *Mathematical Modeling of Coastal and Bay Currents and an Application to İzmir Bay* (in Turkish: Kıyı ve Körfez Akıntılarının Matematik Modellenmesi ve İzmir Körfezi İçin Bir Uygulama). İzmir, Ph.D. Thesis, Dokuz Eylül University, Institute of Science, 1988.

[7] KARAHAN, H., Comparison of Implicit and Explicit Difference Methods in the Mathematical Modeling of Sea Currents (in Turkish: Deniz Akıntılarının Matematik Modellenmesinde Implicit ve Explicit Sonlu Fark Yöntemlerinin Karşılaştırılması). *Doğa, Turkish Journal of Engineering and Environmental Sciences*, vol. 17, No. 2, pp 101-110, 1993.

[8] KARAHAN, H., Mathematical Modeling of Dispersion of Organic Pollutants in Sea Media (in Turkish: Deniz Ortamında Kirletici Madde Yayılımının Matematik Modellenmesi). Araştırma Projesi Raporu, İzmir/Türkiye, Dokuz Eylül University, Investigation Project No: 0902-90-01-02, 1993.

[9] LEENDERTSE, J. J., *A Water Quality Simulation Model for Well-Mixed Estuaries and Coastal Seas:* Vol I, Principles of Computation, The Rand Corp., R-2298-RC.,1978.

Quantification of WwTW UV dose rates for long sea outfalls by means of coastal hydrodynamic modelling

R.A.Crowder[1], A. Nisbet[1], A. Harrington[2] & R. Freestone[3]
[1]Bullen Consultants, UK
[2]Yorkshire Water, Science and Regulation, UK
[3]Environment Agency, North East Region, UK

Abstract

Both the EC Bathing Water Directive and the EC Urban Wastewater Treatment Directive have greatly influenced the treatment levels and requirements of Wastewater Treatment Works (WwTW) that have intermittent and continuous discharges to coastal areas. Over the past decade the requirements of these Directives have been taken into account by Yorkshire Waters' Asset Management Plans and 'Coastcare' programme and have resulted in the upgrading and implementation of new WwTW.

Coastal hydrodynamic modelling has played a key role in developing appropriate WwTW schemes. Calibrated and validated models have typically been used to model the dispersion of nutrients and viral and bacterial determinants in coastal areas. The hydrodynamic models have been used as an effective and economical management tool to evaluate the appropriate level of ultraviolet (UV) treatment required at the WwTW and to ensure that the EC bathing water directives requirements are achieved over a complete ten year bathing period. The study illustrates methods that have been developed to undertake such an evaluation and demonstrates how flow data from the sewer and WwTW modelling can be efficiently incorporated into the process, used in coastal model simulations and then evaluated by analytical methods. The application of the method developed has been applied to the DIVAST hydrodynamic and dispersion model of Whitby, UK, and is presented as a case study of the methodology developed. The benefits of applying this method to ensure compliance with similar EC Directives, as an integrated management and support system, are also illustrated.

1 Introduction

The disinfection policy of the Environment Agency for England and Wales, hereafter referred to as the Agency, requires that a reduction in concentration of target micro-organisms by a nominal factor of 2.5 x 10^4 must be achieved between the raw effluent to the STW and the diluted effluent by the time it reaches the bathing water monitoring point [1]. It is preferred that this reduction is achieved by dispersion within the receiving waters. However, due to the location of the outfall and monitoring point and the characteristic hydrodynamic conditions of receiving waters this can not always be achieved.

The specific micro-organisms that are to be considered are faecal coliforms, faecal streptococci and entroviruses. The evaluation of the required UV dose to achieve the required reduction in concentration of target microorganisms may be considered as four stages:

1. Computation of the total available reduction attributable to all steps in the treatment process other than the UV disinfection
2. Computation of the reduction required from the UV disinfection
3. Identification of the group of micro-organisms for which the received UV dose, to achieve the required reduction, will be greatest (critical group)
4. Computation of the required UV dose for the critical group (critical dose)

Through the use of methodical and repeatable procedure the reduction in micro-organisms, through dispersion for a given outfall load (ppt), can be evaluated by the use of a calibrated and validated coastal hydrodynamic and dispersion model. Furthermore, the period of time that virus levels meet a predefined standard at the monitoring point can also be evaluated i.e. the level of (percentile) compliance.

2 Coastal Model

2.1 Whitby Model

The DIVAST hydrodynamic and dispersion model used for this study is the 75m by 75m grid based around Whitby [2], which is located on the north east coastline of England as illustrated in Figure 1. The model has an input flow from the River Esk, values of which were based upon summer flow conditions modelled by an existing ISIS hydraulic model of the River Esk [3].

Figure 1 illustrates the location of the Whitby outfall, which lies to the east of Whitby harbour and the location of the monitoring point, which is to the west of Whitby harbour. The outfall has a design flow of $0.161m^3/s$ based upon a combination of Mean Daily Probable Flow (MDPF) and three times Dry Weather Flow (3DWF) entering the treatment works from the surrounding catchments.

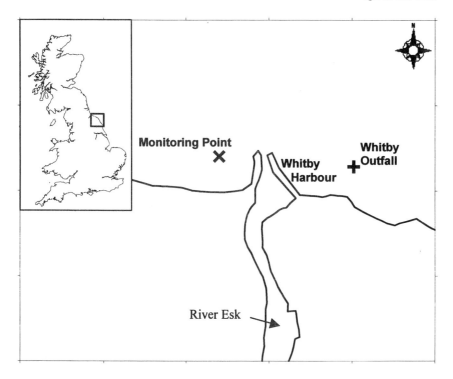

Figure 1: Whitby model domain, outfall location and monitoring point

The monitoring point site, which is the designated benchmarking location for the purpose of compliance, lies to the west of Whitby Harbour, as illustrated in Figure 1.

2.2   DIVAST

The numerical model used in this study is a modified form of the model DIVAST (Depth Integrated Velocities And Solute Transport), originally developed by Professor R A Falconer [4]. The governing differential equations used within the hydrodynamic module to determine the water elevation and depth averaged velocity fields in the horizontal plane, are based on integrating the three-dimensional Navier-Stokes equations [5] over the water column depth. They include the effects of:- local and advective accelerations, the earths rotation, barotropic and free surface pressure gradients, wind action, bed resistance and simple mixing length turbulence model. Assuming that the vertical accelerations are negligible compared with gravity and that Reynolds stresses in the vertical plane can be represented by a Boussinesq approximation [6], then the depth integrated continuity and x- and y-direction momentum equations can respectively be shown to be the following form [4]:

$$\frac{\partial \eta}{\partial t} + \frac{\partial q_x}{\partial x} + \frac{\partial q_y}{\partial y} = 0 \tag{1}$$

$$\frac{\partial q_x}{\partial t} + \beta \left[\frac{\partial U q_x}{\partial x} + \frac{\partial V q_x}{\partial y} \right] = f q_y - gH \frac{\partial \zeta}{\partial x} + \frac{\tau_{xx}}{\rho} - \frac{\tau_{xb}}{\rho},$$

$$+ 2 \frac{\partial}{\partial x} \left[\overline{\varepsilon} H \frac{\partial U}{\partial x} \right] + \frac{\partial}{\partial y} \left[\overline{\varepsilon} H \left[\frac{\partial U}{\partial y} + \frac{\partial V}{\partial x} \right] \right] \tag{2}$$

$$\frac{\partial q_y}{\partial t} + \beta \left[\frac{\partial U q_y}{\partial x} + \frac{\partial V q_y}{\partial y} \right] = f q_x - gH \frac{\partial \zeta}{\partial y} + \frac{\tau_{yx}}{\rho} - \frac{\tau_{yb}}{\rho},$$

$$+ 2 \frac{\partial}{\partial y} \left[\overline{\varepsilon} H \frac{\partial V}{\partial y} \right] + \frac{\partial}{\partial x} \left[\overline{\varepsilon} H \left[\frac{\partial U}{\partial y} + \frac{\partial V}{\partial x} \right] \right] \tag{3}$$

where ζ = water surface elevation above datum, t = time, $q_x q_x$ = depth integrated velocity components in x,y directions, β = momentum correction factor for non-uniform vertical velocity profile, U,V = depth averaged velocity components in x,y directions, f = Coriolis parameter, g = gravitational acceleration, H = total depth of water column, τ_{xw}, τ_{yw} = surface wind shear stress components in x,y directions, τ_{xb}, τ_{yb} = bed shear stress components in x, y directions. ρ = fluid density, and ε = depth averaged eddy viscosity.

3 Method

3.1 Evaluating reduction through dispersion

In determining the level of (percentile) compliance, of a given concentration from the outfall, it is first necessary to know the required compliance level that is to be set. Assuming that the virus concentration at entry to the treatment works is 1×10^7 ppt and knowing that the target reduction in concentration of micro-organisms is a nominal factor of 2.5×10^4, then the required level of compliance at the monitoring point can be set at 400 ppt (1×10^7 ppt / 2.5×10^4). Levels at the monitoring point are recorded along a line that lies at right angles to the coast to ensure that the virus level at the monitoring point takes into account the variation in water level.

3.1.1 Hydrodynamic Conditions
Prior to modelling the virus levels, appropriate hydrodynamic conditions need to be generated, which cover a typical range of flow conditions. At Whitby four hydrodynamic conditions were modelled as shown in Table 1. These conditions cover both a spring and neap tide with either an onshore or offshore wind. The conditions describe the best and worst conditions affecting the outfall plume.

Wind speeds are determined from wind data at Bridlington, recorded over a three year period.

The flows from the River Esk, entering the coastal waters at Whitby Harbour, were modelled for summer conditions only. This criterion is set by the requirement that compliance occurs during the bathing season only, which runs from April to September. To assess compliance of the outfall there is no need to include the contribution of viruses to the model from the River Esk.

Table 1: Hydrodynamic Conditions

Hydrodynamic Condition	Tide	Wind		River Esk Flow
		Speed (m/s)	Direction (deg. True North)	
1	Spring	4.97	39.58 (On-shore)	Summer
2	Spring	4.92	232.56 (Off-shore)	Summer
3	Neap	4.97	39.58 (On-shore)	Summer
4	Neap	4.92	232.56 (Off-shore)	Summer

3.1.2 Procedures

In order to determine the percentile compliance for the outfall the following procedures are adopted.

STEP 1: A known concentration of virus is selected for discharge from the outfall.

STEP 2: Modelling is undertaken using the concentration from STEP 1 for the four different hydrodynamic conditions described in Table 1. Values of virus levels at the monitoring point are recorded and results collected for each condition during the 2nd and 3rd tidal cycles and at a time interval of 0.4 hrs.

STEP 3: For the four different conditions, the maximum level at each time step is determined for the monitoring point.

STEP 4: The percentage of time that the virus levels meet the required compliance level of 400 ppt is calculated for each hydrodynamic condition.

STEP 5: The percentage for each hydrodynamic condition is factored to take into account the probability of the hydrodynamic condition occurring during the bathing season.

The levels of probability for Whitby are shown in Table 2. These values have been determined from wind records for Bridlington, with onshore winds occurring between 330° and 120° and offshore winds occurring between 150° and 300°. The probability of a spring or neap tide occurring has been spilt 50:50.

The individual percentages from each condition summed give the overall compliance for the defined concentration from the outfall. STEPS 1 through to 3 are repeated for different concentrations from the outfall to give a table of compliance levels for the monitoring point. From STEPS 1 through to 4 it is possible to identify the particular regions required to meet the compliance level. A simple procedure of factorising one of the concentrations used in the previous steps allows the selection of a refined range of concentrations which can then be tested in the same fashion as STEPS 1 through to 3 giving a new set of compliance levels.

It is possible to use a simple factoring of the results since there is a natural log reduction between successive results. For example the results from a simulation using a concentration of 1×10^6 are a log reduction from the results using a value of 1×10^7, a factor of 0.1.

Compliance levels are then plotted for each outfall concentration simulated, as illustrated in Figure 2. From this plot and its associated table it is then possible to determine the required concentration from the outfall that would meet the required level of compliance.

Table 2: Probability of Event

Hydrodynamic Condition			Probability of Event	
Case	Tide	Wind	Occurrence	(%)
1	Spring	Onshore	(0.50 x 42)	21
2	Spring	Offshore	(0.50 x 58)	29
3	Neap	Onshore	(0.50 x 42)	21
4	Neap	Offshore	(0.50 x 58)	29
			Sum	100

3.2   Percentile compliance

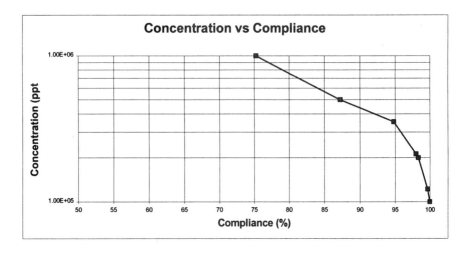

Figure 2: Levels of compliance

The percentile (%ile) compliance is based upon the period of time over a complete bathing season that virus levels meet a predefined standard at the monitoring point. The results from the simulations for Whitby are illustrated in Figure 2 and detailed in Table 3.

Table 3: Levels of Compliance

Outfall Conc. (ppt)	Hydrodynamic Event				
	Spring Onshore (%)	Neap Onshore (%)	Spring Offshore (%)	Neap Offshore (%)	Overall Compliance (%ile)
1.00×10^6	10.38	14.39	21.57	28.92	75.25
5.00×10^5	17.07	17.40	23.87	28.92	87.25
3.53×10^5	20.42	18.41	27.08	28.92	94.82
2.12×10^5	21.09	19.08	28.92	28.92	97.99
2.00×10^5	21.09	19.41	28.92	28.92	98.33
1.22×10^5	21.09	20.75	28.92	28.92	99.67
1.00×10^5	21.09	21.09	28.92	28.92	100.00

3.3 Virus Reduction at treatment works

From the results shown in Table 3 the required outfall concentrations to meet a compliance with a 98.2%ile at the monitoring point can be determined by factoring the results between results. Having determined the required outfall concentrations and hence the available reduction in micro-organisms through dispersion, for the requisite level of compliance, the required UV reduction through the treatment works is now known, as shown in Table 4.

Table 4: Outfall Concentrations and Virus Reduction Ratios for a 98.2% ile

Compliance Location	WwTW Inflow Concentration (ppt)	Required Outfall Concentration (ppt)	Virus Reduction (WwTW Inflow: Outfall)
Monitoring Point	1.00×10^7	2.11×10^5	48:1

4 Evaluation of UV dose

4.1 Theory

For the evaluation of the dosage requirements equations (4) through to (6), which have been derived from the United States Environment Protection Agency [7], have been adopted. Within this approach there is the fundamental concept that there is decay in micro-organism concentration with UV irradiation [8]. This is quantified by a rate of decay K (the proportion of the micro-organisms killed per unit time) and an inactive rate constant k (the proportion of the micro-organisms killed per unit time per unit intensity). These two constants are related since K=kI, where I is the received intensity in $(mW/cm^2)^{-1}$.

$$N = N_o \exp(-kIt) + N_p \qquad (4)$$

$$N_p = eSS^m \qquad (5)$$

$$k = al^b \qquad (6)$$

where N = resultant concentration (No./100ml), N_o = initial concentration (No./100ml), t = time of exposure to the UV radiation (s), N_p = concentration associated with particulates present in the flow (No./100ml), SS suspended solids concentration (mg/l), c = empirically derived constant (No./100μg), m = empirically derived coefficient, a = empirically derived constant $((mW/cm^2)^{-2}s^{-1})$, and b = empirically derived coefficient.

The computation of dosage requires the determination of parameters that are particular to the effluent or hardware setup within the WwTW. For these parameters typical values have been assumed in accordance with current guidelines set by the Agency [9][10].

4.2 Determination of Critical Dosage

The Agency has developed a spreadsheet formulation to enable the determination of the required level of UV dosage, with respect to intensity and contact time, at the treatment works. The development of such a spreadsheet enables repeatable and controlled evaluation of these specific values given that the reduction through dispersion is known. Values used within the spreadsheet analysis are specific to a site.

The required UV dose and time of exposure to the UV radiation to achieve the reduction for 98.2%ile compliance is shown in Table 5 for viruses. Bacteria values are also determined to enable the identification of the critical group and critical dose.

Table 5: Required UV Doses for Indicators and Targets

Required Percentile (%)	Flow (tcmd)	Required Outfall Conc. (ppt)	Log Dispersion	Received Dose (mWs/cm2)	Contact Time (s)
98.20	13.91	2.11Ex105	2.72	20.21	4.04

5 Conclusions

A methodology has been developed to determine the critical UV dose rate required to achieve a reduction of 2.5 x 10^4 between the input to the treatment works and the monitoring point, for a given percentile compliance. Through the use of a computational model it has been determined for the Whitby case study that a 48:1 level of reduction is required through the treatment works in order to obtain a percentile compliance of 98.2%ile. Through the use of the Agency's UV methodology, the critical dose rate for the case study has been calculated as 20.21 mW/cm^2, with a contact time of 4.04 seconds.

Although not addressed here the sensitivity of the proposed dose and the effect/performance of proposed equipment need to be considered in the evaluation process. For the purpose of consenting, the Agency not only requires the specification of UV dose and UV intensity but also the suspended solids concentration, which in some instances may be the limiting factor in overall performance of the process. The effect of dose on micro-organism concentration has a limiting result below a threshold of 500 (No./100ml) and hence UV dose may not provide sufficient Virus reduction to achieve compliance.

6 References

[1] Environment Agency, Discharge Consents Manual, Vol 024A, 1997.
[2] Bullen Consultants Ltd., Whitby Hydrodynamic Model Calibration Report, September 1995.
[3] Bullen Consultants Ltd, Esk River Model Report, October 1994.
[4] Falconer, R.A, An Introduction to Nearly Horizontal Flows, in the Coastal, Estural and Harbour Engineers Reference Book, (Ed. M B Abbott and W A Price), E & F N Spon Ltd, London, Chapter 2, 1993.
[5] Schlichting, Boundary Layer Theory, McGraw-Hill Book Co., New York, Sixth Edition, pp742, 1979.
[6] Rodi, W., Turbulence Models and their Application in Hydraulics, International Association for Hydraulic Research, Delft, Second Edition, pp104, 1984.
[7] United States Environment Agency, Design Manual, Municiple Wastewater Disinfection, 1986.
[8] Havelaar. A.H., F-specific Bacteriophages as Model Viruses in Water Treatment Processes, PhD Thesis, 1986.
[9] National Rivers Authority, Computation of the UV dose required for compliance with NRA disinfection policy for proposed discharges, Internal NRA publication.
[10] Design Guide for Marine Treatment Schemes, Vol I & II, WRc plc, 1990.

Optimisation of Split/Solin sewerage system

N. Ravlic
Civil Engineering Institute of Croatia, Regional Unit in Rijeka, Croatia

Abstract

This paper presents selected results of a techno-economical optimisation analysis conducted prior to the preparation of the Split/Solin (Croatia) sewerage system's backbone final design. Construction of the most important facilities of the system has already started in 1999 as a part of the MEIP Split/Solin (Municipal Environmental Infrastructure Programme) implementation. The project is aimed to improve the non-satisfactory environmental situation in one of the most populated coastal regions in southern Croatia (Kaštela Bay). Modern simulation tools were used to analyse both the impacts of various technical solutions on the environmental situation in the region and to reveal possibilities for improvements of existing technical proposals. The sanitary impact of submarine discharge of treated waste water is modeled by means of hydrodynamic and advection/dispersion models; different schemes of WWTP solution are analysed and presented by means of modern dynamic computer visualization tools. The final result of the optimisation process consists of a proposal to build a cheaper system with satisfactory operational performances and environmental safety, not conditioning the implementation of further steps of the project in the future.

1 Introduction

The towns of Split and Solin are situated in the central part of the eastern Adriatic coast bordered by the Kaštela Bay and the Split and Brač Channels (Figure 1). The towns have a resident population of approximately 240.000 persons. They are located in an area of scenic beauty and have rich archaeological remains which make the cities an attractive tourist destination. Some fifty years ago, the whole area of Kaštela Bay was a popular swimming area but increased pollution, coming from large industrial facilities along the shores of the Bay and from surrounding settlements has lead to a noticeable de-

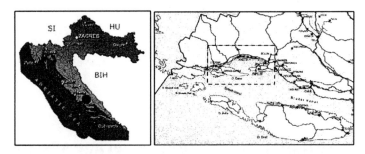

Figure 1: Geographical location of the Kaštela Bay and the Towns of Split and Solin

terioration of water quality. Today raw sewage and industrial effluents are discharged untreated into this semi-enclosed embayment where noticeable ecological problems and occurence of eutrophic conditions were registered in the past [1]. In order to reduce the pollution of Kaštela Bay, a scheme shown in Figure 2 is to be implemented in the near future. According to the Terms of Reference [2], the scheme involves the construction of a collector system to collect wastewater from the Northern catchment and transfer the flow through a hydrotechnical tunnel to a location of planned wastewater treatment works in the area of Stupe (135.000 p.e. in 2010, 360.000 p.e. in 2025), eastward to the town of Split. Treated effluent will be discharged through a long submarine outfall into the Brač Channel.

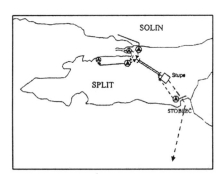

Figure 2: 1st phase of the MEIP Split/Solin project implementation

In 1998, an IBRD Loan Agreement has been signed with the main scope to provide a budget for the 1st phase of the Project implementation. The Client and Investor for the Project is Split Water Supply and Sewerage Company ("ViK" Split). The consulting engineers for final design and international tender documentation preparation are Civil Engineering Institute of Croatia (IGH) – Regional unit in Rijeka, with their sub-consultants: SNC-Lavalin Quebec-Canada and Faculty of Civil Engineering Split. The Project Agency "Eco-Kaštela Bay", which is responsible for the professional and technical aspects of the Project, has been set up to implement the Wastewater Disposal Project.

2 Terms of reference

2.1 Discharge consent standards

Following discharge consent standards were defined by the TOR document [2].

2.1.1 Wastewater treatment works
The wastewater treatment works is to be designed to meet either the values for concentration or the values for the percentage reduction as given in Table 1.

Table 1. Effluent discharge standards

phase parameter	concentration (mg/l)	minimum reduction (%)
1st degree treatment		
TSS		50
2nd degree treatment		
BOD at 20°C	20	70-90
COD	125	75
TSS	35	90

2.1.2 Marine discharge
The Terms of Reference stated that the marine outfall is to be designed to maintain the seawater quality standard on the limit of the protected zone (shoreline to 300 m offshore). The criteria are as follows:
- total colif. <500/100 ml (80 % samples) and <1000/100 ml (20 % samples)
- faecal colif. <100/100 ml (80 % samples) and <200/100 ml (20 % samples)

2.2 Design parameters

The project is to be designed for a planning horizon of 2010 (treatment works and submarine outfall – development phase I) and 2025 (hydrotechnical tunnel and sewer network with pumping stations). Wastewater will include residential, tourist, industrial, public facilities and infiltration flows. Basic wastewater flow design parameters in the Terms of Reference are sumamrized in Table 2, while Table 3 summarizes the design loads at the treatment works for the horizons 2010 and 2025.

Table 2. Basic wastewater flow design parameters (Terms of Reference)

flow		2010	2025
average daily flow	(m³/day)	34630	78400
max. daily flow - summer season	(m³/day)	52000	117600
max. hourly flow - dry period	(m³/s)	0,9	2,04
max. hourly flow – rain period	(m³/s)	2,2	3,0

Table 3. Design loads at the treatment works (Terms of Reference)

loads (kg/day)	2010	2025
BOD$_5$	14600	28200
COD	38615	58800
TSS	22125	28590
Total Nitrogen	1870	5394
Total Phosphorus	442	902

3 Optimisation study

The main objectives of the Optimisation study [3] were as follows: to integrate all relevant technical and economic parameters gathered from the careful analysis and assessment of the existing documents (TOR included) and to propose and select the preference implementation option for the major system facilities in the 1st phase of Project implementation. Several criteria are to be considered: overall economic effects in respective design horisons, construction costs, operational costs, land use and acquisition costs, flexibility for future extensions, environmental impact, phased implementation, safety and reliability.

3.1 Design parameters review

First step of optimisation consisted in review of the TOR designed parameters shown in paragraph 2.2. This assessment was necessary for several reasons: residential and tourist population water consumption decrease was registered in the past, prospective water usage should be reviewed accordingly; population predictions provided in the TOR have been modified in some areas; further decrease in industrial wastewater from the existing industries may be expected in the future; analytical data on industrial and urban wastewater showed significant discrepancy between measured and expected typical loads. Comparison between reviewed and TOR design parameters is shown in Figures 3 and 4.

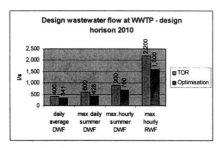 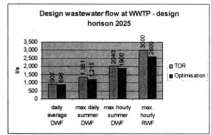

Figure 3: Comparison between the TOR and reviewed design wastewater flow at Split/Solin WWTP (design horisons 2010 and 2025)

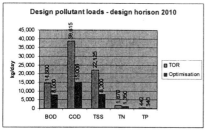 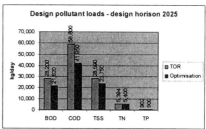

Figure 4: Comparison between the TOR and reviewed design pollutant load at Split/Solin WWTP for design horisons 2010 and 2025

It can be seen from above Figures that significant reduction of both wastewater flow and pollutant loads design parameters was proposed by the Optimization study [3]. Bigger differences between the TOR and optimised values are assessed for the design horison 2010 while somewhat smaller differences are expected for the design horison 2025. Optimisation of short-term design parameters seems reasonable in view of overall noticeable decrease of water consumption in the area, relatively low starting percentage of coverage by the sewerage system and drastic reduction of actual industrial pollution loads. Further, from today's stand-point it seems reasonable to assume that in long-term planning period the cost of water usage (and intention to limit it by either reducing the consumption or reducing the leakages and/or recycling process water) will play an important role in reducing the final expected hydraulic load at the end of the system.

3.2   Submarine outfall optimisation

The concept of the sewer system in Split/Solin region envisages the construction of the central treatment plant at the Stupe-Stobreč location, where all the waste waters will be treated and discharged into the Brač Channel. The "appropriate" treatment level at this plant should be determined by the legislative regulation on one hand, and, on the other, by the conditions in accordance with the carrying capacity of the Split and Brač Channels. Hence, disposal-treatment system optimisation should propose such treatment level, which after discharge allows the receiving waters to meet the relevant quality objectives.

Documents [4] and [5] were carefully analysed and assessed during the submarine outfall optimisation phase. Main conclusions of the report [4], which eventually were shown to be crucial for the Optimisation study, can be summarised as follows: the sea in the Brač-Split Channel as a whole, is mainly clean and is in an oligotrophic state; the simulation results point to a high carrying capacity of the Brač-Split Channel considering the pollution by organic matter; quantities of these pollutants, exceeding the planned quantity several times, do not endanger the sea quality and the required sea standard; according to the results obtained by the simulation model and computation for the planned loading by organic pollutants it is not necessary to treat the effluent at all, but only to dilute it initially; starting from the demand to satisfy aesthetic criteria and the necessity to prevent the settling of organic matter around the outfall, the wastewater should undergo minimum mechanical treatment. Obtained results fully justify the planned concept of a gradual improvement of the treatment level starting from the mechanical treatment up to biological one, according to the actual needs in due course of time.

On the other hand, document [5] provides useful updated information on typical summer/winter oceanographic conditions in the area of submarine discharge in the Brač/Split Channel. Above conclusions regarding the environmental state of the recipient were confirmed once again by the 1997/98-research campaign. However, as a slight possibility of an anticyclonic vortex occurrence in the area of future submarine discharge was revealed, it was suggested to plan the location of outfall diffuser section 3800-m away from the shore.

Conclusions drawn from [4] and [5], in synergy with the results of design parameter review (revealing a significant reduction of both expected wastewater flow and pollutant loads in the next 10-years period), induced IGH to set up a non-stationary hydrodynamic mathematical model and to examine the possibility of outfall length optimisation (reduction) in the 1st phase of project implementation. The model bathymetry is shown in Figure 5.

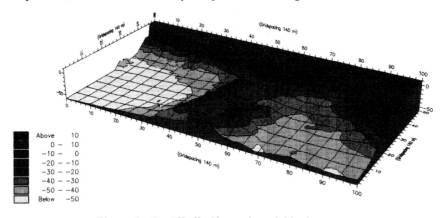

Figure 5: Brač/Split Channel model bathymetry

Different boundary conditions were applied at model boundaries in order to generate the most frequent (90% frequency) E-W current fields of different intensity. Closed steady circulation patterns in superficial layers were observed in extremely slow flow conditions only (Figure 6a). As expected, simulated flow field in superficial layers was influenced by the shallow reefs westward to the Stobreč Bay (Fig. 5), causing the main superficial current turning toward the shore. However, in such conditions no vortices or onshore currents in intermediate and bottom layers were observed (Figure 6b). Those results confirmed that, unlike superficial layers, predominant flow pattern in bottom and intermediate layers is oriented southward, away from the shore.

Further advection/dispersion analysis (with various outfall lengths and in different hydrodynamic conditions) have shown that even in critical, slightly onshore-turned E-W current field conditions (mean current velocity at intermediate level 15 cm/s, $T_{90}=3$ hours, wastewater discharge rate Q=340 l/s, coliform bacteria concentration after initial dilution $c=10^6$ col/100ml, diffuser depth d=36 m) bacteriological pollution field was transported almost parallel to the shoreline, far beyond the 300 m wide coastal strip. Figure 7 shows the extent of bacteriological pollution field after 48-hours release of designed mean daily wastewater flow (year 2010) via a 1600 m long submarine outfall. Obtained modelled results, in synergy with conclusions derived from [4], and supported by the fact that under-thermocline discharge at a depth of 36 m into off-shore oriented current field would take place even with 1600 m long submarine outfall, formed the basis for the outfall optimisation proposal: outfall diffuser section position in the 1st phase of project implementation can be much closer to the shore than suggested by the oceanographers in [5], without any real bacteriological threat to the coastal recreational zones. Being oligotrophic, the sea environment will absorb all organic pollutants expected in the foreseeable period. Preliminary treatment of wastewater, consisting of coarse (10 mm), fine (2 mm) screening and grit/grease removal seems to be an "appropriate" wastewater treatment level in the near future, until either increased hydraulic and pollutant load and/or legal requirements for plants with more than 150.000 p.e. and/or sea quality monitoring data impose higher treatment levels.

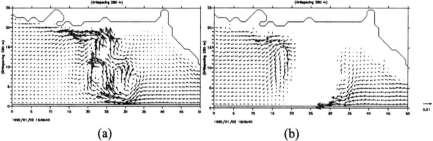

(a)   (b)

Figure 6: (a) modelled on-shore flow field in superficial layers of Brač/Split Channel in slow E-W current conditions
(b) modelled off-shore current field in intermediate and bottom layers of Brač/Split Channel in slow E-W current conditions

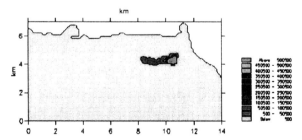

Figure 7: Characteristic bacteriological pollution field dimensions produced by continuous 48-hours release of designed mean daily wastewater flow (year 2010) via a 1600 m long submarine outfall

3.3   Treatment plant optimisation

3.3.1. Plant layout optimisation

Existing studies [6] and [7] have shown that required effluent criteria shown in Table 1 can be met with different process schemes (conventional activated sludge process, two stage A-B process and SBR process) with cost difference margin of only 15%.

Hence, in respect with the expected raw wastewater characteristics, the effluent criteria, the objectives of the optimization study and the reliability/robustness of different treatment technologies, the process scheme with conventional activated sludge is considered to be the most suitable in this particular case. The space reserved for this solution is large enough to receive any other technology, if any change is required for the implementation of the future phases

After preliminary mass balances have been prepared and the equipment sized, the first step optimization of the general layout has been performed and the plant layout shown in Figure 8 has been proposed.

Figure 8: Split/Solin treatment plant aerial view (computer simulation)

3.3.2. Hydraulic profile optimisation

Parallel to the plant layout optimisation, hydraulic profile optimisation was performed in order to define the most suitable leveling of the elements based on technical and economical considerations. General level adjustment has been performed considering the existing ground level and the accessibility of the different elements. Three alternatives have been considered as most appropriate for the optimization study (Figure 9). Alternative 1 (gravity flow, standard head losses) implies significant increase in pumping costs to a higher level of treatment plant inlet channel. Considering that almost 80 % of wastewater will be pumped to the treatment plant, pumping costs play an important role in economic considerations. In alternative 2 (gravity flow, minimum head losses), whole plant would be completely below adjacent ground levels and would require major excavations. Alternative 3 (intermediate pumping required, standard head losses) would tremendously reduce excavations and optimizes the

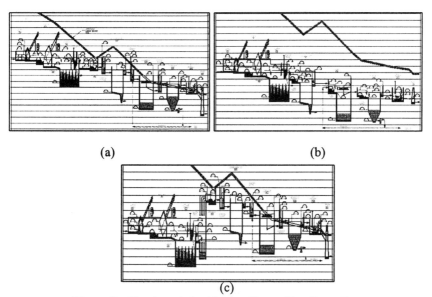

(a)   (b)

(c)

Figure 9. Treatment plant hydraulic profile alternatives

hydraulic profile as a function of level of the influent at the incoming hydrotechnical tunnel. If 2010 were assumed as design horison for pretreatment facilities, this alternative would be the absolute optimum in terms of overall economic efficiency.

3.3.3. First phase treatment scheme selection

TOR effluent criteria for the 1st degree treatment (Table 1) require 50% removal of total suspended solids. In order to provide for a 50 % suspended solids removal in the 1st phase of plant implementation, primary treatment processes should be provided (e.g. very fine screens or conventional primary clarification). A complete sludge treatment line must be implemented for both alternatives. The construction cost estimates for those two alternatives have shown that 50% total suspended solids removal require double investment costs with respect to the proposed preliminary treatment (with TSS removal in the lower range of around 15 % to 20 %). Going to finer screens in the 1st phase of plant development would increase the organic material content and might jeopardize the disposal of screenings in a landfill site. For all these reasons, it was proposed to install 2-mm fine screens, which is thought to be sufficient to prevent visible slicks at sea and settling of organic matter around the outfall diffuser section.

Due to the fact that the design loads to the Split-Solin wastewater treatment plant have been estimated and that actual observed values might differ, it was recommended to provide 2 mm fine screens and grit removal as the first step of treatment. Once the sewer network will be constructed to an acceptable degree, wastewater characterization could be carried out and it will be possible to choose further treatment based on sound technical analysis and, if required, pilot plants.

For these reasons, it was deemed preferable to leave the final decision on further process to a later date.

4 Conclusions

In general, stepwise approach in wastewater treatment and disposal facilities development enables not only to characterize the wastewater received at the plant (and determine the best treatment processes on a well-known basis to meet the effluent criteria and/or recipient standards), but forms an economically and technically reasonable solution. This is especially true in coastal areas where submarine discharge of "appropriately" treated wastewater takes place in oligotrophic, less sensitive deep-sea conditions.

By relying on the proven recipient autopurification capacity and favourable hydrodynamic conditions, the construction of much cheaper treatment and disposal facilities than required by stringent and non-selective legislative requirements can help to bridge the gap between legal issues and the need to efficiently protect the environmental quality in coastal areas. This approach seems reasonable especially in the early phases of system development and/or limited project budget conditions, similar to the one presented in this paper.

References

[1] University of Split, Croatian Academy of Science and Arts in Zagreb, UNEP Regional Activity Centre (PAP-RAC) in Split. *Environmental Management of the Kaštela Bay*, University of Split and CIMIS Split: Split, pp. 90-102, 1993.

[2] ECO Kaštela Bay Agency. *Terms of Reference for Split/Solin WasteWater Disposal Project Final Design Preparation*. ECO Kaštela Bay Agency: Split, 1998.

[3] Civil Engineering Institute of Croatia – Regional Unit in Rijeka, SNC-Lavalin, Faculty of Civil Engineering in Split. *Optimisation Study for the 1st Phase of Split/Solin WasteWater Disposal Project*. Civil Engineering Institute of Croatia, 1999.

[4] Margeta, J., Barić A., Gačić M. *Selection of the Optimum Treatment Level for the Central Treatment Plant – Final Report*, UNEP Regional Activity Centre (PAP-RAC) in Split, 1992.

[5] Institute for Oceanography and Fishery, State Hydrography Institute. *Results of the Research Works for Split-Stobreč Submarine Outfall Final Design Preparation*, Institute for Oceanography and Fishery: Split, pp. 113-136, 1998.

[6] Civil Engineering Institute Split, Schueffl & Forsthuber Consulting. *Report on Comparative Conceptual Design of Treatment Plant Stupe – Split*, Civil Engineering Institute Split: Split, 1991.

[7] Southern Waters Global. *Integrated Environmental Project Kaštela Bay - Split/Solin WasteWater Disposal – WasteWater Treatment Plant Feasibility Study*, Southern Waters Global: Split, 1996.

Mechanisms of water exchange at the platform edge of oceanic islands

J.L. Pelegrí, A. Antoranz, C. Gordo, A. Hernández-Guerra,
A. Martínez & P. Sangrà
*Departamento de Física, Facultad de Ciencias del Mar,
Universidad de Las Palmas de Gran Canaria, Spain*

Abstract

Islands in deep oceanic waters have rather vulnerable ecosystems due both to
the narrowness of their coastal platforms and the presence of very steep
slopes. We examine the different mechanisms capable of exchanging water at
the platform edge of oceanic islands, and conclude that the principal ones are
turbulent transport in the boundary layers generated by intense deep ocean
currents and wind-induced cross-slope transport at selected coastal sites.

1 Introduction

A common acquiescence is that coastal areas in oceanic islands are in good
health because whatever is thrown to the water is carried to the deep ocean by
the currents. The only "evidence" behind this puerile idea, however, is that
xenobiotic compounds are not visible along other coastlines or under the sea
surface, but this is simply because there are not such coastlines and the water
bottom is not discernible. The truth is that coastal waters of oceanic islands
are rather isolated and this makes their coastal ecosystems very vulnerable.

Islands in oceanic deep waters are, up to a high degree, separated from the
deep ocean waters because of the narrowness of their coastal platforms and
the presence of very steep slopes. On one hand the narrowness of oceanic
island platforms causes that the condition of non-normal flux at the coast is
still rather valid at the shelf edge. On the other hand, the very steep slope of
these islands prevents permanent currents to cross the isobaths at the shelf
edge. Both factors inhibit water exchange between the coastal and the deep

oceans, resulting in slow water renewal at the platform edge and undesirably high residence time of these waters.

The main physical processes responsible for the circulation in the coastal ocean and, in particular, for water exchange at the platform edge of oceanic islands, are tides, river discharge, deep-water forcing, and winds. Tidal currents, however, because of their periodic character and orientation almost parallel to the coast, usually cannot provide such exchange. Only internal tidal waves, generated at the slope, may propagate into the platform. River discharge creates a water plume that moves along the coast as a Kelvin wave but, unless the discharge is very large, will hardly penetrate into deep waters.

Turbulence generated near the islands is a very important mechanism for water exchange. This occurs in islands found on the path of relatively strong oceanic currents, characterized by large Reynolds numbers: water import into the platform is then associated to mesoscalar structures generated at the lee of the island, and water export usually takes place further upstream. The residence time of platform waters is linked to the period of generation of the mesoscalar structures, typically several days. In this work we will illustrate this type of situations with data from the island of Gran Canaria.

For islands in the verge of rather weak permanent currents, exchange is essentially controlled by wind. If the platform is very narrow the condition of non-normal flux at the coast still holds at the platform edge and exchange is highly restricted. If the water column is stratified an upwelling-type vertical circulation cell could develop, which needs not to verify an integrated non-normal flux condition. For many small islands, however, upwelling is not significant. Even more, the water column in many instances is well mixed and the velocities show little depth dependence. In this case the pattern of water exchange, as well as the width of the boundary layer between the coastal and deep oceans, may be investigated using Csanady's [1] model of the arrested topographic wave. This model suggests the existence of horizontal cells of recirculation and determines the natural lengths of such cells. In practice, however, the coastal morphology will largely control the generation of coastal filaments that stretch out into the deep ocean.

Our aim here is to examine the limited pathways for water, and other substances, to be exchanged at the platform edge of oceanic islands. The survival of coastal ecosystems in these islands depends, first of all, on our capacity to understand and quantify how this exchange takes place, and second, but not less important, on our skill to convince the local administration that the island platforms are not an unlimited dumping site.

2 Tidal currents on the island platform

Tidal currents are very susceptible to the presence of continental masses, with the tidal ellipses orientating grossly parallel to the predominant orientation of the coastline up to several hundred kilometers away [2]. The oceanic islands, because of their relatively small size as compared with the wavelength of the tide, have little effect on the surrounding deep waters but do control the characteristics of the tidal currents on the platform. Figure 1 illustrates the bathymetry around Gran Canaria and indicates the locations of stations where currents were simultaneously measured, and Table 1 presents the tidal ellipse information for the M_2 component obtained from these current-meters [3]. The results clearly illustrate how the current orientates parallel to the coast over the island platform.

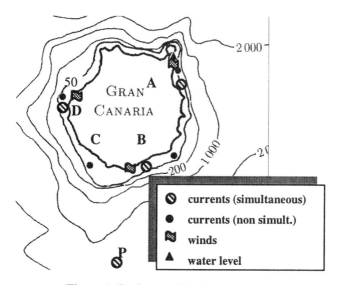

Figure 1. Stations and bathymetry.

Table 1. Semi-major axis, M, semi-minor axis, m, orientation relative to the east, O, and Greenwich phase lag, G, for the M_2 ellipses.

Current-meter	M (cm/s)	m (cm/s)	O (°)	G (°)
A	9.6	-0.2	143.9	265.7
B	3.78	-0.4	178.5	313.8
C	2.9	0.1	140.3	127.6
D	26.91	1.69	7.42	314.9

The modelization of tidal currents in the presence of topography verifies the importance of the coastal morphology. Figure 2 shows the axes of the M_2 ellipses around Gran Canaria, as obtained from a numerical model using real bathymetry [4]. The currents are always approximately parallel to the coastline over the platform and flow along the isobaths over the slope.

So far we have considered the barotropic tidal wave. A different situation arises with the internal tidal wave, which has a wavelength of only a few kilometers, depending on the vertical stratification. The barotropic tidal wave, when running into the coastal slope, generates an internal tide that is either reflected back into the deep ocean or forth into the platform [5]. In this last case there will be cross-slope water movements which, when associated to the different along-shore regimes in the coastal and deep-ocean, will cause permanent exchange. This may be important in those regions exposed to the incoming tidal wave, e.g., along the southern portion of Gran Canaria. This is the situation illustrated in Figure 3, with a rather strong cross-slope tidal current south of Gran Canaria at the 100 m isobath [6].

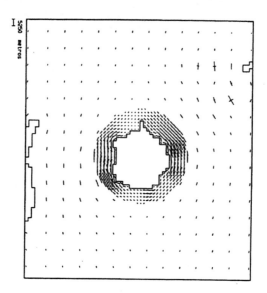

Figure 2. Numerical simulation, using real bathymetry,
of the M_2 ellipses around Gran Canaria.

3 The insulating effect of the island slope

Let us consider the simplest possible model, a homogeneous ocean in approximate geostrophic balance, and consider the current flowing into a region where the water depth shows substantial horizontal changes (as compared with changes in the latitude):

$$fv = g\frac{\partial \eta}{\partial x}, \qquad (1)$$

$$fu = -g\frac{\partial \eta}{\partial y}, \qquad (2)$$

$$\frac{\partial(uh)}{\partial x} + \frac{\partial(vh)}{\partial y} = 0, \qquad (3)$$

YOY2

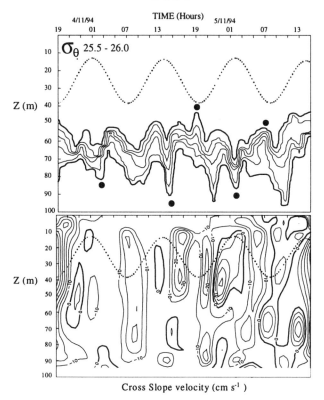

Figure 3. Density stratification and cross-slope velocities near the shelf edge south of Gran Canaria, in the top figure the dotted line illustrates the barotropic tide.

with the usual notation, η being the surface elevation, h the water depth, and x, y the across-shore and along-shore coordinates. It is easy to show that the current will flow along isolines of constant surface elevation, which coincide with the isobaths. Cross diferentiation of eqns (1) and (2) and substitution into (3) leads to the desired condition $\vec{u} \cdot \nabla h = 0$. The deep currents will never flow over a sea mountain or penetrate into the slope of an oceanic island, what we may call the fundamental insulating effect of a steep slope. The presence of horizontal stratification can break this constraint, i.e., if the ocean is stratified along-shore then there may be a geostrophic cross-slope flux. However, stratification is usually normal to the shelf edge [7] and the dominant geostrophic flow is along-slope.

Notice that it is not really necessary to assume geostrophy to derive the condition of along-isobath flow. Consider the case of a homogeneous coastal ocean, with the isobaths parallel to a long straight coast. Then the coastal constraint $U \equiv uh = 0$ at the coast, together with eqn (3), implies $V \equiv vh = V(x)$, so that between upstream and downstream transects V remains constant along isobaths, the isobaths behaving as streamlines.

For real applications the model should include other effects such as horizontal stratification, wind stress, and bottom friction acting on the along-slope direction. Csanady and Shaw [8] considered this problem and found that even in this case there is no significant bottom flow escaping from the platform. We will see below, however, that the presence of wind may indeed produce depth-integrated cross-slope currents different from zero, providing a mechanism of water transfer out of the shelf at localized sites.

4 Islands in the path of strong currents

The transition to turbulence in homogeneous flows is assessed by means of the Reynolds number, which compares the stabilizing effect of viscosity against the destabilizing role of inertia. The Reynolds number, $Re = UL/v$, is expressed in terms of the horizontal and velocity scales of the flow, L and U, and the horizontal viscosity v. A classical problem in fluid mechanics is the development of turbulence in a flow past a circular cylinder. For $Re < 4$ the flow is laminar and perfectly conforms to the cylinder. For $4 < Re < 40$ there is a wake behind the cylinder, characterized by two rotating vortices, over which the main stream flows. For $40 < Re < 200$ the wake becomes unstable and detaches from the cylinder as an oscillation in space and time. As Re increases the oscillations turn into two row of vortices, periodically shed from the two sides of the cylinder and rotating with opposite senses, which has been called a Karman vortex street. When $Re > 200$ the vortex street becomes unstable and irregular, and the flow within the vortices chaotic. For Re beyond 5000 the periodicity is lost and the flow is fully turbulent.

The geophysical problem of an oceanic island immersed in the path of a strong current resembles the Karman street vortex problem, aside of the important role played by the earth's rotation [9]. In this case L is given by the size of the island, U by the velocity of the incoming current, and v by the characteristic oceanic value of horizontal viscosity. As in the Karman street vortex problem, the characteristics of the leeward flow depend on Re. Hence, given L and v, the leeward flow depends on the impinging velocity U.

Table 2. Relation between U, Re, and the vortex shedding period.

U (m s^{-1})	Re	T (days)
0.04	20	no shedding
0.07	40	no shedding
0.1	60	36
0.19	100	16.2
0.47	250	5.4

Table 2 summarizes the numerical experiments carried out by Sangrà [9] for a range of realistic flows impinging onto Gran Canaria. The Reynolds number is calculated using v =100 m^2 s^{-1}, L=54 km, and the value of the impinging velocity for each numerical run. The vortex shedding period T is obtained from the fluctuations of the drag coefficient on the island.

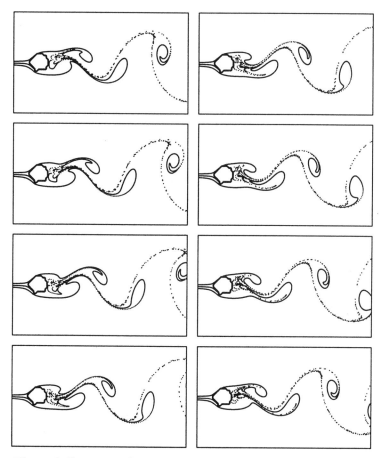

Figure 4. Sequence of tracer lines over one full period for a flow with Re=100 impinging onto a near-circular island which resembles Gran Canaria.

Figure 4 illustrates a full cycle of streak lines using Sangrà's [9] model for Re=100. The sequence goes from top to bottom, starting with the left column, each box separated by time intervals of 1/8 of a period. The flow enters the domain from the left (northeast of Gran Canaria) and leaves it at the right (southwest of Gran Canaria), the shedding of successive cyclonic and anticyclonic vortices is very clear. In this simulation we have chosen three streak lines, originated at the upstream end of the domain, which pass very close to the island. The dot points illustrate the position of different particles that have passed through the origin of the streak line at successive time intervals. Where the dot points become closer it means that the current slows down, where they separate apart it means that the current speeds up.

One very important feature is the separation of the flow associated to small capes at both sides of the island, particularly by the one in the lower part (northwest of the island). Another remarkable aspect is the reincorporation of flow, from vortices sometimes quite away from the island, at the lee side of the island. These results are confirmed with satellite images from the region, such as the one in Figure 5 that shows the surface temperature distribution around the island on August 31st, 1999.

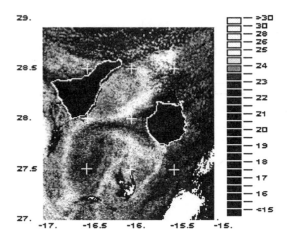

Figure 5. Surface temperature distribution around Gran
Canaria on August 31st, 1999.

5 Wind induced exchange

We will briefly examine now whether wind may provide a mechanism for water exchange across the shelf edge of oceanic islands. The coastal constraint, together with the vertical homogeneity of the water over the shelf, inhibits the existence of significant cross-shelf flow near the coastline. In the far deep ocean, however, the surface Ekman transport is perpendicular to the wind. Hence, if we have a situation where the wind blows parallel to the coast we need to understand how, and over which distance, the transition between the coastal constraint and the Ekman flux takes place. This problem was examined by Csanady [1]. Depth-integration of the eqns (1) and (2), including the surface and bottom stresses, gives:

$$fV = gh\frac{\partial \eta}{\partial x} + F_x, \tag{4}$$

$$fU = -gh\frac{\partial \eta}{\partial y} + F_y - B_y. \tag{5}$$

F_x and F_y are the wind-stress components, $B_y \equiv -rv$ is the along-slope bottom stress, inversely proportional to the velocity with a friction coefficient r typically of about 10^{-4} m s^{-1} [10], and the across-slope bottom stress is neglected because of the low cross-shelf velocity.

By taking the curl of eqns (4) and (5), with $h=h(x)$, the following depth-integrated vorticity tendency balance is obtained:

$$\frac{\partial^2 \eta}{\partial x^2} + \frac{f}{r}\frac{dh}{dx}\frac{\partial \eta}{\partial y} = \frac{f}{rg}\left(\frac{\partial F_y}{\partial x} - \frac{\partial F_x}{\partial y}\right). \tag{6}$$

The solution of eqn (6), subject to appropriate boundary conditions, determines the transition between the coastal constraint and the Ekman offshore transport. Consider the case of constant wind, parallel to the coast, which makes the right hand side of eqn (6) zero. The coastal constraint and eqn (5) imply $F_y = rv$ at $x=0$, which using (4) becomes the first boundary condition for (6), $\partial \eta/\partial x=(fF)/(rg)$ at $x=0$. A second boundary condition corresponds to zero surface elevation far away from the coast, $\eta=0$ as $x\rightarrow\infty$.

Under these conditions eqn (6) has the form of the heat conduction equation in one dimension, with surface elevation playing the role of temperature and negative y the role of time. The longshore velocity, proportional to the elevation gradient, is the analog of heat flux, so the first boundary condition by the along-shore wind is like prescribing a heat flux at the coast. The boundary condition at large x is like requiring a constant temperature far from the end of the semi-infinite slab. Csanady [1,11] solved this equation for different wind forcing or initial surface distributions. He noted that surface perturbations, such as those produced by river discharge, propagate along the shelf in the same direction as Kelvin waves. He also found that wind forcing is responsible of exchange at rather localized positions of the shelf-edge. In real cases, where the coastal morphology is quite different from an idealized straight coast, it seem reasonable to expect that the exchange at the shelf edge will be related to features such as capes.

We may estimate the distance over which the coastal constraint vanishes by simply asking the two terms on the left-hand side of eqn (6) to be of the same order, as $l_x \approx (rl_y/f\alpha)^{1/2}$. This means that the size of the island, characterized by l_y, controls the width l_x over which the flow adapts to the deep-ocean Ekman transport. For small islands, say about 10 km long, this width is going to be approximately 1km. This suggests that small islands with narrow shelves will have very limited exchange at the shelf-edge.

Acknowledgements

This work has been supported by the Spanish government through CICYT's grant number MAR96-1893.

References

[1] Csanady, G.T. The arrested topographic wave. *Journal of Physical Oceanography*, **8**, 47-62, 1978.

[2] Siedler, G. & Paul, U. Barotropic and baroclinic tidal currents in the eastern basins of the North Atlantic. *Journal of Geophysical Research,* **96**, 22259-22271, 1991.

[3] Martínez, A., Marrero, M., Pelegrí, J. L., Cisneros, J., Hernández-Guerra, A., López-Laatzen, F. & Ratsimandresy A. W. Análisis de las corrientes mareales y submareales en la plataforma de Gran Canaria. *Memorias de la 2ª Asamblea Hispano Portuguesa de Geodesia y Geofísica*, pp. 593-594, Lagos, Portugal, 2000.

[4] Martínez, A. *Estudio sobre la propagación de ondas de marea en canales: aplicación al análisis de niveles y corrientes en la plataforma de Gran Canaria.* Doctoral Thesis, Universidad de Las Palmas de Gran Canaria, 144 pp, 1995.

[5] Pingree, R. D., Mardell. G. T. & New, A. L. Propagation of internal tides from the upper slope of the Bay of Biscay. *Nature*, **321**, 154-158, 1986.

[6] Sangrà, P., Basterretxea, G., Pelegrí, J. L. & Arístegui, J. Observational evidences of internal tides in the shelf break of Gran Canaria. *Scientia Marina*, in press.

[7] Csanady, G. T. Mixing in coastal regions (Chapter 17). *The Sea: Ocean Engineering Science*, eds. B. Le Mehaute & D. M. Hanes, John Wiley & Sons, Inc., New York, pp. 593-629, 1990.

[8] Csanady, G. T. & Shaw, P. T. The insulating effect of a steep continental slope. *Journal of Geophysical Research*, **88**, 7519-7524, 1983.

[9] Sangrà, P. *Perturbación de un flujo geofísico por un obstáculo: aplicación a la Isla de Gran Canaria.* Doctoral Thesis, Universidad de Las Palmas de Gran Canaria, 201 pp, 1995.

[10] Weatherly, G. L. A study of the bottom boundary layer of the Florida Current. *Journal of Physical Oceanography*, **2**, 54-72, 1972.

[11] Csanady, G. T. Ocean currents over the continental slope. *Advances in Geophysics*, **30**, 95-203, 1988.

A simplified approach to photochemical box model meteorological data input

G. Latini, R. Cocci Grifoni, G. Passerini, S. Tascini
Dipartimento di Energetica, Università di Ancona, Italy

Abstract

One of the most important meteorological input parameter for photochemical air pollution models is the mixing height h_{mix}, which has a strong influence on the shape and intensity of the vertical diffusivity and, as a consequence, on ground level air concentration of primary and secondary pollutants. In this paper the analytical model of the boundary layer height proposed by Gryning and Batchvarova has been taken into consideration. A new meteorological input, developed according to specific requirements, has been developed to supply input data to a simple numerical air quality model (Photochemical Box Model) that simulates urban scale photochemical smog evolution.

1 Introduction

Photochemical Box Models (PBM) are possibly the simplest numerical models for air quality modelling. The region to be modelled is treated as a single cell, or box, bounded by the ground at the bottom, by the mixing height at the top, and the east-west and north-south boundaries on the sides. Inside the box volume pollutant concentrations are spatially homogeneous and directly mixed. Under these conditions, the pollutant concentrations can be described by a balance among the net amount carried in and out the volume, the rate of emissions, and the rate at which pollutants react chemically or decay.

The role of meteorological conditions in a box model can be a determining factor to predict the severity of pollution episodes. The most dramatic effects of meteorological conditions on pollutant evolution, simulated by a Photochemical Box Model, are caused by the diurnal growth of the Mixed Layer. The description of the growth of the Mixed Layer depends on some meteorological variables which are not commonly available. In the PBM a standard approach to Mixed Layer rise is considered: the algorithm uses a daily minimum and maximum Mixed Layer depth.

In this paper a more flexible solution to the problem is presented. It is based on the analytical model of the boundary layer height proposed by Gryning and Batchvarova. A simple algorithm is described, that estimates diurnal evolution of the height of the Planetary Boundary Layer. The input information for the model are therefore modified. The model has been applied over a Coastal Area, in the middle of Italy (Marche), that comprises valleys, hills, urban zones like Ancona and Jesi, and the industrial zone of Falconara. The box-approach to air quality simulation modelling considers the Coastal Area as a single cell 20 Km wide both in length and in width and with a variable height representing the changing depth of the Mixed Layer.

2 Planetary Boundary Layer Parameterisation

The vertical structure of the Planetary Boundary Layer (PBL) can be described through a reduced set of scale parameters and the result is a simple one-dimensional description of the turbulent PBL. Several routines have been already proposed and implemented to obtain turbulent surface fluxes and typical scale parameters of the PBL from data of wind speed, temperature and radiation (Van Ulden and Holstlag [1]). In this paper we present a new simple routine for the calculation of PBL height, as a function of time, based on the Batchvarova-Gryning model (Batchvarova-Gryning [2]).

2.1 The Mixed Layer

During the daytime, surface heating leads to convective motions in the PBL. Heat transfer from the surface rises warm air. Radiative cooling from clouds forms sinking cool air. Convective motion also leads to significant turbulence which mixes the air within this layer. Because of the convective motion and significant mixing of air, this sub-layer is called the Convective Layer or Mixed Layer. Above the Mixed Layer is a stable layer which prevents a further upward motion of thermals. This stable layer also restricts turbulence, preventing frictional influences from shifting out of the upper PBL; it is called the Entrainment Zone, since in such layer air from above the PBL entrains into the Mixed Layer (Stull [3]). During the day, the Mixed Layer reaches heights over 1 km and make up the entire layer of the PBL above the Surface Layer. However, the Mixed Layer vanishes with the sun as the thermally driven convection ceases.

2.1.1 The Mixing Height

Because turbulent fluxes vary based on surface heating and other factors, the height of the PBL also varies. At night, the height of the PBL decreases dramatically as the Stable Layer forms. During daytime, the height of the PBL is called the Mixing Height, because it is the height up to which the air is well-mixed. The Mixing Height is an important, often critical, input for air quality experts when determining air pollutant dispersion.

2.2 The Batchvarova-Gryning model

The Batchvarova-Gryning approach is used to model the evolution of the Internal Boundary Layer during near neutral and unstable atmospheric conditions over the land. A reasonable parameterization of the top-layer temperature flux - as a function of the surface flux, the friction velocity and the Convective Boundary Layer height- allowed the Batchvarova-Gryning model to obtain a simple equation for the evolution of the mixing height h_{mix}:

$$\left\{\left[\frac{h^2}{(1+2A)h-2BkL}\right]+\left[\frac{Cu_*^2 T}{\gamma g\left[(1+A)h-BkL\right]}\right]\right\}\frac{\Delta h}{\Delta t}=\frac{\overline{(w'\theta')}_s}{\gamma} \qquad (1)$$

where

$$L = -u*^3 T\big/Kg\overline{(w\theta')} \qquad (2)$$

is the Monin-Obukhov length, T is a representative Convective Boundary Layer temperature, g is gravity acceleration and K the Von Karman constant.

The parameters A, B and C are experimentally determined (A=0.2, B=5 and C=8), and a typical value of 0.007 K/m is used for the temperature gradient γ above the Planetary Boundary Layer.

The aim is to implementing a simple running code to be used in convective situations (Figure 1) in which only data series are available and vertical sounding is not possible.

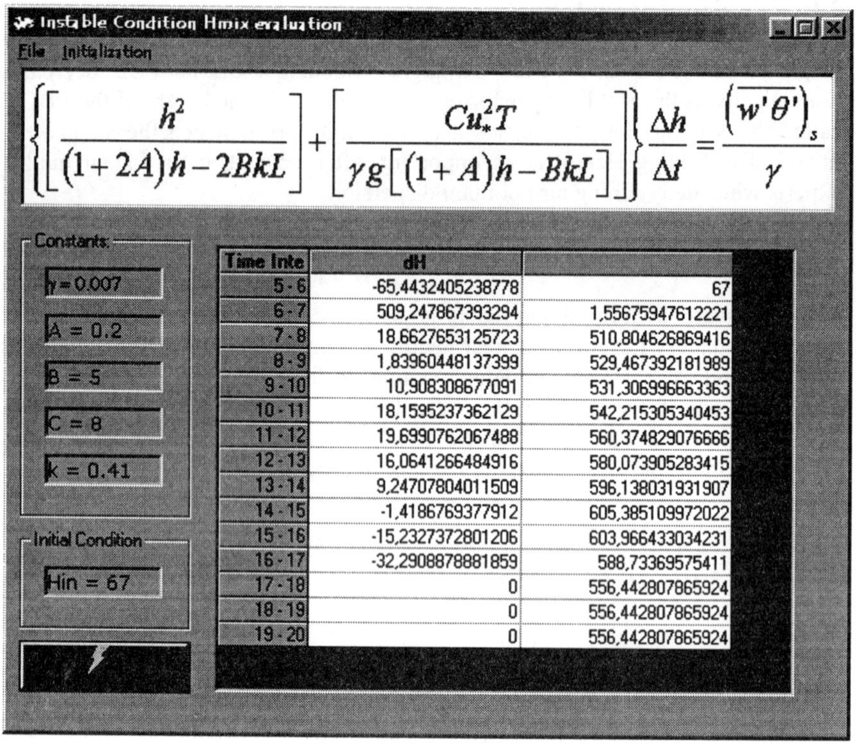

Figure 1: Main window showing mixing height values.

3 The study area

We initially tested our approach in lower valley of Esino (Figure 2). This area covers 20 km inland from the coast and is 20 km wide. The Esino valley (in the province of Ancona) is characterised, like most Italian valleys, by a riverbed that is roughly perpendicular to the coast and in a NE direction. The valley is surrounded by hillsides in increasing height as they distance from the coast. The first hills rise close to the coast and at this distance the height does not exceed 100 m. A further 20 km inland, the height does not exceed 200 m, but at about 30 km the valley undergoes a sharp narrowing near the gorge "Della Rossa" where the height exceeds 1000 m.

The climate in this area is classified *subcoastal*. There is an all year round sea breeze although of different intensity and influenced by a heavy component from NW. The sea breeze is caused by the different temperatures in the air bodies above the sea on one side and the coast on the other. This gives way to a meandering current along the coast with a component parallel to the coast caused by the synoptic winds.

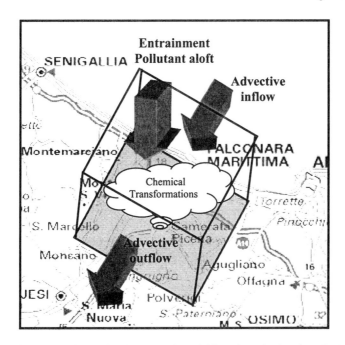

Figure 2: Schematic diagram of modelling domain for the PBM.

The component perpendicular to the coast, according to the direction, is called *coastal breeze* or *sea breeze* based on the origin of the winds. In presence of discontinuity as in the hillsides situated at the entrance of the valley, the sea breeze forces the breeze coming from the valley. This phenomenon is due to a difference in temperature between the land and the air around slopes. The highest intensity of the sea/valley breeze occurs during the warmest hours (about 3 PM) and it starts decreasing after the sunset. The temporal evolution of the Planetary Boundary Layer over Esino Valley on 16th of July was simulated starting at sunrise when the heat flux becomes positive. The simulation was based on hourly averages of the model input parameters (temperature, Monin-Obukhov length, friction velocity and the vertical kinematic heat flux). The friction velocity was taken as a fraction of wind speed, the fraction being 0.1 for agricultural areas and 0.13 for urban areas.

The meteorological output data here in analysed, is related to the measurements performed by the Italian Air Force Weather Station in Falconara Marittima (ISTAT), located in Chiaravalle airport, 2 km from the coast and more or less in the middle of the Esino valley (43.37 Lat, 13.22 Long, 12 m height). The output data comes from the observations performed hourly and gives information in both frequencies and monthly/seasonal distribution of speed, direction and directional persistence of wind, atmospheric stability and air temperature.

4 The Photochemical Box Model

The main task of Photochemical Box Model is to establish causal relations among emission levels, meteorological conditions and air quality (Figure 2). The numerical model simulates the transport and the chemical transformations of pollutants in the atmosphere and has been applied to local, urban and regional scales.

When applying the PBM, ozone is one of the pollutants of main interest. Ozone is a secondary pollutant formed through non-linear chemical reactions between precursor species, nitric oxide (NO_x) and volatile organic compounds (VOCs) that are emitted by a variety of anthropogenic and natural (biogenic) emission sources. Certain meteorological conditions can significantly enhance the concentration of ozone in specific areas, leading to substantial health risks. The PBM is also employed to study the effectiveness of alternative strategies to reduce primary emissions and, therefore, secondary pollutants with a special attention paid to ozone levels. The atmospheric diffusion equation is given by:

$$\frac{\partial c_i}{\partial t} = u \frac{\partial c_i}{\partial x} + \frac{\partial z}{\partial t} \frac{\partial c_i}{\partial z} + \frac{Q_i}{z} + R_i(c_i, \ldots \ldots, c_n) \tag{3}$$

where c_i denotes the mean concentration of species i within the domain, u is the mean wind speed within the domain, Q_i is the source emission flux of species i into the domain and R_i is the rate of production and/or destruction of species i from chemical species. This equation expresses the mass conservation of each pollutant in a fluid in which chemical reactions occur.

Certain assumptions are made in order to facilitate the use of eqn (3) in the boundary layer. Among all, we assume that the emissions rates are averaged over all sources within the box, that they do not change substantially over one hour period and that the sources are uniformly distributed over the horizontal surface of the box.

The term $\partial z / \partial t$ represents the diurnal growth of the Mixed Layer and the effect of the rising top of the box is the dilution of pollutant species contained within the volume. The description of the growth of the Mixed Layer is dependent upon many meteorological variables, being atmospheric stability and insolation two of the more important. The prediction of this growth is a separate and rigorous problem in and of itself. Several models have been developed to treat the Mixed Layer evolution (Tennekes [4], Carson [5], Gryning-Batchvarova [2], Latini [6]).

4.1 Meteorological data input

The PBM requires values of specific meteorological parameters in order to solve eqn (3). Required data are:

- date (day, month, year; starting time is fixed at 05.00 local time);
- location (latitude, longitude);
- wind speed (hourly, meter/second; the wind speed is assumed to be measured at 10 m above ground);
- mixing height (hourly, meters above the ground) computed with Gryning-Batchvarova model;
- total solar (TSR) or ultra-violet radiation (ly/minute, optional);
- ambient air temperature (hourly, degree Centigrade, optional);
- cloudiness (coverage, height, optional).

The meteorological data related to July 1998 (temperature and wind speed) have been supplied by IRPEM, department of National Centre of Research (CNR), Ancona.

5 Application of Model

During the first validation of the PBM in a valley-coastal area we have considered the Esino Valley (Latini [7]). The results of the comparison between ozone observations in the Esino Valley and the ozone levels predicted by the PBM were encouraging. Further work was necessary to assess the confidence of the model and to establish the uncertainty and the capabilities of the model in a coastal area. Therefore, in this paper we have applied the model with a different meteorological input. The modelling domain is described by a 20 Km x 20 Km x h_{mix} volume (where h_{mix} is the temporally varying depth of the Mixed Layer) and includes most of the major emissions sources.

Air quality data are required to compute values for some c_i terms in (eqn 3). Initial conditions are necessary for $\partial c_i/\partial t$ terms, lateral boundary conditions for $\partial c_i/\partial x$ terms, and boundary conditions at the top of the domain for $\partial c_i/\partial z$. Air quality data were supplied by IRPEM department of National Centre of Research (CNR), Ancona.

Data related to source-specific emission inventories or continuous emission observations are not always available and, when available, they might not be representative of the actual variability of emissions over time. Thus, emission factors, despite their limitations, are applied as the choice method for estimating emissions.

Ozone pollution is mainly a daytime problem during summer months since sunlight plays a primary role in its formation. Nitrogen oxides and hydrocarbons are known as the chief "precursors" of ozone. These compounds react in the presence of sunlight to produce ozone. The sources of these precursor pollutants include cars, trucks, power plants, factories, and any other activity involving natural gas, gasoline, diesel fuel, kerosene and oil combustion. Large industrial areas and cities with heavy summer traffic are the main contributors to ozone formation. When temperatures are high and the mixing of air currents is limited, ozone can accumulate to reach unhealthy levels.

6 Results and Discussion

July 16nd, 1998 was the case study day, being representative of typical high ambient ozone day. The meteorological conditions of this day also allow the PBM software to work at its best, inside its intrinsic limits. In fact it is a sunny day with low wind and therefore characterised by unstable conditions. Figure 3 shows the result of the described routine, evaluating the mixing height, based on a data set obtained from surface measurements of wind speed and temperature.

In Figure 4 the ozone prediction is in good agreement with monitored values although the magnitudes of the predictions are slightly higher than the observed data (figure 5).

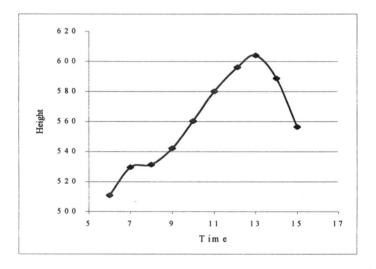

Figure 3: Mixing height behaviour

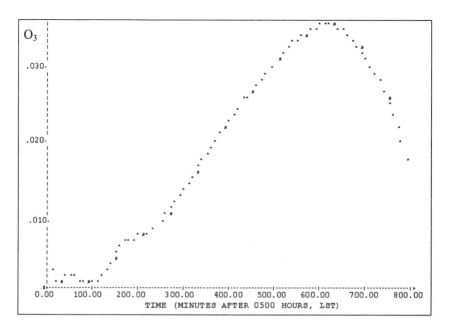

Figure 4: Ozone day-time evolution

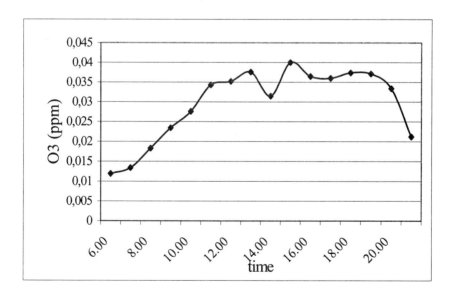

Figure 5: Monitored Ozone evolution

7 Conclusions

A photochemical box model for air quality simulation in conjunction with a different meteorological input has been applied over a coastal area. A new algorithm for the parameterisation of the growth of the Mixed Layer has been developed. The model has been slightly changed from the version being used in the St. Louis model validation studies, and is now demonstrated to be generally applicable to a complex valley-coastal terrain as Esino valley.

Model results showed generally good agreement between prediction and spatially–averaged observed data for Ozone.

References

1. Holtslag, A.A.M. & Van Ulden, A.P., A Simple Scheme for Daytime Estimates of Surface Fluxes from Routine Weather Data, *Journal of Climate and Applied Meteorology,*22, pp. 517- 529, 1983.
2. Batchvarova, E. & Gryning, S.E., Applied Model for the Growth of the daytime Mixed Layer, *Boundary Layer Meteorology,*56, pp.261-274, 1991.
3. Stull. R.B., *An Introduction to Boundary Layer Meteorology*, Kluwer Dordrect, 1994.
4. Tennekes, H., A Model for the Dynamics of the Inversion Above a Convective Boundary Layer, *Journal Atmos. Sci.,*30, pp.558-567, 1973.
5. Carson, D.J., The Development of a Dry, Inversion-Capped Convectively Unstable Boundary Layer, Quart.J.Roy.
6. Latini, G., Cocci Grifoni, R., Passerini, G. & Tirabassi, T., Evaluation of Mixing Height over Complex Coastal Terrain. *Air Pollution 99,* eds. C.A. Brebbia, M. Jacobson, H. Power, Wit Press: Southampton and Boston, pp.1073-1081, 1999.
7. Latini, G., Cocci Grifoni, R., Passerini, G. & Fava, G., Applicability of a Photochemical Box Model over Complex Coastal Areas . Air Pollution *99,* eds. C.A. Brebbia, M. Jacobson, H. Power, Wit Press: Southampton and Boston, pp.697-706, 1999.

Sedimentological analysis of deposits due to combination of strong rain, floods and wave erosion: Case of Albuñol Delta, Granada, Spain

J.S. Lopez-Gutierrez & V. Negro-Valdecantos
Coastal and Harbour Department, School of Civil Engineering,
Politechnic University of Madrid, Spain

Abstract

A catastrophic flood due to extraordinary rain in a very short time period, over 600 mm in five hours, produces important changes in the evolution of the shoreline, because it generates higher deposits than the natural evolution of the longshore transport caused by normal wave action. The high magnitude, low frequency flood called 'spark freshet' (avenida relámpago) is the opposite of the natural slanted wave transport of sediments and it represents the dynamic nature of the alluvial geomorphic changes with significant accretion occurring during hours, and limited erosion over years. The potential for dramatic changes in hydrological events and small variations in oceanographical aspects are the topics of this paper. We try to establish the procedure to recover the coast with improved coastal management. The Albuñol and Huarea natural riverbeds, Granada, South East of Spain are two cases studied to understand the whole process.

1   Introduction. Development of the work

The catastrophic flood of October 19th, 1973, in Albuñol and Huarea deltas, Granada, Spain, was caused by an extraordinary storm longer than five hours and with over 600 mm of precipitation.

The hydrological event generated a delta strongly dominated by river sediments with small wave action (see figure 1). The return period of this flood was over 700 years, with a peak flow of 2,500 m^3/s. The results were more than 15 Hm^3 solid deposits and a new shoreline was created 1 Km far from the old one. The river basin is approximately 100 Km^2.

After this event (which caused over 100 deaths), the dynamic action of the wave climate produced the natural erosion and distribution of the grain size all over the beach profile along the shoreline. This was the procedure of the hazard evolution.

This study tries to analyze the procedure of the small erosion and strong accretion caused by different actions: wave transport and high magnitude flow in the very short time period. The calculations were carried out using several skills: one-line model for shoreline evolution (GENESIS, GEneralized model for SImulating Shoreline change, Hanson y Kraus, 1989), a 3D numerical model (MOUSLE, MOdified Universal Soil Loss Equation), and Wave Climate records from the Spanish Network of Waves Recording and Measuring, (REMRO).

The permanent purpose of a Coastal Engineering Project is to strengthen the positive aspects and solve problems outstanding within the complex system making up the littoral, wave climate, hydrological aspects and, finally, the consequent urban development and agricultural uses of the land.

In this particular case, the Integrated Coastal Zone Management (ICZM) tries to achieve wave action compatible with the use of the area and to analyze the procedure caused by the strong accretion and the fast response of the high magnitude flood as well as the slow response and the effect of the small erosion due to the wave climate action.

2 Morphological behaviour of deltas

The zone is characterized by a seasonal torrential fluvial regime on a dry bed such as the Spanish Mediterranean Coast. The spark freshet presents flows over 2,500 m^3/s. Two examples are: one at the end of 1970, with one hundred years return period, and the other, the catastrophic flood of October of 1973 with seventy hundred years return period.

The extraordinary event in a hydrological dry basin creates a deltaic cone with a bottom drag transport regime. The wave energy patterns the fluvial sediments with a grain size distribution and a very high suspension transport item.

Galloway comparative deltas classification, 1975, [figure 1], shows the different actions: fluvial dominated, wave dominated and tidal dominated of this outstanding shoreline configuration.

Albuñol and Huarea deltas are comparable with Po, Danube and Missisipi with a very long periods of inactivity and wave action processes in between. The South East part of Spain is fully of torrents and gullies like Adra, Castillo de Baños, Punta de los Hornicos, Cuevas de Almazora, Vélez, etc.

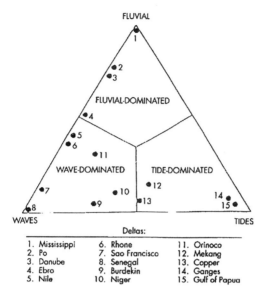

Figure 1: Comparative classification of deltas, Galloway, 1975

According to Suarez Bores scientific notes, the behaviour of the deltas can be classified in the following steps:

Agents
1	Bottom drag transport	Medium - high
2	Wave energy	Low - medium
3	Delta type	Arch

Nature of the basin
1	Basin	Dry most of time
2	Transport type	Bottom drag
3	Deposits type	Deltaic cone

The figure 2 shows the natural coastline evolution after the catastrophic flood to nowadays.

The natural erosion of the coast is one of the reason of this project because the adjacent land to the Coastal Public Property (protection easement, passage easement and zone of influence, Spanish Shore Act, 1988) and the Coastal Public Property is occupied by farming uses in illegal form, setting randomly rubble mound protection avoiding the wave actions during

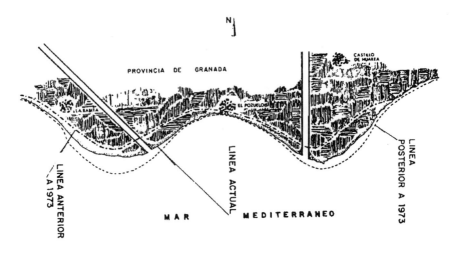

Figure 2: Coastline evolution, Albuñol and Huarea deltas, 1973-1999

the storms.

3 Sediment transport rate

Coastal engineers are now routinely called upon to make predictions of beach profile response due to seasonal wave conditions. The beach will erode or accrete by wave induced cross shore sand transport and the natural evolution with the storm erosion, post-storm recovery and combination of hydrological process with wave energy. The traditional equation by Inman examines the capability to predict whether a beach will erode or accrete:

$$Q_1 + q \cdot \Delta l \left\{ \begin{array}{c} < \\ = \\ > \end{array} \right\} Q_2 \qquad (1)$$

where,

Q_1 Inner transport rate of the cell
Q_2 Outer transport rate of the cell
q Fluvial mass rate
Δl Length of the cross section

where = means equilibrium situation; >, accretion situation; <, erosion situation.

The simple equation can easily reproduce the natural state and evolution of the shoreline, because the sediment transport balance depends on the hydrological episodes (rare, fast response) and the wave climate ones (common, slow response).

The longshore transport rate magnitude is discussed using the CERC and Bagnold formulae, getting the following results:

Table 1: Transport rate using CERC and Bagnold formulae

Significant wave height $H_{\frac{1}{3}}$, m	Wave direction, E Transport rate, Q_E	Wave direction, W Transport rate, Q_W
0.50	19,000	-11,000
1.00	90,000	-62,000
1.50	112,000	-89,000
2.00	90,000	-47,000
2.50	53,000	-37,000
3.00	0.00	-55,000
TOTAL	364,000	-301,000

$$Q = 1.042 \cdot 10^8 \cdot \frac{1}{(\theta_2 - \theta_1)} \cdot f \cdot H_0^{\frac{5}{2}} \cdot \left(\cos^{\frac{9}{4}} \theta_2 - \cos^{\frac{9}{4}} \theta_1 \right) \qquad (2)$$

Q	net transport rate	$m^3/year$
f	frequency of wave energy	s^{-1}
H_0	wave height in deep water	m
θ	offshore wave angle	$^\circ$

Using the CERC formula, (Shore Protection Manual), table 1 shows the different balance of the transport rate in the main directions of the study area, East one and West one, because the wave climate is absolutely bidirectional and this fact is represented in the transport components [figure 3]

Wave Rose in Alboran Sea (Swell)

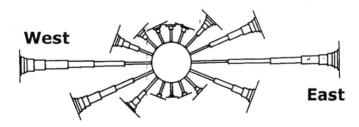

Figure 3: Wave rose in Alboran Sea. ROM 0.3-91

The theoretical longshore transport rate is 63,000 m^3 *per year*, direction East from West. The real situation according different researchers in Spain, $Q_{net} = 0.75 \cdot Q_{calculated}$, that is the net transport rate becomes approximately 47,250 $m^3/year$.

The coefficient 0.75 is the result of several surveys in the Mediterranean Spanish Coast (Maresme zone, NE of Spain). This term is published in *'Evaluación del transporte de sedimentos paralelo a la orilla empleando observaciones visuales de oleaje y teoría cnoidal de ondas'*, J.M. Grassa, *Ingeniería Civil, 66 (1988)*.

The paper processed the drag material from the hydrological episode in storm conditions to compare the results derived from the wave action in common situation. In a basin over 100 Km^2 and with a rain of 600 mm in five hours time, a simulation using the MOdified Universal Soil Loss Equation (MOUSLE) was carried out to determine the specific basin demotion in $Tonne/Ha/year$, with the climatic and topographic items, type of soil, plantations, for obtaining the deposits evolution in different return periods.

The model architecture was founded in the following design concepts to develop a 3D numerical model for wide application use. The first parameter was to combine the climate module and the effect of the high storm on the soil (R); the second one is the behavior of the ground such as its components and history (edaphology), the property for being eroded (K); the third one, the topographical factor (T); the slope of the natural material and the friction angle (m, φ); and the last one, the effect of the vegetation on the soil and the capability against the erosion process (P). The MOUSLE equation (Muskingum, 1936; William and Berndt, 1977) interacts and represents the profile shape using the Gumbel distribution for rain in the basin and the flow regime, the storm current speed, the Froude number, the specific weight of the material, the suspension and drag profile and finally the transport rate in combination with the return period of the event. This equation analyzes the essential features for calculating storm induced erosion and long term change in shoreline position with the suspension and drag rates from the ground to the sea. These calculations are disposed in tables 2 to 6 including in the paper.

Table 2: Deposits evolution in five years return period

Froude Number	Speed	Rain time	Flow	Rubble mound	Specific weight	Alluvium	$D_{n,50}$
	m/s	h	m^3/s	*tonne*	t/m^3	*tonne*	m
1.05	2.73	5.00	92.20	422.575	1.14	50.984	0.394
0.96	3.61	5.67	274.3	$1,42 \cdot 10^6$	1.13	108.738	0.567

Table 3: Deposits evolution in twenty years return period

Froude Number	Speed m/s	Rain time h	Flow m^3/s	Rubble mound tonne	Specific weight t/m^3	Alluvium tonne	$D_{n,50}$ m
1.12	3.60	4.92	139.50	932.519	1.15	104.933	0.614
1.01	4.60	5.67	519.8	$2,9 \cdot 10^6$	1.14	210.287	0.832

Table 4: Deposits evolution in fifty years return period

Froude Number	Speed m/s	Rain time h	Flow m^3/s	Rubble mound tonne	Specific weight t/m^3	Alluvium tonne	$D_{n,50}$ m
1.15	4.06	4.75	258.1	$1,31 \cdot 10^6$	1.15	143.300	0.743
1.05	5.12	5.67	686.2	$3,96 \cdot 10^6$	1.14	280.335	0.976

Table 5: Deposits evolution in one hundred years return period

Froude Number	Speed m/s	Rain time h	Flow m^3/s	Rubble mound tonne	Specific weight t/m^3	Alluvium tonne	$D_{n,50}$ m
1.18	4.38	4.67	312.0	$1,62 \cdot 10^6$	1.15	173.420	0.830
1.06	5.44	5,67	813.9	$4,81 \cdot 10^6$	1.14	334.985	1.081

Table 6: Deposits evolution in five hundred years return period

Froude Number	Speed m/s	Rain time h	Flow m^3/s	Rubble mound tonne	Specific weight t/m^3	Alluvium tonne	$D_{n,50}$ m
1.22	5.00	4.58	441.8	$2,4 \cdot 10^6$	1.165	247.061	1.027
1.10	6.11	5.67	1114.9	$6,87 \cdot 10^6$	1.149	466.082	1.29

The changes of the Public Property Domain shows the following values in the contrast profiles using the different data sources and the evolution of the shoreline with aerial photography.

There are various factors which have negatively affected the preservation of this natural environmental which value has increased as a result of changes in human behavior and the mass phenomenon of the leisure civilization. The decrease of solid materials in rivers and streams, the construction of dams, groynes and breakwaters, on the other hand, the development of high walls of buildings on the very short front, are the most important reasons for the disappearance of dunes and the environmental degradation of the seashore Public Property.

This fact can be observed using several aerial photographic surveys from 1947 to 1998, describing the natural evolution of the shoreline. The shoreline evolution is shown in tables [7, 8, 9, 10, 11 y 12], where the symbol "+" means accretion and "-" means erosion of the coast.

Table 7: Relative shoreline locations in Albuñol and Huarea deltas,1

La Rabita beach, western part of Albuñol delta

1947	1957	1973	1977	1981	1995	1998
310 m	285 m	363 m	418 m	348 m	339 m	332 m
	(-25)	(+78)	(+55)	(-70)	(-9)	(-7)

Table 8: Relative shoreline locations in Albuñol and Huarea deltas,2

Albuñol delta, western part

1947	1957	1973	1977	1981	1995	1998
189 m	193 m	334 m	388 m	275 m	265 m	262 m
	(+4)	(+141)	(+54)	(-113)	(-10)	(-3)

Table 9: Relative shoreline locations in Albuñol and Huarea deltas,3

Albuñol delta, eastern part

1947	1957	1973	1977	1981	1995	1998
68 m	68 m	234 m	269 m	215 m	204 m	204 m
	(-)	(+166)	(+35)	(-54)	(-11)	(0)

Table 10: Relative shoreline locations in Albuñol and Huarea deltas,4

El Pozuelo beach, between both deltas

1947	1957	1973	1977	1981	1995	1998
329 m	333 m	350 m	395 m	486 m	450 m	430 m
	(+4)	(+17)	(+45)	(+53)	(+2)	(-20)

Table 11: Relative shoreline locations in Albuñol and Huarea deltas,5

Huarea delta, western part

1947	1957	1973	1977	1981	1995	1998
273 m	263 m	472 m	415 m	404 m	400 m	395 m
	(-10)	(+209)	(-57)	(-11)	(-4)	(-5)

Table 12: Relative shoreline locations in Albuñol and Huarea deltas,6

Huarea delta, eastern part

1947	1957	1973	1977	1981	1995	1998
96 m	92 m	348 m	354 m	312 m	310 m	303 m
	(-4)	(+256)	(+6)	(-42)	(-2)	(-7)

4 Rehabilitation proposal

The researchers propose the confidence bands according to the hydrological regime and the natural wave energy history, in order to determine the boundaries of the Coastal Public Property, the use of it, free and public for the common and natural uses, the territorial planning and zoning of land.

This protection easement can be enlarged by the Central Coastal Authority upon prior agreement with the corresponding Regional Government and Town Council, up to maximum of 40 to 50 meters.

Figure 4: Aerial view of Albuñol delta before and after flooding, 1973-1999

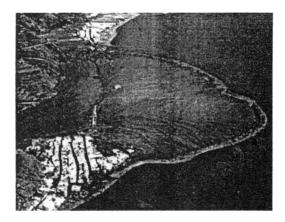

Figure 5: Aerial view of Albuñol delta after flooding, 1973

The authorization for the carrying out of farming and plantations of land in these confidence bands is denied. The shoreline must be clear without illegal protection and indiscriminated uses of the Public Property Domain, as the natural evolution of the longshore transport by waves and hydrological response events.

5 Conclusions

The Central Government, without prejudice to the powers of the Regional Government or Town Council, issues general and specific regulations for certain coastal stretches on the protection and use of the Coastal Public Property in compliance with the terms of the Shore Act, July 28th, 1988, Title III, "Use of the Coastal Public Property", Chapter I, sections 31 to 46.

By this reason, the shoreline shall never be for private uses. In this case, agricultural uses, farming and plantations in the land is absolutely denied.

The shoreline shows important changes in the natural evolution, hydrological processes with a fast answer increase the delta with deposits, and the wave action produces the erosion of the coast with the common energy.

The territorial planning and zoning is one of the purpose of the Spanish Shore Act (1988), joined to the protection and regulation of the Coastal Public Property.

In addition, the use of this domain and, in any event, of the sea and its seashore shall be free and public for the common and natural uses. The occupation of the Coastal Public Property shall only be permitted for those activities or installations which by their nature may not be located elsewhere.

The Central Government and the Granada Coastal Agency can require the total or partial expropriation –compulsory purchase– or the temporary occupation of the necessary land for agricultural facilities or uses. Albuñol represents this study case.

The use of illegal rubble mound protection in the public domain must be avoided.

The territorial planning and zoning of land recommends a confidence bands of forty to fifty meters for the natural and free evolution of the shoreline. This proposal is agreed with the hydrological regime and the wave climate.

Albuñol and Huarea, natural river beds, Granada, South East of Spain, are two cases study to understand the whole process.

6 References

[1] Galloway, W.E., Process framework for describing the morphologic and stratigraphic evolution of deltaic depositional systems. Deltas. Models for exploration, Ed. Martha Lou Broussard, Houston Geological Society, pp. 87-98, 1975.

[2] Hanson, H. & Kraus, N., GENESIS: GENEralised model for SImulating Shoreline changes. CERC, TR-CERC-89-19, 1989.

[3] De la Peña, J.M. & López, J.S., Dinámica litoral del delta del Ebro. CEPYC-CEDEX, 1996.

[4] Moreno, L.J., Modelo de evolución de playas. CEPYC-CEDEX, 1992.

[5] Zenkovich, V.P., Processes of Coastal Development. Ed. Oliver & Boyd, London, 1967.

[6] The Shore Act, Ministry of Public Works, 1988

[7] Grassa J.M., Evaluación del transporte de sedimentos paralelo a la orilla empleando observaciones visuales de oleaje y teoría cnoidal de ondas, Ingeniería Civil, 66, 1988

Acknowledgment

The major part of the work reported here was carried out in INTECSA, Consulting Company in Madrid, Spain. The authors greatly appreciate Mr. Severiano Benavides and Mr. Miguel Calleja from the General Directorate of Coasts in Granada (Central Coastal Authority - Ministry of Environment), for their sincere assistance and help during the study.

Environmental impact assessment of Marennes–Oléron bay: operational mathematical modelling and GIS

C. Freissinet[1], P. Sauvaget[1] & C. Bacher[2]
[1] SOGREAH, France
[2] CREMA, France

Abstract

Throughout the world, coastal zones are the most heavily populated and ecologically vulnerable areas. Many different and often conflicting interests exist in such areas which have to be (at least) properly addressed and (hopefully) carefully managed. For this purpose it is absolutely necessary to enhance the forecasting capacity regarding the medium- to long-term changes in the coastal marine system in the light of human activities and natural changes.

In this framework, the primary objective of the European OPCOM project (Operational Modelling for Coastal Zone Management - Mast 3 - website: *http://www.hydromod.de/projects/OPCOM/OPCOM.html*) is to improve methods and techniques for continuous monitoring and operational forecasting in coastal waters. In this respect, operational models and monitoring infrastructure are used in four European coastal areas: the Elbe Estuary (Germany), the Bay of Marennes-Oleron (France), the Tagus Estuary (Portugal), the Archipelago Sea (Finland).

Consequently, quality standards and the needs of interested first users in coastal zone management are being considered in the project work. Marennes-Oléron Bay was chosen as the French site for implementation of a hydrodynamical model developed by SOGREAH, coupled with the Geographical Information System built by IFREMER and implemented at the CREMA laboratory.

1 Introduction

Marennes-Oléron Bay is the largest oyster farming area in France. Both inter tidal areas and salt marshes are important for this aquaculture activity and for the ecosystem equilibrium. Tourism has also been developing and is giving additional value to the coastal zone as a recreational area. Industrial growth has occurred as well in some locations and, combined with urban activities, contributes to the increasing awareness of the multiple interests of the coastal zone. Agricultural activity in the catchment areas of the rivers, bringing fresh water and nutrients to the coastal zone, plays a role in the functioning of the bay ecosystem. Several institutions are therefore concerned by impact assessments, environmental protection and economic development of the bay. They are interested in the long-term variability of the hydrodynamical and environmental parameters of the bay – e.g. changes in water quality in the bay (salinity, temperature, dissolved oxygen) depending on changes in river discharge, meteorological parameters, tide, etc. They are also interested in sea-level prediction and sediment transport.

Some of the achievements obtained for this bay are presented here, including mathematical modelling and GIS, as well as the requirements that the operational modelling will have to meet.

2 Coastal zone management areas: Marennes-Oléron Bay

Marennes-Oléron Bay is a coastal inlet, roughly 40 km long. It is oriented north-south and located in the central part of the French Atlantic coast. It is limited by the Island of Oléron to the west, the Island of Ré to the north and the Charente Maritime coast to the east.

The bay is connected to the ocean through three channels: Pertuis de Maumusson in the south, Pertuis d'Antioche and Pertuis Breton in the north. Pertuis d'Antioche and Pertuis Breton converge into the Coureau d'Oléron (see figure 1).

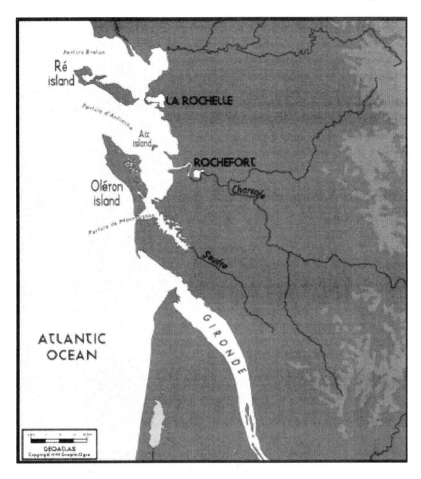

Figure 1: Marénnes d'Oleron Bay

The bay is fed with fresh water from three sources:

- The Gironde estuary, located south of the island of Oléron, which receives the discharges of the rivers Garonne and Dordogne ,
- The river Charente, between the Island Madame and Fouras,
- The river Seudre, between Cayenne and La Tremblade.

The main characteristics of these systems are summarised in Table 1.

Table 1: Marennes-Oléron Bay fresh water sources

Upper limit of the model	Curvilinear distance (km)	Discharge / year			Yearly fresh water flow (m^3)
		Average discharge (m^3/s)	High flow (m^3/s)	Low flow (m^3/s)	
Charente river	St Savinien	45	36.1	600	13	1.1 x 10^9
Seudre river	Saujon	20	1.01	8.9	0.037	3.2 x 10^7
Gironde Estuary	Bec d'Ambes	70	Garonne river: 631 Dordogne river: 284	Garonne river: 5700 Dordogne river: 2600	Garonne river: 39 Dordogne river: 8.45	2.9 x 10^{10}

The yearly flow of the Seudre represents only 3 % of the yearly flow of the Charente [1]. Only a fraction of the water from the Gironde estuary enters the Marennes-Oléron Bay through the Pertuis de Maumusson [2] and possibly through the Pertuis d'Antioche. This fraction depends on the tidal currents and has to be determined with the model.

3 The mathematical model

The numerical software used in this work is TELEMAC-2D developed by the French Electricity Board (EDF-LNH). TELEMAC-2D is based on finite-element numerical schemes solving the two-dimensional shallow water equations on triangular unstructured meshes [3]. Another software embedded in the system enables salinity and temperature dispersion to be computed, represented as a passive tracer (SUBIEF-2D).

3.1 The hydrodynamical nested model

The study is divided into two parts:
- Construction and operation of a large model of the Bay of Biscay which extends from the Brittany peninsula (Pointe de Penmarch) to the north of the Iberian Peninsula.
- Construction and operation of the Marennes-Oleron Bay refined nested model. The model extends from the south of the Gironde Estuary up to the town of Les Sables d'Olonne in the north (about 120 km long by 80 km wide).

In Marennes-Oléron Bay application, a large-scale model was calibrated using measurement data, so that the model could be used in generating the boundary conditions for a refined model of the area of interest.

3.1.1 Model with large geographical area
Bathymetric data sources
GEBCO 97 (General Bathymetric Chart of the Oceans from the British Oceanographic Data Centre) data were used to define the coastline and the bathymetric values between -200 m and –5000 m including data on the continental shelf.
SHOM data (Service Hydrographique et Océanographique de la Marine) were used to describe the bottom between –200 m and the coast.
Boundary conditions
The boundary conditions are the tidal constituents (amplitude and phase) of the Atlantic Ocean free surface elevation. These conditions were obtained from the LPAL 94 database [4] developed by LEGI (Laboratoire des Ecoulements Geophysiques et Industriels, Grenoble University, France).
Thirteen main tidal constituents are imposed on the maritime boundary. These main constituents are diurnal or semi-diurnal.
Model calibration
To calibrate the large model, two sets of tide gauge data were used (see figure 2). However, the validity of each tide gauge must be checked, considering: the duration of the time recording, its situation at sea, the comparison of the amplitude and phase between close tide gauges

To determine whether a tide gauge is coherent, the vectorial RMSE (Root Mean Square Error) can be calculated from the harmonic analysis of the tide signal on the large model. This technique is used to compare the amplitudes and phases of the computed and measured data. If a tide gauge has a high vectorial RMSE value while its neighbours have quite small values, this means that the tide gauge is in a zone where local phenomena occur (often a coastal zone), or that it is not reliable.

There are some differences between the calculated model results and the observed tide gauge data (see figure 3). The calculated amplitudes are increased in comparison with the tide gauge amplitudes.

Before any modifications of the model parameters, the increase in height was:
- + 7 cm in the north (near the Brittany coast),
- + 3 cm near Marennes-Oléron Bay,
- + 4 cm in the south (near the Spanish coast)

To improve these results, it was necessary to modify different physical parameters:
- modification of bed friction (Chezy coefficient)
- integration of the astral potential that takes into account the generator power of the tide
- use of the available information on velocities for forcing the model at its open boundaries,
- Bathymetry sensitivity analysis (in the continental shelf margin, i.e. between -200m and -100m),
The bathymetry is the most sensitive parameter affecting model results.

Figure 2 – Location of the two sets of tide gauges

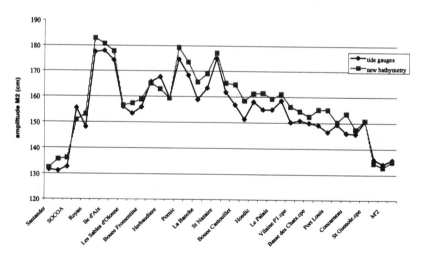

Figure 3 – Model calibration

3.1.2 Marennes-Oléron regional model

Bathymetric data sources

Different data sources were used to build the model:

- SHOM data (Service Hydrolographique et Oceanographique de la Marine) for details on the bathymetry near the coast, in particular in Marennes-Oléron Bay

- DDE data (Direction Départementale de l'Equipement) for scattered bathymetric data of Marennes-Oléron Bay and for cross-sections of the river Charente.

- IGN data (Institut Géographique National) for the description of oyster beds in the river Charente and river Seudre.

Boundary conditions

The open boundary conditions must be specified as accurately as possible. It is better to use not just the water levels at the boundaries but also the flow velocities. The results are then closer to reality. Unfortunately, it was impossible to do so for the regional model as the mesh of the large model (from which the boundary conditions of the regional model are extracted) was not fine enough compared to the mesh of the refined model along its boundaries. For this reason, the values of water velocities and levels were interpolated at certain nodes and were not exactly coherent from one computation point to the next. This induced small wiggles in the model near the maritime boundaries.

The boundary conditions are :

- Amplitude and phase of the 13 main tide constituents. The large model determined these conditions.

- **Banque Hydro** (national data bank for hydrometry and hydrology) to obtain the discharges of the three fresh water systems: river Charente, river Seudre, Gironde Estuary.

Model calibration

Most of the parameters that had been modified to calibrate the large model were also changed in the regional model. The astral potential was not taken into account, the friction coefficient was considered with the Strickler formula to obtain a better fit in the rivers and was taken equal to 60, the boundary conditions consisted only in water levels (no velocities) as there is an incompatibility when these data are extracted from the large model.

To calibrate the model, comparisons with water levels were made at different points. SHOM data were used at several tide gauges in the bay (Sables d'Olonnes, La Rochelle, Ile d'Aix, Pointe de Gatseau). Some water levels in the river Charente, were drawn from a former study carried out by Sogreah.

The assimilation of the real river discharges is quite important in order to have a good fit to measured data at the considered date. This can be achieved thanks to the Banque Hydro, which contains the real discharges of the rivers for given dates.

The model results in the bay were quite similar to the amplitudes and phases of the 13 main constituents from the SHOM[5].

3.2 Dispersion model

The model results for salinity and temperature dispersion have been compared to available data at different sites (e.g. in the Charente river mouth and inside the bay). It is thus possible to see whether the model results are accurate or not.

4   Case Studies

4.1 The different scenarios

For Marennes-Oléron bay, the set of sensitivity studies focused on the hydrodynamics of the bay and on the variation in salinity and temperature when parameters are changed (river inflow, tide amplitude, wind). To evaluate the influence of each parameter on the hydrodynamics of the bay and the diffusion of fresh water from the rivers, different scenarios have been considered: spring tide with river mean flow, neap tide with river mean flow, spring tide with flood discharge in rivers, neap tide with flood discharge in rivers, spring tide with wind from west and river mean flow, spring tide with wind from north-west and river mean flow.

The best way of analysing the results is to compare the different scenario results. This approach only gives qualitative and not a quantitative information. To be able to give the exact concentrations, it would be necessary to know the initial salinity everywhere in the model at the initial time (or to let the model run for a very long period in order to reach an established regime). This requires to simulate real tide cycles (spring tide and neap tide) and would ask for very long computational time.

4.2 Results

4.2.1 Tide influence
The results showing tide influence are very conclusive. When comparing the results during a neap tide cycle with those during a spring tide cycle, fresh water dispersion is found to be much greater during the spring tide cycle (cf. figure 4). This is due to the fact that the amplitude of water movements is greater during the spring tide.

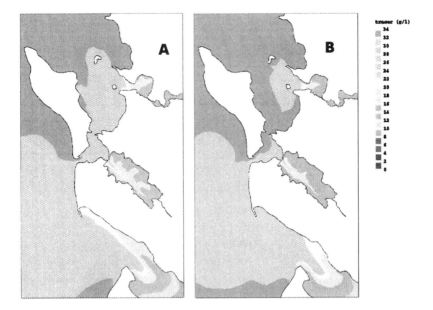

Figure 4 : Comparison between spring (A) and neap tide (B) influence in normal
river discharge conditions for 1 month of simulation

In normal conditions, as shown above, the fresh water from the river Charente
mainly remains south of Aix island. It flows mainly southwards, crosses Le
Chapus channel and comes out through the Pertuis de Maumusson.

4.2.2 Influence of river discharge
In river flood conditions, the fresh water flows mainly southwards as shown in
figure 5. The diffusion in the different directions is proportional to the fresh
water discharge.

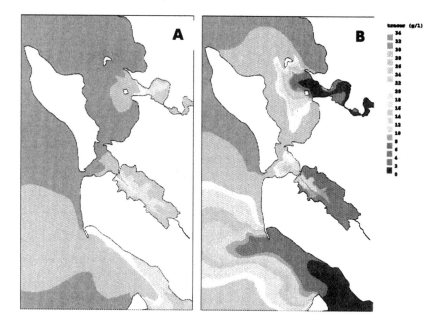

Figure 5: Comparison between normal (A) and flood discharge (B) conditions
during spring tide for 14 days of simulation

4.2.3 Wind effects
The wind also has a major effect on fresh water dispersion. Two cases have been
considered:
- wind of 20 m/s from west
- wind of 12 m/s from north-west

The westerly wind prevents the fresh water from flowing out of the river
mouths. The fresh water is pushed towards the eastern coasts of the model.

The fresh water from the river Charente cannot reach the central channel of
the bay and go southwards. It flows through the exit channel and follows the
coast northwards.

The fresh water from the river Seudre follows the coast southwards and does
not cross the central channel either. The fresh water from the Gironde estuary is
pushed southwards

The north-westerly wind has less impact. Its main effect is to prevent fresh
water from going north of Aix island. The fresh water is logically pushed south-
eastwards (*see figure 6*).

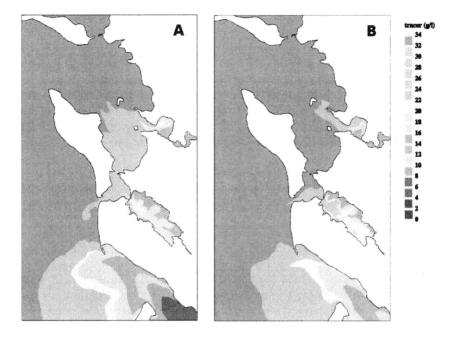

Figure 6: Fresh water dispersion for westerly (A) and north-westerly (B) wind
conditions (1 month)

Fresh water diffusion is mainly driven by tides. However, exceptional
weather conditions such as strong wind or river floods have a major impact on
water movements.

4.2.4 Conclusions on the scenarios
The tide is the main parameter determining the water level in the bay. Wind also
has to be taken into consideration as it has an impact on water level and
hydrodynamics and, as a consequence, an influence on fresh water dispersion.
The main winds to be considered are westerly winds as they are the most
frequent and strongest winds. North-westerly winds are also fairly frequent but
weaker.

The salinity variability is important because of the influence of the river
Charente and the river Seudre. The salinity generally goes from 30‰ to 36‰,
but fresh water inputs may affect this salinity value, principally at the mouth of
the river Charente. At this site, both the inter annual and seasonal variability are
more pronounced and the salinity can drop below 10‰.

4.2.5 Residual velocities

The residual velocities have been calculated for one tidal cycle in order to have an indicator of the overall water movement. A residual circulation from north to south can be observed inside the bay (see figure 7). In the northern part of the bay (above Le Chapus), this circulation follows the central channel. In the southern part of the bay, the residual current becomes stronger and flows towards the Pointe de Gatseau in the Pertuis de Maumusson. This indicates that most of the water situated in this zone is rapidly evacuated to the ocean through the Pertuis de Maumusson. North of the Island of Oléron, two eddies, one cyclonic and the other anticyclonic, have been noticed. These results are quite similar to those presented in the thesis of Cédric Bacher [6].

This general north-to-south movement was confirmed by the release of floats at different points in the bay. Excursions due to the tide also appear to be quite considerable.

The wind impact on bay hydrodynamics is quite considerable. Westerly winds were found to have a major influence on water level in the bay. Indeed, the increase in water level ranges from 10 to 25 cm. Moreover, the flood tide occurs a little in advance and the ebb tide is a little late. The westerly wind induces a current towards the north in the northern part of the bay and a current towards the river Seudre in the southern part of the bay. This is due to the geometry of the bay.

The influence of the north-westerly wind on water level is not so significant. This is partly due to the wind velocity modulus which is smaller than for the westerly wind. However, there is a small difference in phase.

Outside the bay, flow velocities increase considerably and a general southerly movement was noted. In addition, velocities tend to rotate southwards compared to the situation without wind. This rotation is noticeable in relatively deep water zones, thus indicating that rotation is due to the Coriolis force.

The model results of salinity variations due to tide depend on the location in the bay. The salinity variations in the Charente river mouth are greater than elsewhere. The smallest salinity variations occur on the tidal flats near Oléron island.

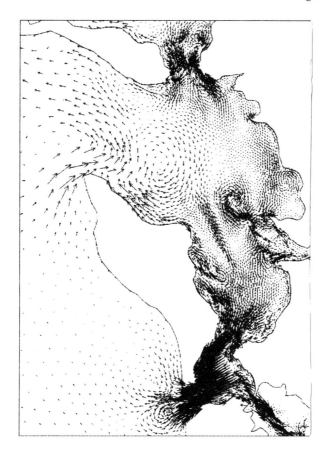

Figure 7: Residual velocity

5 Integration in the GIS

End-user inquiries have shown considerable demand for integration of the modelling results in decision-making tools for the overall water quality management of the Bay. For instance, it is of great interest for oyster farmers to know under which hydrometeorological conditions the plume of fresh water flowing from the river Charente into the bay is pushed to the south-east part of the bay, along the coastline, and therefore enters the Seudre estuary where a lot of oyster farms are located. Freshwater inflow is an important factor for the sustainability of oyster farming (survival, growth and reproduction).

The combination of different parameters (meteorological factors, tides, discharge of the rivers, wave conditions) in the model was established in order to highlight the sensitivity of water quality to individual parameters and to build a databank of contrasted scenarios representing typical hydrometeorological conditions in the Bay.

The interface between the TELEMAC results and the GIS was established in order to select, visualise and superimpose on existing geographical information the outputs of the hydrodynamic and transport computations. This interface is based on the ARCVIEW® capabilities, including the database structure, and the AVENUE object-oriented language. The model output files are converted into direct access binary files thereby guaranteeing a quick means of selecting and retrieving data. Basically, the following steps were defined:

▪ Convert the computation grid to geographic coordinates in the Lambert system.
▪ Build a regular grid based on user selection of node density and geographic area to visualise. This includes spatial interpolation and masking of areas not taken into account within the calculation.
▪ Display the model output characteristics (type of output, time step, number of temporal outputs) contained in the result files.
▪ Select and import a series of vector (current velocity) or scalar fields (concentration, bathymetry, water level) into the GIS. In the first case, the data are imported as a SHAPE file, in the other case as a GRID.
▪ Export the GIS views as GIF or JPG files in order to build an animated sequence

The analysis of the model outputs in the GIS will be further developed in order to allow statistical analysis of the results within a given area or a temporal window – e.g. mean current velocity, mean water level, emersion time. At the present stage, simple queries can be made – e.g. the area with a given salinity concentration – and exported as a table to other applications (see figure 8).

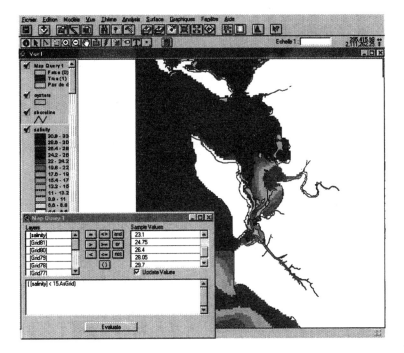

Figure 8 : Importation of the salinity concentration field computed in the high freshwater discharge case, superimposition on the cultivated areas and results of a query on the area with low salinity values.

6 Conclusions

The immediate objective is to provide an easy way of visualising the results of the hydrological parameters which are sensitive to environmental variations and have some potential impact on coastal zone management. As outlined by the end-users interviews, the water quality of specific areas, the sediment uptake and release due to dredging and the water level on the coastline are some of the key parameters. The model is able to show the sensitivity of the transport of substances and the water level to local and regional natural factors – wind, tidal coefficient, freshwater inputs of the Charente and the Gironde estuary. Case studies were defined to assess and to illustrate the importance of such factors for the salinity gradient.

The model results cover the short-term changes in physical descriptors of the environment – e.g. current velocity, bathymetry, water height in the coastal zone over periods of less than 1 month, taking into account the meteorological forcing variables – e.g. wind, freshwater inflows from the rivers, tidal components. The forcing variables are derived from the existing summaries of the statistical characteristics over long term periods – e.g. statistical distribution of wind speed and direction, river maximum and minimum flows.

The results of the model are imported in a GIS and are superimposed on other types of information - e.g. oyster farming areas and salt marshes. By integrating the mathematical results in the GIS an operational tool is obtained which can be used to improve the decision-making process concerning the coastal zone management projects supported by local end-users.

Acknowledgements

The work presented here was partly funded by the European Commission under the DGXII MAST Programme (MAS3-CT97-0089).

References

[1] SOGREAH, *Estimation des débits d'étiage de la Charente à Saint-Savinien*, étude hydraulique, Février 1990.

[2] L.H.F. (Laboratoire d'Hydraulique de France), Programme Charente-Marennes-Oléron, *Contribution au modèle global de l'écosystème d'un bassin versant et de la zone littorale associée ; Modélisation numérique des influences météorologiques sur l'hydrodynamique d'un bassin littoral peu profond.*, Décembre 1990.

[3] Hervouet J.M. and Van Haren L. *Recent advances in numerical methods for fluid flow*, in Anderson M.G., Walling D.E. and Bates P.D. (editors) Floodplain processes, John Wiley, Chichesteer, 1997.

[4] Le Provost, C., Genco, M.L., Lyard, F., Vincent, P. and Canceill, P. *Spectroscopy of the world ocean tides from a finite element hydrodynamic model*, Journal of Geophysical Research, Vol. 99, n°C12, pp.24,777-24,797., 1994

[5] Rapport SHOM 540-MJA, *Table des marées des grands ports du monde*, 186p.

[6] Bacher, C., *Capacité trophique du bassin de Marennes-Oléron : couplage d'un modèle de transport particulaire et d'un modèle de croissance de l'huître Crassostrea Gigas* , Aquat. Living Resour., 2, 199-214, 1989.

Using bioassays for testing seawater quality in Greece

A. Kungolos[1], P. Samaras[2], E. Koutseris[1] & G.P. Sakellaropoulos[2]
[1]*Department of Planning and Regional Development and Department of Civil Engineering, University of Thessaly, Volos, Greece*
[2]*Chemical Process Engineering Research Institute, Thessaloniki, Greece*

Abstract

The purpose of this study is to show the seawater quality in the Thermaikos Gulf, the Pagassitikos Gulf and Skiathos island in the Northern Aegean Sea. The assessment of coastal water quality presented here is based on two bioassays that use marine organisms as indicators of seawater quality, the invertebrate *Artemia franciscana* and the marine bioluminescent bacterium *Vibrio fischeri*. Bioassays are necessary in water pollution evaluations as physical and chemical tests alone are not sufficient to assess potential effects on aquatic organisms. According to our results, there was an improvement in coastal water quality in the Thermaikos Gulf between September 1997 and April- May 2000. In the Pagassitikos Gulf the coastal water quality was generally good in April- May 2000, while in October 1999 it was generally poor. Between the two bioassays that we used in this study, the Microtox test, which uses the marine bacterium *Vibrio fischeri* as a test organism, was more sensitive in detecting toxicity in seawater.

1 Introduction

In Greece, the assessment of seawater quality contaminated with chemicals has been based on chemical analyses and the compliance or not of the measured concentrations with limit concentrations imposed by the legislation. This approach has quite some drawbacks which could be summarised as follows: The cost of a complete chemical analysis is very high, therefore only few chemical tests are performed, investigating the presence and concentration of a limited number of substances. On the other hand, with the number of new chemical substances entering the market increasing every year, one can never be sure that the chemical analysis can cover all potentially hazardous chemicals. Finally, chemical tests alone do no give information concerning potential interactions of the chemical substances existing in the seawater. It has been shown, on more than one occasions, that mixtures of chemicals in water present at the level of current water quality criteria have caused severe toxic effects to aquatic organisms [1].

Because of all the above mentioned disadvantages associated with a monitoring strategy of water quality which depends exclusively on physical and chemical parameters, many countries have initiated including biological parameters in freshwater and seawater monitoring [2]. The European Union Commission has made a draft proposal in 1999, suggesting that the water quality criteria should not be only chemical but also ecological. In USA, EPA (Environmental Protection Agency) regulations require the use of an integrated strategy to achieve and maintain water quality standards. This strategy includes chemical specific analysis, biosurveys (organisms in receiving water body) and bioassays [2]. Bioassays do not measure the cause of the pollution, but they measure directly the polluted seawater effect on test organisms. In this study two bioassays have been used, the *Artemia* test, using the brine shrimp *Artemia franciscana* as a test organism, and the Microtox test, using the marine bacteria *Vibrio fischeri* as a test organism. Two coastal areas were investigated. Thermaikos gulf in Northern Greece, where the port city of Thessaloniki is located, and Pagassitikos gulf in Central Greece, where the port city of Volos is located.

The *Artemia* [3,4] and Microtox [5] bioassays have been used extensively in the past for toxicity assessment of substances and mixtures of substances. The literature covering toxicity of environmental samples on these two organisms is more limited [6,7]. The authors of the present study had used *Artemia franciscana* in 1997 [6] for assessment of coastal water quality in Thessaloniki area. The purpose of this study includes the investigation of the use of marine organism bioassays for testing the ecological quality of coastal water, the comparison of coastal water quality at various time periods and the sensitivity comparison between *Artemia* and Microtox tests.

2 Materials and methods

Coastal water samples were collected in glass containers. Sampling was done in shallow waters from piers or docks. The coastal water samples were transferred to the laboratory within 6 hours from the sampling time. The bioassays were performed either the same day or after 24 hours. In the latter case coastal water samples were maintained at 4 °C.

The bioassays Artoxkit M test [3], and Microtox were used in this study. Hatching of *Artemia franciscana* was done with the help of a NOVITAL 504 incubator. Larvae of the brine shrimp *Artemia franciscana* were hatched from cysts. Cyst hatching was initiated 48 hours prior to the start of toxicity tests. Three replicates were done for each sample and 10 organisms were transferred into each well. The multiwell plate was put in the incubator at 25±1 °C in darkness for 24 hours for the actual test. The inhibitions caused by the coastal water on the bioluminescence of *Vibrio fischeri* were measured with a Microtox 500 analyser (Azur Environmental, USA).

3 Results and Discussion

A map of Thessaloniki area is shown in Figure 1. Coastal waters samples were taken from 7 different sites, which are also shown in Figure 1. Sampling site A was in Potamos, Epanomi, which is after Epanomi cape, where Thermaikos Gulf

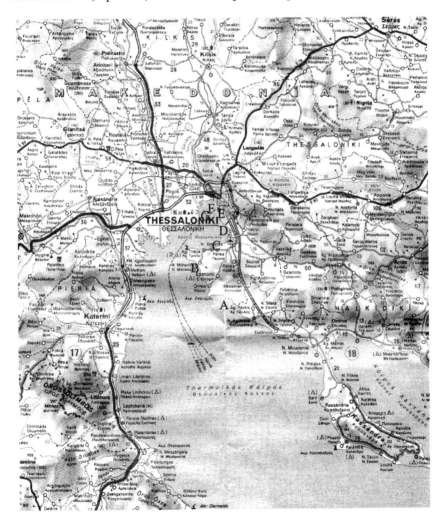

Figure 1. Map of Thessaloniki area indicating the sampling sites for coastal water samples.

becomes very wide and opens to the Aegean Sea. Potamos beach is very popular for swimming within Thessaloniki residents. Sampling site two was in the fishing village of Mihaniona, between the capes of Megalo Emvolo and Epanomi. The other 5 sampling sites were in the inner Thermaikos Gulf (Gulf of Thessaloniki). Sampling site C was in Perea and sampling site D in Aretsu, Kalamaria, between the capes of Megalo Emvolo and Micro Emvolo. The other three sampling sites were in the inner Gulf of Thessaloniki (Thessaloniki Bay), in the area which is supposed to be most polluted from anthropogenic activities. Sampling site E was in Analipsi, Thessaloniki in the heart of the east part of the city. Sampling site F was in the port of Thessaloniki. Sampling site G was in Kalohori, close to the industrial area of Thessaloniki.

The effect of coastal water from the 7 different sampling sites on the survival of *Artemia franciscana* is given in Table 1. As it is presentedin this table, toxicity of natural seawater on *Artemia franciscana* was generally low, with the exception of sampling site F, the port of the city. The authors of the present study had tested the toxicity of coastal water on *Artemia franciscana*, with samples from the same sampling sites during September 1997 [6].

Table 1. Toxicity assessment of Thermaikos Gulf with *Artemia franciscana* as test organism.

Symbol	Sampling site	Tests performed	Toxicity range (%)
A	Potamos, Epanomi	2	0
B	Mihaniona	3	0-10
C	Perea	3	0-20
D	Aretsu, Thessaloniki	3	0-5
E	Analipsi, Thessaloniki	3	0
F	Port, Thessaloniki	3	10-30
G	Kalohori	3	0-10

Comparison between the two tests can be seen in Figure 2. Figure 2 shows a clear improvement in coastal water quality between September 1997 and April-May 2000. One possible reason for this improvement is the extension of the treating capacity of Thessaloniki city municipal wastewater treatment plant between 1997 and 2000. The improvement in coastal water quality is very prominent in Analipsi area (sampling site E). This result can be attributed to the fact that domestic effluents are no more discharged untreated to Thessaloniki Bay in that area [6]. Another possible explanation for the improvement of coastal water quality between September 1997 and May 2000 is the fact that rainfall height in Greece is generally high in the period from January to May, but very low from July to October. Therefore, dilution of toxicants is very low in September and potential toxicity of seawater higher.

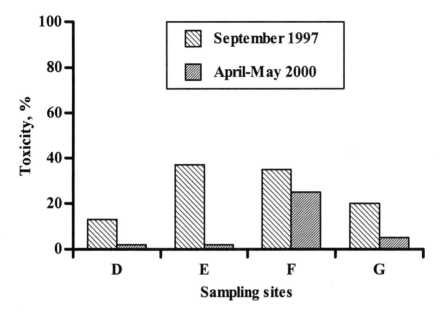

Figure 2: Comparison of coastal water toxicity between September 1997 and April - May 2000 in Thessaloniki area, using *Artemia franciscana* as test organism (D: Aretsu, E: Analipsi, F: Port, Thessaloniki, G: Kalohori).

The effect of coastal water from the 7 different sampling sites on the bioluminescence of *Vibrio fischeri* is shown in Table 2. These results were deduced by the Microtox 500 analyser. The Microtox test uses a sublethal effect, the inhibition on bioluminescence as an endpoint and it is, therefore, more sensitive than *Artemia* test. Figure 3 shows the comparison of *Artemia franciscana* and *Vibrio fischeri* sensitivity in detecting coastal water toxicity in Thessaloniki area during April - May 2000. It is clear that Microtox test is more

Table 2. Toxicity assessment of Thermaikos Gulf with *Vibrio fischeri* as test organism.

Symbol	Sampling site	Tests performed	Toxicity range (%)
A	Potamos	3	0
B	Mihaniona	3	0
C	Perea	3	0-10
D	Aretsu, Thessaloniki	3	0-10
E	Analipsi, Thessaloniki	3	5-10
F	Port, Thessaloniki	3	0-40
G	Kalohori	3	0

sensitive than *Artemia* test. *Artemia franciscana* has been found to be an organism tolerant to certain toxicants [8]. In addition, *Artemia* test is a very useful test for seawater toxicity assessment, because it is a simple and it can be performed with the naked eye, without need for expensive analysers. Microtox test is a reliable test for detecting seawater toxicity [5]. Still, when detecting seawater toxicity with *Vibrio fischeri*, one should be careful to use standard of same salinity with the sample to be tested. The salinity in our coastal water samples was generally high, between 38 and 40 ppt.

Figure 4 shows the 13 sampling sites in Magnesia prefecture in Central Greece. Two sampling sites A and B were in Skiathos island in the open Aegean Sea. Nine sampling sites were in Pagassitikos Gulf. Sampling sites L and M are in a small bay in the opening of Pagassitikos Gulf. All these beaches are used for swimming, with the exception of sampling site F which is the port of Volos. Among these 13 sampling sites, only sites E (Agria) and F (Volos Port) are close to industrial areas. The other sampling sites are affected only by municipal sewage and effluents from foodstuff processing small units, mainly olive oil mills. There are more than 40 operating olive oil mills in Magnesia prefecture. Their season for operation is usually between October and January.

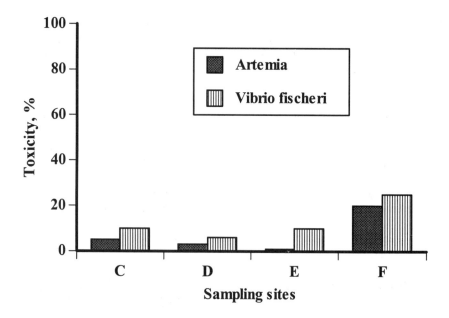

Figure 3: Comparison of *Artemia franciscana* and *Vibrio fischeri* sensitivity in detecting coastal water toxicity in Thessaloniki area during April - May 2000 (C: Perea, D: Aretsu, E: Analipsi, F: Port, Thessaloniki).

Figure 4. Map of Volos area indicating the sampling sites for coastal water samples.

It can be seen in Table 3 that in October 1999 coastal water toxicity was generally higher than one would expect in areas which are not much affected by heavy industries. On the other hand in May 2000, the coastal water toxicity was generally very low, in most cases it was zero.

Figure 5 shows the comparison of coastal water toxicity between October 1999 and May 2000 in Magnesia prefecture coastal areas. It is clear that coastal water toxicity was generally low in May 2000, but relatively high in October 1999. It was also observed in Thessaloniki area (see Figure 2) that toxicity was lower in spring than in early autumn. One possible explanation could be the fact that October comes after a long dry season, therefore toxic pollutants are not so much diluted. On the other hand, the operation start-up of many olive mills in October could have affected the coastal water toxicity. Another possible

Table 3. Toxicity assessment of Pagassitikos Gulf with *Artemia franciscana* as test organism.

Symbol	Sampling site	May 1999	October 1999	May 2000
A	Skiathos	-	10-50	0-20
B	Koukounaries	-	10-50	0
C	Kala Nera	-	20	30
D	Kato Gatzea	-	0	0
E	Agria	-	20-40	0
F	Port, Volos	0-5	20	0
G	Alykes beach	0-5	10	0-10
H	Aghialos	-	40	0
I	Almyros beach	-	0-20	0-30
J	Amaliapoli	-	0-10	0-10
K	Sourpi beach	-	0	0
L	Rodia beach	-	0	0
M	Lihoura beach	-	10-40	0

explanation is the fact that a lot of tourists use to visit these beaches in the season from June to September, increasing the amount of untreated sewage discharging to the sea, since there is no municipal wastewater treatment plant in the tourist beaches of Magnesia prefecture.

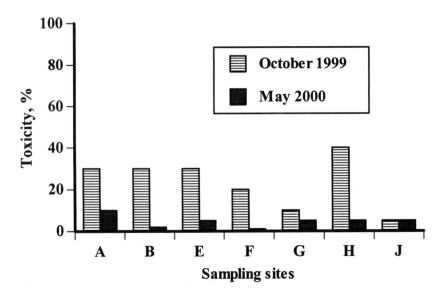

Figure 5: Comparison of coastal water toxicity between October 1999 and May 2000 in Volos area, using *Artemia franciscana* as test organism (A: Skiathos, B: Koukounaries, E: Agria, F: Port, Volos, G: Alykes, H: Aghialos, J: Amaliapoli).

4 Conclusions

Bioassays can be a very useful tool in assessing the coastal water quality. In this study, two bioassays, *Artemia* test and Microtox test were used for assessing the coastal water quality from 7 different sites in Thessaloniki area and 13 different sites in Magnesia prefecture. It was found that there was a clear improvement in coastal water quality in Thessaloniki between September 1997 and April- May 2000. Microtox test was found to be more sensitive than Artemia test in detecting seawater toxicity. In addition, due to the fact that no single marine organism is sensitive to all potential toxicants, it is essential that a battery of bioassays should be used for a proper assessment of seawater quality.

Coastal water quality was generally very good in Magnesia prefecture in May 2000. On the other hand, it was not so good in October 1999. It is suggested that there should be more tests in future with bioassays covering the whole year and that bioassays should be combined with chemical tests for a proper assessment of seawater quality.

Acknowledgement

The authors of this study express their gratitude to Sewerage Authority of Thessaloniki S.A. and Prefecture of Magnesia Development Company S.A. for financial support on this study.

References

[1] Van der Gaag, M.A. Combined effects of chemicals: An essential element in risk extrapolation for aquatic ecosystems. *Proc. of the 1st IAWPRC International Symposium on Hazard Assessment and Control of Environmental Contaminants in Water*, ed. S. Matsui, TIC Ltd, Osaka, Japan, pp. 662-668, 1991.

[2] Environment Agency. *Regulatory whole sample ecotoxicological testing: Comparison of approaches*, Environmental Agency, UK, pp. 21- 45, 1995.

[3] Persoone, G. and Castritsi - Catharios, M. A simple bioassay with Artemia larvae to determine the acute toxicity of antifouling agents. *Water Res.*, **23** (7), pp. 893- 897, 1989.

[4] Pandey, A.S. and MacRae T.H. Toxicity of organic mercury compounds to the developing brine shrimp, Artemia. *Ecotox. and Environ. Saf.*, **21**, pp. 68- 79, 1991.

[5] Kaiser, K.L.E. and Palabrica V.S. *Photobacterium phosphoreum* toxicity data index. *Water Poll. Res. J. Canada*, **26 (3)**, pp. 361- 431, 1991.

[6] Samaras P., Sakellaropoulos G.P., Kungolos A. and Dermissi S. Toxicity assessment assays in Greece. *Fresenius Envir. Bull.*, **7**, pp. 623- 630, 1998.

[7] Cook N.H. and Wells, P.G. Toxicity of Halifax harbour sediments: An evaluation of the Microtox solid phase test, *Water Qual. Res. J. Canada*, **31 (4)**, pp. 673- 708, 1996.
[8] Trief N. M. Toxicity and heavy metals, oils and other organics on *Artemia. The Brine Shrimp Artemia Vol. I. Morphology, Genetics, Radiobiology, Toxicology*, eds. G. Persoone, P. Sorgeloos, O. Roels and E. Jaspers: Universal Press, Wetteren, Belgium, pp. 253-262, 1980.

Section 10
Water Quality Issues

How to protect our coasts – main results from the OPCOM project

K. Duwe & I. Nöhren
HYDROMOD Scientific Consulting, Germany

Abstract

Within the MAST project OPCOM general criteria were derived for the design of operational model systems to improve integrated coastal management. This comprised a comprehensive data and information basis, state-of-the-art but easy-to-handle and cheap information technology, sophisticated data assimilation techniques as well as long-term stability and the ability of the model systems to describe the hydrographic variability of the respective area.

The improvement of long-term operability and reliability of operational and pre-operational coastal zone models has been addressed by means of sensitivity analyses of selected test calculations. Furthermore, the evaluation of sensitivity studies of the input from meteorological and large-scale hydrographic models resulted in requirements on data quality and coverage and guidelines on judgement on larger scale model data resolution and quality have been formulated. The forecast of boundary values is also of high importance for coastal models. Therefore statistically significant trends in the differences between forecasts and measurements have been established by means of analysis of residual errors of operational models and algorithms both by classical statistical methods (multi-regression techniques) and by new-generation predictive filtering methods.

Finally a corresponding coastal management database was defined. Besides the development of a database structure incorporating essential information derived from (pre-) operational model systems and monitoring networks a user interaction structure for model management has been established. The application of the whole system is mainly targeted at model related data access and visualisation and comprises tasks like the display of map information from the model area, display of selected model results as e.g. time series or 2D fields and animations, and the comparison of modelled and measured data using visual comparison and statistical analyses.

1 Introduction

More than half of the world's population lives within 60 km of the shoreline. The coasts are unique in a very real economic sense as sites for port and harbour facilities but they are also rich and diverse ecosystems. The combination of freshwater and salt water in coastal estuaries creates some of the most productive and richest habitats on earth. Furthermore the coasts are highly valued and greatly attractive as sites for resorts and as vacation destinations.

The existing different interests comprising sustainable protection and the economic use of the coastal and shelf environments require the long-term predictive capability of coastal zone evolution. An important prerequisite are general improvements of forecast ability of operational systems.

In the framework of the OPCOM project methods and techniques for continuous monitoring and operational forecasting for coastal waters were applied based upon existing operational model systems (Duwe et al. [2]).

These models could provide important improvements to the current state of affairs in continuous monitoring, operational forecasting, and integrated coastal management (including risk assessment), in short: helping to protect our precious coastal environment.

2 Minimum User Requirements

The most important conclusions and user requirements derived in the OPCOM project regarding the usefulness and importance of local operational models in integrated coastal zone management are described in this chapter.

In this context experiences were used from either currently running operational model systems of coastal zones (e.g. Elbe estuary) or pre-operational models (e.g. Tagus and Gironde estuaries). Operational model results were analysed for coastal zone management purposes (Freissinet et al. [3], Lauri et al. [4], Neves et al. [5]), after the respective model systems were modified accordingly taking into account specific local demands for coastal management in close cooperation with possible first users.

2.1 Operability of operational models

All four project applications covered a wide range of user interests and hydrographic conditions. Some of the major experiences drawn from many years of operation show a considerable potential for local small-scale operational models if they could be operated in a smooth and rather inexpensive way.

There are, however, some conditions to be met to enable a successful operation:

1. A comprehensive data and information basis.
2. State-of-the-art but easy-to-handle and cheap information technology.
3. Sophisticated data assimilation techniques.

4. Long-term stability for coastal management and planning purposes.
5. Ability to describe the hydrographic variability of the area to enable meaningful user investigations.

A number of important suggestions are outlined below which could improve the operability of local operational models in the coastal environment:

❏ Data and information basis: Important criterion is the accuracy of the bathymetry and the boundary conditions. This information has to be easily available for long periods of time and continuously updated.

❏ Requirements on information technology: The necessary information technology becomes more and more easily available. Field instrumentation technologies have been improved, so that the implementation of monitoring plans with lower cost and better coverage of the areas of interest are possible. In general, the models that are being used within the OPCOM project are running on low price workstations, and there is a generalised use of high performance software to analyse and interpret the model results. The increase of measuring and modelling capacity implies the need of efficient data management systems capable to take advantage of such amount of data and enable the users to easily search and analyse this available knowledge.

❏ Data assimilation topics: For management purposes data assimilation must be carried out on two perspectives: (a) to prove that the model is functioning properly and (b) to improve model results.

❏ Long-term management topics: The basic infra-structures of a region are built to last for long periods of time. Their design and construction must be based on long-term management options. Even considering the uncertainty of the boundary conditions, operational modelling can still be useful. They can predict evolution in different scenarios and, as the time passes and we move from several scenarios to a single reality, models can still be used to refine previous predictions.

❏ Reproduction of coastal variability: A local operational model has to take into account the spatial and temporal variability of parameters and their governing processes. The ability of the model in this respect will have to be ascertained by thorough checks with available field data over a long period (preferably several years to cover also some extreme situations and long-term trends).

2.2 Improving coastal management

Generally there are quite a number of different objectives of coastal zone management where local operational models could be used to its advantage. These include

❏ Long-term routine monitoring and assessment activities in coastal waters to complement existing field measurements and observations. High-resolving (in time and space) model information can provide a far more consistent picture than single point time series or accidental observations. Updating of extension of tidal flats, in-situ measurement probes for nutrients and dissolved oxygen or the vast amount of data from ferry box operations are expected to be feasible in the short term.

❑ Future planning activities and necessary assistance in decision support by providing impact assessment information. The very existence of an operational system could help to avoid considerable efforts and costs for the set-up and qualification of another version of consulting tool (preventing the everlasting reinvention of the wheel).

❑ Needs of short-term information provision like nowadays weather forecasts. Such applications could include early warning systems (pollution control or hazardous surge predictions) and on-line nowcast information on current and wave conditions. However, especially the necessary on-line data flow from field stations and larger-scale hydrographic and meteorological models prove to be essential and (in most cases) very expensive.

2.3 End user requirements

From an end user point of view, a generalised use of local operational models could provide positive additions to coastal zone management activities.

However, here the generalisation of past and present experiences almost reach their limitation. This is due to the very nature of end users: They are in essence unique. These species may comprise members of the general public, routine clients from local authorities, adhoc clients like engineering consultants or members of the scientific community.

Combined with the day-to-day routine user contacts of the project partners interviews have been performed within the project. This interview information has been put into perspective to elaborate a meaningful list of minimum requirements. Generally the end user requirements may be categorised according to the following rough distinction:

❑ Availability of information: Information produced by a management tool must be easily and rapidly available. This includes the capability to process large amounts of information in a structured way to facilitate easy search and selection operation. For the combination of information with information from other (external) sources a database must be used. It must be capable to deal, besides time series, with 2D and 3D array data. For the distribution of information a client-server architecture via internet is becoming more and more customary.

❑ Handling of the system: A very costly part of the system is the time required learning it. Therefore pre- and post-processing software must be as much as possible the same already in use by the final user. Standard software should be preferred. Graphical Information Systems (GIS) are becoming standard tools for visualisation of model results and provision of input data and should be used where possible.

❑ Costs: In general the costs for a modelling tool are split into the development, maintenance, learning/training and exploitation. Minimisation of the costs depends on each case. For small institutions, minimisation of the costs can be done through a gradual transfer of the tasks from the developer team to the end user. Using the internet facilities the model can be installed and maintained by this team and exploited remotely by the final user, using a client-server philosophy or similar. Overall there has to be a compromise between the marketing effect of low cost or even free overall information and the cost-covering activities in providing specific detail (incl. data banking etc.).

❑ Reliability: Quality assurance must be done for the code of the model and for results. To obtain good quality results good data must be provided to the model.

❑ Actuality of information: Results must be available in real time and short and long time forecasts, too. For emergency purposes the quality of the results (including their error bars) must be guaranteed. To allow these objectives to be met both the model and data acquisition must be done in real time.

3 Requirements on operability

A combination of experience and selected model investigations in all OPCOM sites and subsequent generalisations of results were used to arrive at some guidelines for the improvement of operational coastal models regarding forecast ability, field data assimilation, and long-term stability and reliability in continuous operation.

3.1 Operability and reliability

The improvement of long-term operability and reliability of coastal operational models was one of the main aims of the project. By means of user interviews, the collection of already existing experiences from running local models, and selected sensitivity studies associated guidelines are being elaborated in the closing stages of the project.

3.2 Input from larger-scale models

An important basis for the long-term reliability of a local operational model as an added value for successful coastal zone management is the availability of larger-scale hydrographic and meteorological information. This is normally only feasible by using already existing meso-scale hydrographic models of the adjacent sea area and data from regional weather forecast models. The elaboration of ground-rules and guidelines on the necessary resolution and quality of larger scale model data by means of user interviews and selected sensitivity studies in the individual OPCOM application areas was performed. Results of these activities will be published in a separate publication.

3.3 Forecast of boundary conditions

A major step towards improving the forecast ability of a local model is a better prognosis of hydrodynamic variables at the open boundaries. Accordingly specific combinations between deterministic and statistical forecasting and calculation of residual errors of model boundary conditions were developed in a special project task. Thereby classical and non-classical statistical methods (see e.g. Aver & Strichartz [1], Mandelbrot [6]) were applied.

Different case studies to improve waterlevel and current information have been undertaken within OPCOM. In the case of the Bay of Marennes-Oléron the classical statistical analysis used the mean and standard deviation of some parameters in order to estimate the low, average and high flow of the three main rivers (Charente, Seudre and Gironde). The exact evaluation of the discharges of these three fresh water inflows then gives the boundary conditions of the rivers and provides the salinity variation near the oyster farms. The non-classical statistical method used the harmonic analysis to validate the results of a coarser grid coastal model. These results give the boundary conditions for the refined model of Marennes-Oléron. This method is restricted to a minimum duration of a tide record in order to allow the separation of couples of tide waves. Investigations show that for the separation of the largest number of tide waves a record of one year duration is necessary. Since such long waves records are generally not available it is not possible to separate some waves of specific frequencies. The technique commonly used is to group these non-separable waves and to carry out the harmonic analysis on only the dominating wave within the group. The analysis will give information on amplitude and phase for the whole group. In order to deduce each tide wave characteristics a coefficient of proportionality between the waves inside one group is generally adopted (this assumption is acceptable in local areas). The problem is then to define this coefficient of proportionality from a sufficiently long field data record measured close to the considered area or deduce it from a world ocean numerical model which takes into account these individual waves.

Regarding the Elbe estuary error time series obtained as differences between observed measurement data and values predicted by a numerical model were investigated. Statistically significant (seasonal) trends were identified and eliminated with regression procedures. After removing the trend classical linear time series models (ARMA, ARIMA) were fitted to the remaining time series. To measure the performance of these models the one-step-ahead prediction errors were computed and its standard deviations were compared with the original ones. In the statistical average over the given error time series waterlevel prediction errors at low tide can be reduced by 5 cm (approx. 18 % error reduction). The waterlevel prediction error for high tide can be reduced by 3 cm (approx. 16 % error reduction). In case of the non-classical statistical analysis wavelet transformations were applied before extracting the seasonal trend.

The Archipelago Sea data set was obtained from the differences of observed and predicted values. It contains the residual errors of flow velocities in four selected straits of the Archipelago Sea. The velocities are measured in longitudinal (along-channel) and transversal (cross-channel) directions with respect to the position of the corresponding channel. A significant error reduction of more than 95 % can be achieved for the along-channel flow and more than 90 % in cross direction if wavelet decomposition of flow velocities is performed and the ARMA method applied afterwards. The major reason for the much better performance compared to the Elbe results is the different size of the time step of the prediction: 24 hours for the Elbe and 10 minutes for the Archipelago Sea.

After the application of the wavelet transformation the ARMA model fitted to the synthesised or synthesised plus de-trended (removal of the seasonal trend) time series reduces the standard deviation of the prediction error only to a very small amount compared with the result obtained by using the ARMA model for the original time series.

4 Database management system

A wide range of data storage techniques and accompanying expertise has been identified among the OPCOM partners: Starting at a "low level" storage of plain ASCII files within a special directory structure over small local data bases for special applications up to a sophisticated numerical model system bound to a commercial RDBMS (Relational Database Management System).

Due to the huge amount of e.g. numerical model results even modern and powerful database systems will be overloaded with managing single numbers of these data. A solution could either be the use of an object oriented DBMS capable of managing BLOBs (Binary Large Objects) or just storing header- or meta information within the DB together with a pointer to the location of affiliated numerical data, or a mixture of both. Storing just meta information (like e.g. parameter, time and date, measurement device, model run) has the advantage to be independent from the physical storage device on which numerical data reside on, i.e. a DB system itself will remain operational without numerical data.

A "mixed mode" operation may be appropriate at sites where special needs occur, i.e. different storage for large size data (outside a DBMS) and for small size data (inside a DBMS), where instant access by the DB systems itself is required (online monitoring data, for example).

A definition of general rules is not possible since every single site has special requirements in order to maximise performance.

Different file formats and storing techniques for DB-external data (or BLOBs) are available, some of them for free. HDF (Hierarchical Data Format) and netCDF (Common Data Format) are the most wide spread and common used ones, very well documentation exist, and – concerning netCDF – several coding standards for it's application to oceanographic purposes.

Typical tasks in (operational) modelling consists of manipulating data and parameter files using editing tools, running the model, and the viewing of model results with specific tools. The model data editing and viewing tools are diverse reaching from a simple text editor to graphical map-based parameter editor and 3D visualisation tools. In addition to this Input – Compute – Output sequence a set of different software tools are used to move and convert input data to be available in the correct format, and then to publish the results. A model management browser may support the model management system.

The model browser requires separate tools for model input data handling, model parameter setting and model result visualisation. These tools are handled directly by the user with a graphical user interface (GUI). It enables the user to access the data in an easy-to-handle way and represent the data illustrative and clearly.

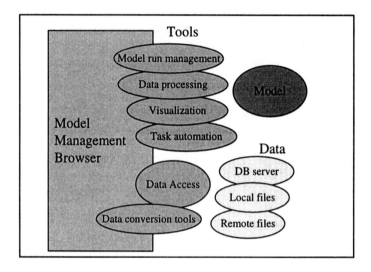

Figure 1: Model management toolbox

Typical tasks done by the user with the system include at least the following:

- displaying map information from the model area,
- selecting model data to be displayed,
- displaying selected model results, for example, as time-series, 2D fields, or 2D animations,
- displaying measurement data from the model area,
- computing statistical information from the model and measurements data,
- exporting modelled and measured data as number or figures for further processing
- comparing modelled and measured data using visual comparison and statistical analyses.

The integration of the database and user interaction structure leads to the definition of an open application structure where an entire application entity is based on open application standards, such as SQL, SGML, TCP/IP-protocols and Java script. A demonstration version of a OPCOM database prototype will be completed within 2000 and a respective CD-ROM will be available via the European Commission early in 2001 for the general public.

Conclusions

Although in the early stages of reflections about the achievements of the OP-COM (Operational Modelling for Coastal Zone Management) project it can be stated that local models of coastal waters can provide useful, sometimes even crucial information for a large variety of coastal management objectives. It is hoped that this initiative will help to understand that operational modelling in coastal zones is no more a subject only of pure science or large hydrographic offices. Possibilities are real and chances plentiful.

Acknowledgements

The OPCOM project is partially being funded by the European Commission under the DG Research MAST III Programme (contract no. MAS3-CT97-0089). The work described in this publications was done in co-operation with the project partners SOGREAH, Echirolles (France), Environmental Impact Assessment Centre of Finland Ltd., Espoo (Finland), ARGOS Ltd., Vaasa (Finland), Hidromod, Lisbon (Portugal), CREMA-CNRS, L'Houmeau (France), and Instituto Superior Técnico, Lisbon (Portugal).

References

[1] Aver, E. & Strichartz, R.S., Exact Hausdorff measure and intervals of maximum density for Cantor sets. *Ttrans. Amer. Math. Soc.*, **351**, pp. 3725–3741, 1999.

[2] Duwe, K., Nöhren, I., & Post, J., Operational modelling for coastal zone management (OPCOM) – Goals, results, recommendations and still open questions. *Proc. of the Int. Conf. COASTAL ENVIRONMENT 2000*, WIT Press: Southampton, 2000 (in this volume).

[3] Freissinet, C., Sauvaget, P., & Bacher, C., Environmental Impact Assessment on the Marennes-Oleron Bay: Operational Mathematical Modelling and GIS. *Proc. of the Int. Conf. COASTAL ENVIRONMENT 2000*, WIT Press: Southampton, 2000 (in this volume).

[4] Lauri, H., Ylinen, H., Koponen, J., Helminen, H. & Laihonen, P., Combating nutrient spillage in the Archipelago Sea – a model system for coastal management support. *Proc. of the Int. Conf. COASTAL ENVIRONMENT 2000*, WIT Press: Southampton, 2000 (in this volume).

[5] Neves, R., Delfino, J., Silva, A., Leitão, P., Leitão, J., Pina, P., Braunschweig, F. & Miranda, R., Coastal Management Supported by Modelling. Optimising the Level of Treatment of Urban Discharges into Coastal Waters. *Proc. of the Int. Conf. COASTAL ENVIRONMENT 2000*, WIT Press: Southampton, 2000 (in this volume).

[6] Mandelbrot, B., Fractals and Scaling in Finance., Springer Press, Berlin-Heidelberg-New York, 1997.

Operational modelling for coastal zone management (OPCOM) – goals, results, recommendations and still open questions

K. Duwe, I. Nöhren & J. Post
HYDROMOD Scientific Consulting, Germany

Abstract

OPCOM is an EU funded research & development project which aims at improving methods and techniques for continuous monitoring and forecasting of coastal waters. OPCOM takes advantage of the experience acquired by individual project partners using existing coastal zone models (namely for the Elbe estuary, the Bay of Marennes-Oléron, the Tagus estuary and the Finnish Archipelago Sea). The project strives also to incorporate advice from the user community and looks at such diverse management objectives like environmental monitoring for the protection of oyster farms or the provision of up-to-date hydrographic forecasting for ship routing. In this paper special details are provided concerning the implication of project results for local monitoring practice. Also, experiences from a local modelling application are reported which was concerned with navigational support for pilots in a narrow waterway.

1 Introduction

Between 1997 and 2000 a European research project under the framework of the MAST-III Programme was carried out to look into the possibilities of improving coastal zone management by considering information from local operational hydrodynamic models. This paper outlines project goals and presents general conclusions from the actual project activities. The OPCOM consortium consists of seven organisations from Germany, France, Portugal, and Finland.

2 Goals

The primary objective of the project is to improve methods and techniques for continuous monitoring and operational forecasting in coastal waters by using already existing operational model and monitoring infrastructure to address needs and quality standards of interested first users. The individual technical and scientific tasks of the project may be outlined as:

❑ A. Site-specific investigations on coastal management aspects to define user requirements: Improvement of coastal management and monitoring practice by analysing operational data and model results and defining user requirements on model and monitoring systems in four OPCOM key pilot studies for different hydrographic regimes and coastal management objectives.

❑ B. Model investigations on operational aspects and forecast ability: Improvement of operational and pre-operational coastal zone models regarding forecast-ability, field data assimilation, and long-term stability and reliability in continuous operation.

❑ C. Integration of modelling, coastal monitoring, and user specifications: Goal is a dedicated database structure as basic operational forecast segment of coastal zone management information systems also suitable for GOOS (Global Ocean Observation System) purposes.

There are many different hydrographic situations and coastal management problems relevant in European waters. OPCOM intends to address a wide range of this spectrum. Thereby it is convenient to pay special attention to suitable areas where a profound infrastructure of in-situ data acquisition and validated model systems of coastal circulation and transport already exist (e.g. Nöhren and Duwe [6]). Especially the problems of pollutant transport (or the environmental impact assessment based on transport forecasts in general) are of great importance and need a thorough modelling technique (Post [7], Duwe and Pfeiffer [1]). Within OPCOM the following geographical areas were investigated:

- Bay of Marennes-Oléron (French Atlantic Coast)
- Elbe estuary (German North Sea Coast)
- Tagus estuary (Portuguese Atlantic Coast)
- Archipelago Sea and Luodonjärvi Lake (Bothnian Sea)

Starting with the MAST-project OPCOM in late 1997 the main goals were already defined. Especially these goals and the planned approach to reach them were reasons enough for the EU to fund this project. The highlights were:

- Four highly interesting target areas with first users eager to test the new OPCOM-products to find solutions for their problems.

- A number of different model approaches were proposed for the application on a wide range of hydrodynamic modelling and forecasting tasks.
- Optimisation of model codes and algorithms to make models faster and more reliable.

Specific local experiences from the OPCOM project are described in Freissinet et al. [3], Lauri et al. [4], and Neves et al. [5]. These activities were also directed towards

- User defined requirements for operational models for the coastal management.
- New approaches to improve the reliability of coastal operational models.
- How to reproduce the coastal variability to improve operational model approaches.

Main objective in these local applications was how to help to protect the coastal environment by means of e.g. expert systems for traffic guidance, sewage outlet control, environmental monitoring of salinity variations on oyster banks, and nutrient transport between islands in a non-tidal sea.

3 Recommendations derived from project results

Generally detailed project results concerned with minimum user requirements on local operational models are described in Duwe and Nöhren [2] in this volume. However, some interesting features often sidelined in the discussions about coastal zone monitoring are mentioned here.

3.1 General recommendations for local monitoring practice

The existing local monitoring practice in coastal waters could also be assisted and considerably optimised (even in an economic sense) by consulting small-scale operational models. Here special attention has to be paid on modelling strategies (Sündermann et al. [9]) and sampling design (Schoer and Duwe [8]). Some very important remarks can be given by drawing experiences both from within the OPCOM project and from the already functioning operational networks.

❑ By matching relevant information 'from both sides of the shoreline' only a comprehensive account of changes and impacts in coastal zones can be achieved. Here a local hydrographic model can be an efficient tool, which could even be extended in future by ecological modules or the connection with a river basin model of the corresponding catchment area.

❑ Readily available model information can be used very intelligently to optimise location or sampling rates of field stations that are costly to be operated and maintained. Moreover dedicated interpretations of model results can be very informative to improve future strategies of measurement and monitoring activities.

❑ The continuous check of field measurement registrations, with the corresponding model results, could be used very efficiently to look at data quality and, accordingly, lead to swift reactions in case of quality deterioration or malfunctions.

The described avenues of progress through operating smaller-scale local models will have to be defined more clearly and tailored to specific individual demands in a case-by-case manner. Here the requirements and resources of the local users become overwhelmingly important.

3.2 Example regarding support of a pilots positioning system

As example is given here an operational model system applied to the tidal estuary of the Elbe river and the dynamically complex port region of the City of Hamburg. The numerical model can be used for navigational support by integrating the results in the pilots positioning system LOPOS®.

LOPOS® is a modular port and waterway control system. It is designed to become an integrated part of the complex port logistics. Planning, engineering design and decision making in ports, waterways and coastal zones highly depend on forecasts and case studies of dynamic situations. Thus a wide range of applications can be treated by utilisation of the circulation model integrated into LOPOS® and supplemented by adequate program components.

Within a VTS applied to the Elbe River the integrated operational hydrodynamic module provides now- and forecasts of the dynamical situation with high accuracy and in high spatial and temporal resolution. These comprise e.g. current speed and direction and water-levels which can be target-related incorporated into the electronic sea chart display (ECDIS, established by 7Cs, Hamburg) of LOPOS® both on board and ashore. Generally this application of a local numerical model system has mainly two different kinds of users: the navigational authorities which are mostly interested in the spatial distribution of parameters and pilots who are keen to know the temporal development of parameters on a defined route.

The public authority responsible for the maintenance of waterways is predominantly concerned with:

- the temporal changes in water-level to safeguard their shipping forecasts,

- the spatial distribution of currents to monitor the development of sedimentation tendencies and the dynamical conditions at dumping sites to improve dredging activities.

All other information from a local model is of minor interest to such institutions. From the pilots (or captains) point of view this is quite different although the major concern is also connected with water-level information. Here two major information sources have to be available apart from an operational model system:

A very accurate bathymetry (admittedly this is not a real-model variable unless a very good morphological forecast model is incorporated which is not state of the art). The accuracy should preferably be in the range of decimetres.

A similarly crucial parameter is the geographical position (e.g. by DGPS) with accuracy in the range of metres (relative to the bathymetric chart referred to).

The following list of model parameters is of considerable or at least some importance for vessel traffic information systems: the actual water depth, as a function of time and ship route taking into account bathymetry, water-level and water density.

The latter is important because of the different draught of ships in freshwater compared with seawater, accordingly a spatially and temporally variable estuarine salt wedge has to be accounted for. Preferred accuracy: < 0.5 m draught.

Wind speed and direction is not only influencing water currents and waterlevels but also have a pronounced direct effect on ship movements. This is especially the case for crosswinds. On very exposed routes a local meteorological model may be necessary to complement the available general circulation model and singular field stations. Preferred accuracy: < 5 knots.

In areas with strong tides information on current speeds is very useful for navigation. Favourable current conditions could save fuel costs along the route, in locations with considerable crosscurrents a safer navigation could be ensured. Preferred accuracy: < 0.2 m/s.

Locally there may also be need to have reasonably good information on waves or ice drift which points to the necessity of introducing additional modules into the system.

According to OPCOM experiences well-adapted and verified operational models could be used very efficiently to optimise ship tracks according to hydrographic and weather conditions to improve safety and minimise fuel costs. Forecasts should extend to 12 hours or 24 hours in advance and should be updated regularly (e.g. every hour). In this way it would provide welcome additional decision making support in a vessel traffic information system.

4 Open Questions

Data acquisition in real time is becoming standard, models are progressing very quickly and pre- and post-processing software is becoming more and more powerful and user-friendly. As research progresses also the availability of modellers in the labour market is growing. Therefore generally future prospects of local operational models for coastal zones are promising.

However, some open questions still loom on the horizon: These include:

- How to improve the operability of coastal operational models in the framework of future developments regarding the water management of river basin scale and activities within GOOS concerned with coastal waters.

428 Environmental Coastal Regions III

- How to improve the models accuracy and efficiency even further which are needed for individual applications and planning necessities.
- How to improve the situation for the environment at European coasts with increasing ship traffic, municipal sewage outlets, tourism, fish and shell-fish farming, etc.
- How to improve the model's forecast ability concerning topographic changes within coastal areas of high spatial and temporal variability.

The easy accessibility of results to a wider audience, in addition, is essential for acceptance and considerable development prospects. An Internet based service with easy to understand 'basic' information would be an ideal start, more sophisticated questions would be posed almost concurrently after going on-line. So far developments have been hampered mostly by the lack of long-term planning possibilities on the side of the service providers and not by lacking user interests on information from local operational models.

Acknowledgements

The OPCOM project is partially being funded by the European Commission under the DG Research MAST III Programme (contract no. MAS3-CT97-0089). The work described in this publication was done in co-operation with the project partners SOGREAH, Echirolles (France), Environmental Impact Assessment Centre of Finland Ltd., Espoo (Finland), ARGOS Ltd., Vaasa (Finland), Hidromod, Lisbon (Portugal), CREMA-CNRS, L'Houmeau (France), and Instituto Superior Técnico, Lisbon (Portugal).

References

[1] Duwe, K. & Pfeiffer, K.D., Three-dimensional Modelling of Transport Processes and its Implications for Water Quality Management. Computer Modelling in Ocean Engineering, eds. B.A. Schrefler & O.C. Zienkiewicz, A.A. Balkema, Rotterdam/Brookfield, pp. 319–425, 1988.
[2] Duwe, K. & Nöhren, I., How to protect our coasts – Main results from the OPCOM project. Proc. of the Int. Conf. COASTAL ENVIRONMENT 2000, WIT Press: Southampton, 2000 (in this volume).
[3] Freissinet, C., Sauvaget, P., & Bacher, C., Environmental Impact Assessment on the Marennes-Oleron Bay: Operational Mathematical Modelling and GIS. Proc. of the Int. Conf. COASTAL ENVIRONMENT 2000, WIT Press: Southampton, 2000 (in this volume).
[4] Lauri, H., Ylinen, H., Koponen, J., Helminen, H. & Laihonen, P., Combating nutrient spillage in the Archipelago Sea – a model system for coastal management support. Proc. of the Int. Conf. COASTAL ENVIRONMENT 2000, WIT Press: Southampton, 2000 (in this volume).
[5] Neves, R., Delfino, J., Silva, A., Leitão, P., Leitão, J., Pina, P., Braunschweig, F. & Miranda, R., Coastal Management Supported by Modelling. Optimising the Level of Treatment of Urban Discharges into Coastal

Waters. *Proc. of the Int. Conf. COASTAL ENVIRONMENT 2000*, WIT Press: Southampton, 2000 (in this volume).

[6] Nöhren, I. & Duwe, K., An Operational Modelling System for Shortterm Forecasting of Transport Processes in Regional Seas and Coastal Waters. *Proc. of the 1ˢᵗ Int. Conf. On Interaction of Computational Methods and Measurements in Hydraulics and Hydrology (HYDROCOMP'92)*. eds. J. Gayer, Ö. Starosolszky & C. Marksimovic, Vituki, Budapest, Hungary, 1992.

[7] Post, J., Simulation models for pollutant transport in the marine environment – capabilities and limitations. *Proc. of the Symposium: Oceanographic Modelling of the Kuwait Action Plan Region*, University of Petroleum & Minerals, Dharan, Saudi Arabia, 1983.

[8] Schoer, J. & Duwe, K., Sampling design for estuarine investigations. *Trends in Analytical Chemistry*, **5(5)**, pp. 20–25, 1986.

[9] Sündermann, J., Duwe, K., Nöhren, I., Pfeiffer, K.D. & Longfei, Y., Strategy of Current and Transport Modelling in Different Estuaries. *Estuarine Water Quality Management (Monitoring, Modelling and Research)*, eds. W. Michaelis, Coastal and Estuarine Studies, Springer Berlin-Heidelberg-New York, pp. 25–34, 1990.

Disturbances: their role in shaping coastal environments

W.W. Schroeder
Marine Science Program, The University of Alabama and Dauphin Island Sea Lab, USA

Abstract

Disturbances such as tropical cyclones, extratropical cyclones, and floods are known to play important roles in shaping coastal environments. These events generate combinations of high winds, large waves, strong currents, storm surges, fluctuations in sea level, flooding river conditions, excessive local run-off, and subfreezing air temperatures. In turn, these processes can produce both short- and long-term alterations to shoreline and basin geomorphology, bottom sediments, water quality, and biological habitats. It has been clearly documented that short-term, local changes can be significant and extensive, particularly in developed areas. On the other hand, available literature indicates that from a long-term, regional perspective, these events do not generally result in detrimental impacts to unmodified areas but present a serious recurring problem for developed regions. Therefore, a thorough understanding of disturbances is essential in efforts attempting to accurately model and manage coastal environments.

Introduction

Coastal environments and their adjacent watersheds are dynamic environments, constantly undergoing alteration. Temporal scales of change range from less than an hour (e.g., intra-tidal) to geological time references (e.g., eustatic sea level fluctuations), while spatial scales of change range from less than a meter to thousands of kilometers. Within the time frame of days-to-weeks and across distances of 100's to 1000's of kilometers (Figure 1) lies the domain of "synoptic" disturbances (e.g., tropical cyclones, extratropical cyclones, and floods) [1]. They can occur as annual events or at multiple-year recurrence frequencies. Along the Gulf of Mexico and Atlantic coastlines of North America, the approach and landfall components of tropical cyclones are usually classified as the most severe disturbances followed by either floods or extratropical cyclones. This paper examines the role disturbances and disturbance related processes play in shaping coastal environments.

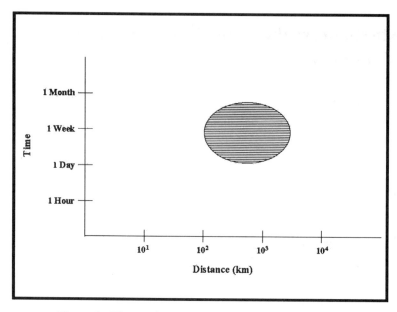

Figure 1. Time and space scales for synoptic disturbances.

Discussion

The effect a disturbance will have on coastal environments (e.g., short- or long-term alterations to shoreline and basin geomorphology, bottom sediment composition and quality, water quality, and/or biological habitats) is principally a function of the dominant characteristics of the disturbance (Table 1) coupled with disturbance related processes (Table 2) and the nature of the coastal region and adjacent watershed (Table 3).

Disturbance characteristics

Table 1 presents a list of disturbance characteristics. Four of these characteristics [size, duration, timing (annual cycle or out-of-season), and recurrence frequency] are common to tropical cyclones, extratropical cyclones, and floods. Other characteristics that are shared by tropical cyclones and extratropical cyclones are track/heading, forward speed, barometric pressure, wind speed and direction, and rate of precipitation. For flooding events rate of precipitation is the other critical characteristic while subfreezing air temperatures are generally only associated with winter extratropical cyclones.

Table 1. Characteristics* of disturbances.

1) Size	6) Barometric pressure
2) Track/Heading	7) Rate of precipitation
3) Forward speed	8) Subfreezing air temperatures
4) Duration	9) Timing (annual cycle or out-of-season)
5) Wind speed and direction	10) Recurrence frequency

* Not all characteristics apply to every disturbance.

Table 2. Disturbance related processes*.

1) Coastal sea level set-up/set-down
2) Large waves
3) Strong currents
4) Storm surge
5) River flooding conditions
6) Excessive local run-off
* Not all processes are generated by every disturbance.

Table 3. Characteristics of coastal regions and adjacent watersheds.

1) Exposure to the open ocean
2) Geology and geomorphology of the offshore and coastal areas
3) Proximity of major rivers
4) Watershed drainage patterns
5) Types of biological habitats
6) Degree of development

Disturbance related processes

Table 2 presents a list of disturbance related processes. All of these processes can be significant components of both tropical cyclones and extratropical cyclones. Along any coastal region, the magnitude of sea level set-up/set-down and the extent of development of waves, currents, and storm surges is principally a function of a combination of the storm's intensity (e.g., central barometric pressure and wind speed), track/heading, location of landfall, and forward speed, and the nature of the coastal region (Table 3); particularly the exposure to the open ocean and the geology and geomorphology of the offshore and coastal areas. One method of classifying tropical cyclones is the Saffir-Simpson Disaster-Potential Scale (Table 4). For rating "northeasters", severe extratropical cyclones that normally occur along the middle U.S. Atlantic seaboard during winter, Dolan and Davis [2] have developed the Dolan/Davis Storm Classes and Coastal Impact classification system (Table 5). The magnitude of river flooding and the extent of excessive local run-off are generally dependent on the size of the area of precipitation, rate of precipitation, duration, watershed drainage pattern, and degree of development. Other important factors are pre-disturbance river and stream levels and watershed soil moisture content.

Table 4. Saffir/Simpson Disaster-Potential Scale.

Storm Category	Central Pressure (millibars)	Wind Speed (knots)	Storm Surge (meters)
1	<980	64-82	1.2 - 1.7
2	965-979	83-95	1.8 - 2.6
3	945-964	96-113	2.7 - 3.9
4	920-944	114-135	4.0 - 5.5
5	<920	>135	>5.5

Case studies

Tropical cyclones
Two excellent case studies of specific tropical cyclones are "The effects of Tropical Storm Agnes on the Chesapeake Bay estuarine system" [3] and "Impacts of Hurricane Hugo: September 10-22, 1989" [4] in the Caribbean Sea and along United States mainland shores adjacent to the South Atlantic Bight. Both of these publications present a broad spectrum of topics in both the natural and social sciences and examine storm effects along both unmodified and developed coastal areas. More recently Young et al. [5] investigated the impacts of Hurricanes Dennis and Floyd on coastal North Carolina and discussed related coastal zone management issues. In addition, Hayes [6] summarizes studies conducted on fourteen separate hurricanes. He focuses on hurricane-generated processes that affect coastal sedimentation including storm surges, strong currents, wave action, and flooding resulting from heavy rainfall. Conner et al. [7] conducted an assessment of the impact of tropical cyclones on the natural environment with a specific objective of synthesizing existing data on the impact of hurricanes on ecosystems in the northern Gulf of Mexico. Nichols [8] analyzes observations from five episodic storm events in the Rappahannock and James River estuaries, Virginia. His focus is on the combined effects of meteorological forcing and extreme river runoff.

Table 5. Storm Classes and Coastal Impacts. (Modified from Dolan & Davis 1992 [2].)

Storm Class	Beach Erosion	Beach Recovery	Dune Erosion	Dune Breaching	Overwash	Inlet Formation	Property Damage
Class 1 (weak)	Minor changes	Full & usually immediate	None	No	No	No	No
Class 2 (moderate)	Modest: confined to lower beach	Full	None	No	No	No	Minor, local
Class 3 (significant)	Erosion: extends across entire beach	Usually recovery over considerable period of time (months)	Can be significant	No	On low profile	No	Loss of many structures at local scale
Class 4 (severe)	Severe beach erosion and recession	Recovery seldom total	Severe dune erosion or destruction	Where beach is narrow	On low profile beaches	Occasionally	Losses of structures at community level
Class 5 (extreme)	Extreme beach erosion (up to 50m in places)	Permanent and clearly notice-able changes	Dunes destroyed over extensive areas	Widespread	Massive in sheets and channels	Common	Extensive regional scale: millions of dollars

Examples of other studies of interest are: McGowen and Scott [9] examine post-hurricane evidence along the northwestern coast of the Gulf of Mexico and then attempt to model the effects on shorelines and bays; Fletcher et al. [10] describe oceanographic and meteorological factors that influenced overwash impacts in the coastal zone of Kaua'i, Hawai'i, during Hurricane Iniki; Halper and Schroeder [11] analyze current meter records to determine the response of shelf waters to the passage of five tropical cyclone systems in the northern Gulf of Mexico; Livingston et al. [12] present a series of field analyses of oyster reefs in Apalachicola Bay system, Florida, to document their recovery following two hurricanes; and Fenner [13], Hubbard et al. [14], Hubbard [15], Van Veghel and Hoetjes [16], and Jordan-Dahlgren and Rodriguez-Martinez [17] all report on various aspects of effects of tropical cyclones on coral reef systems.

Extratropical cyclones
Case study information on U.S. Atlantic coast extratropical cyclones or "northeasters" can be found in Dolan et al. [18] and [19], Davis and Dolan [20], and Jones and Davis [21]. Kocin et al. [22], Uccellini et al. [23], Caplan [24], and Schumann et al. [25] describe various aspects of the March 1993 "Storm of the Century" that affected much of the Gulf and East Coasts of the U.S. The impacts of cold-air outbreaks in the Gulf of Mexico are documented by Huh et al. [26], Roberts et al. [27], and Schroeder et al. [28].

In addition, Kahn [29] investigated the role both extratropical and tropical cyclones play in the post-hurricane recovery of the Chandeleurs barrier island system and Snedden et al. [30] examined a combination of current meter, wave, tide, and wind records from the central Texas shelf during fair-weather periods and extratropical and tropical storm events to assess the kinematics and dynamics of shelf flow and sediment transport. Examples of catastrophic mortality events resulting from subfreezing air temperatures accompanying winter storms are reported in Dahlberg and Smith [31], Taylor [32], and Beddingfield and McClintock [33].

Floods
Ten case studies that examine impacts of flooding events are Chesapeake Research Consortium [3], Hayes [6], McGowen and Scott [9], Zabawa and Schubel [34]; Nichols [35], Schroeder [36], Ingram et al. [37], Schubel and Pritchard [38], Stumpf [39], and Eyre [40]. Tee and Lim [41] report the results from their development of a 2-D numerical model to investigate freshwater pulsing in the St. Lawrence Estuary.

Conclusions

The case studies referred to above, as well as many others (e.g., see references cited in these publications), clearly document that short-term, local changes resulting from tropical cyclones, extratropical cyclones, and floods can be significant and extensive, particularly in developed areas. On the other hand the available literature indicates that from a long-term, regional perspective, these disturbances do not generally result in detrimental impacts to unmodified areas, and often can provide a net benefit [7], [42], [43], and [44]. However, they do present a serious recurring problem for developed regions. Therefore, a thorough understanding of disturbances is essential in efforts attempting to accurately model and manage coastal environments. For example, construction set-back lines and minimal elevations for private homes, commercial enterprises and public facilities should be based on predicted water levels resulting from coastal set-up, storm surge and river flooding while building codes should incorporate climatological statistics for maximum sustained winds and wind gusts, and wave characteristics.

References

[1] Atkinson, B. W. *Meso-scale Atmospheric Circulations*. Academic Press: London, pp. 5-7, 1981.

[2] Dolan, R., & Davis, R. E. An intensity scale for Atlantic coast northeast storms. *Journal of Coastal Research*, **8(4)**, pp. 840-853, 1992.

[3] Chesapeake Research Consortium, Inc. *The Effects of Tropical Storm Agnes on the Chesapeake Bay Estuarine System*, CRC Publication No. 54, The Johns Hopkins University Press: Baltimore and London, 1976.

[4] Finkl, C. W. & Pilkey, O. H. (eds). Impacts of Hurricane Hugo? September 10-22, 1989, *Journal of Coastal Research* Special Issue No. 8, 1991.

[5] Young, R. S., Bush, D. M., Coburn, A. S., Pilkey, O. H., & Cleary, W. J. Hurricanes Dennis and Floyd: Coastal Effects and Policy Implications, *GSA Today*, **9(12)**, pp. 1-6, 1999.

[6] Hayes, M. O. Impact of hurricanes on sedimentation in estuaries, bays, and lagoons. *Estuarine Interactions*, ed. M. L. Wiley, Academic Press: New York, San Francisco, London, pp. 323-346, 1978.

[7] Conner, W. H., Day, J. W., Jr., Baumann, R. H. & Randall, J. M. Influence of hurricanes on coastal ecosystems along the northern Gulf of Mexico. *Wetlands Ecology and Management* **1(1)**, pp. 45-56, 1989.

[8] Nichols, M. M. Response of coastal plain estuaries to episodic events in the Chesapeake Bay region (Chapter 1). *Nearshore and Estuarine Cohesive Sediment Transport*, ed. A. Mehta, Springer-Verlag: New York, Berlin, pp. 1-20, 1992.

[9] McGowen, J. H., & Scott, A. J. Hurricanes as geologic agents on the Texas coast. *Estuarine Research*, ed. L. E. Cronin, Academic Press: New York, pp. 23-46, 1975.

[10] Fletcher, C. H., Richmond, B. M., Barnes, G. M, & Schroeder, T. A. Marine flooding on the coast of Kaua'i during Hurricane Iniki: hindcasting inundation components and delineating washover. *Journal of Coastal Research*, **11(1)**, pp. 188-204, 1995.

[11] Halper, F. B., & Schroeder, W. W. The response of shelf waters to the passage of tropical cyclones - observations from the Gulf of Mexico. *Continental Shelf Research*, **10(8)**, pp. 777-793, 1990.

[12] Livingston, R. J., Howell, R. L., IV, Niu, X., Lewis, F. G., III, & Woodsum, G. C. Recovery of oyster reefs (*Crassostrea virginica*) in a Gulf estuary following disturbance by two hurricanes. *Bulletin of Marine Science*, **64(3)**, pp. 465-483, 1999.

[13] Fenner, D. P. Effects of Hurricane Gilbert on coral reefs, fishes and sponges at Cozumel, Mexico. *Bulletin of Marine Science*, **48(3)**, pp. 719-730, 1991.

[14] Hubbard, D. K., Parsons, K. M., Bythell, J. C., & Walker, N. D. The effects of Hurricane Hugo on the reefs and associated environments of St. Croix, U.S. Virgin Islands - A preliminary assessment. *Journal of Coastal Research*, Special Issue No. 8, pp. 33-48, 1991.

[15] Hubbard, D. K. Hurricane-induced sediment transport in open-shelf tropical systems - an example from St. Croix, U.S. Virgin Islands. *Journal of Sedimentary Petrology*, **62(6)**, pp. 946-960, 1992.

[16] Van Veghel, M. L., & Hoetjes, P. C. Effects of Tropical Storm Bret on Curaçao reefs. *Bulletin of Marine Science*, **56(2)**, pp. 692-694, 1995.

[17] Jordán-Dahlgren, E., & RodrPguez-MartPnez, R. E. Post-hurricane initial recovery of *Acropora palmata* in two reefs of the Yucat<n Peninsula, Mexico. *Bulletin of Marine Science*, **63(1)**, pp. 213-228, 1998.

[18] Dolan, R., Lins, H., & Hayden, B. Mid-Atlantic coastal storms. *Journal of Coastal Research*, **4(3)**, pp. 417-433, 1988.

[19] Dolan, R., Inman, D. L., & Hayden, B. The Atlantic coast storm of March 1989. *Journal of Coastal Research*, **6(3)**, pp. 721-725, 1990.

[20] Davis, R. E., & Dolan, R. The "All Hallows' Eve" coastal storm - October 1991. *Journal of Coastal Research*, **8(4)**, pp. 978-983, 1992.

[21] Jones, G. V., & Davis, R. E. Climatology of Nor'easters and the 30 kPa jet. *Journal of Coastal Research*, **11(4)**, pp. 1210-1220, 1995.

[22] Kocin, P. J., Schumacher, P. N., Morales, R. F., Jr., & Uccellini, L. W. Overview of the 12-14 March 1993 superstorm. *Bulletin of the American Meteorological Society*, **76(2)**, pp. 165-182, 1995.

[23] Uccellini, L. W., Kocin, P. J., Schneider, R. S., Stokols, P. M., & Dorr, R. A. Forecasting the 12-14 March 1993 superstorm. *Bulletin of the American Meteorological Society*, **76(2)**, pp. 183-200, 1995.

[24] Caplan, P. M. The 12-14 March 1993 superstorm: Performance of the NMC Global Medium-Range Model. *Bulletin of the American Meteorological Society*, **76(2)**, pp. 201-212, 1995.

[25] Schumann, S. A., Moser, J., Johnson, G. A., Walker, N. D., & Hsu, S. A. An overview of a strong winter low in the Gulf of Mexico 12-13 March 1993. *National Weather Digest*, **20(1)**, pp. 11-25, 1995.

[26] Huh, O. K., Wiseman, W. J., Jr., & Rouse, L. J., Jr. Winter cycle of sea surface thermal patterns, northeastern Gulf of Mexico. *Journal of Geophysical Research*, **83(C9)**, pp. 4523-4529, 1978.

[27] Roberts, H. H., Rouse, L. J., Jr., Walker, N. D., & Hudson, J. H. Cold-water stress in Florida Bay and northern Bahamas: A product of winter cold-air outbreaks. *Journal of Sedimentary Petrology*, **52(1)**, pp. 0145-0155, 1982.

[28] Schroeder, W. W., Huh, O. K., Rouse, L. J., Jr., & Wiseman, W. J., Jr. Satellite observations of the circulation east of the Mississippi Delta: Cold-air outbreak conditions. *Remote Sensing of Environment*, **18**, pp. 49-58, 1985.

[29] Kahn, J. H. Geomorphic recovery of the Chandeleur Islands, Louisiana, after a major hurricane. *Journal of Coastal Research*, **2(3)**, pp. 337-344, 1986.

[30] Snedden, J. W., Nummedal, D. & Amos, A. F. Storm- and fair-weather combined flow on the central Texas continental shelf. *Journal of Sedimentary Petrology*, **58(4)**, pp. 580-595, 1988.

[31] Dahlberg, M. D., & Smith, F. G. Mortality of estuarine animals due to cold on the Georgia coast. *Ecology*, **51(5)**, pp. 931-933, 1970.

[32] Taylor, D. S. Notes on the impact of the December 1989 freeze on local populations of *Rivulus marmoratus* in Florida, with additional distribution records in the state. *Florida Scientist*, **56(3)**, pp. 130-134, 1993.

[33] Beddingfield, S. D., & McClintock, J. B. Environmentally-induced catastrophic mortality of the sea urchin *Lytechinus variegatus* in shallow seagrass habitats of Saint Joseph's Bay, Florida. *Bulletin of Marine Science*, **55(1)**, pp. 235-240, 1994.

[34] Zabawa, C. F., & Schubel, J. R. Geologic effects of tropical storm Agnes on upper Chesapeake Bay. *Maritime Sediments* **10(3)**, pp. 79-84, 1974.

[35] Nichols, M. M. Response and recovery of an estuary following a river flood. *Journal of Sedimentary Petrology*, **47(3)**, pp. 1171-1186, 1977.

[36] Schroeder, W. W. The impact of the 1973 flooding of the Mobile River system on the hydrography of Mobile Bay and east Mississippi Sound. *Northeast Gulf Science*, **1(2)**, pp. 68-76, 1977.

[37] Ingram, R. G., D'Anglejan, B. F., & Lepage, S. Changes in current regime and turbidity in response to a freshwater pulse in the Eastmain estuary. *Estuaries*, **9**, pp. 320-325, 1986.

[38] Schubel, J. R., & Pritchard, D. Responses of upper Chesapeake Bay to variations in discharge of the Susquehanna River. *Estuaries*, **9**, pp. 236-249, 1986.

[39] Stumpf, R. P. Sediment transport in Chesapeake Bay during floods: analysis using satellite and surface observations. *Journal of Coastal Research*, **4**, pp. 1-15, 1988.

[40] Eyre, B. Water quality changes in an episodically flushed subtropical Australian estuary: a 50 year perspective. *Marine Chemistry*, **59**, pp. 177-187, 1997.

[41] Tee, K-T., & Lim, T-H. The freshwater pulse - a numerical model with application to the St. Lawrence estuary. *Journal of Marine Research*, **45**, pp. 871-909, 1987.

[42] Isphording, W. C., Imsand, D., & Flowers, G. C. Storm-related rejuvenation of a northern Gulf of Mexico estuary. *Gulf Coast Association of Geological Societies Transactions*, **37**, pp. 357-370, 1987.

[43] Isphording, W. C., Imsand, F. D., & Flowers, G. C. Physical characteristics and aging of Gulf Coast estuaries. *Gulf Coast Association of Geological Societies Transactions*, **39**, pp. 387-401, 1989.

[44] Schroeder, W. W., Cowan, J. L. W., Pennock, J. R., Luker, S. A., & Wiseman, W. J., Jr. Response of resource excavations in Mobile Bay, Alabama, to extreme forcing. *Estuaries*, **21(4A)**, pp. 652-657, 1998.

INDEX OF AUTHORS

WIT*PRESS*

Oil and Hydrocarbon Spills II
Modelling, Analysis and Control
Editors: G. RODRIGUEZ, University of Las Palmas de Gran Canaria, Spain, C.A. BREBBIA, Wessex Institute of Technology, UK and R. GARCIA-MARTINEZ, Universidad Central de Venezuela, Venezuela

The transport and fate of hydrocarbons in the sea is governed by complex oceanographic and physico-chemical interacting processes which makes oil spill research one of the most difficult challenges in present day science. The risk of spills still continues to overwhelm prevention measures and to create potential danger in sensitive natural environments.
Helping to close the gap between theoretical developments and practical applications, this book presents state-of-the-art techniques designed to model, prevent and control oil spills on land and in water. The papers featured come from the Second International Conference on this subject and include discussion of: Spills in Estuaries and Delta Areas; Operational Procedures for Storage, Handling and Transportation of Oil; Biological Impact of Oil Pollution; Biological Control Methods; Use of Dispersants; Oil Spill Treating Agents; In-Situ Burning; Experimental and Laboratory Analysis; Sensitivity Maps; Rehabilitation of Oil-Damaged Resources; and Recovered Oil Management.
Series: Water Studies
ISBN: 1-85312-828-7   2000   apx 400pp apx £120.00/US$189.00

WIT*Press*
Ashurst Lodge, Ashurst, Southampton, SO40 7AA, UK.
Tel: 44 (0) 238 029 3223
Fax: 44 (0) 238 029 2853
E-Mail: witpress@witpress

Air Pollution VIII
Editors: C.A. BREBBIA, Wessex Institute of Technology, UK, J.W.S. LONGHURST, University of the West of England, UK and H. POWER, Wessex Institute of Technology, UK

Air pollution is one of the most challenging problems facing the international community. The human need for transport, manufactured goods and services brings impacts on the atmospheric environment both locally and globally. However, whilst there are examples of regulatory success in minimising such impacts, the rapid development of the global economy brings new pressures and the willingness to regulate is often limited by economic concerns. Science holds the key both to identifying the nature and scale of air pollution, its prediction and assessment, and to providing the necessary information for the regulatory decision making process.
This book contains state-of-the-art contributions from scientists working on various aspects of air pollution. Over 70 papers are included and these were first presented at the eighth in a highly successful series of international conferences focusing on this problem.
The topics covered are as follows: Air Pollution Modelling; Air Quality Management; Urban Air Pollution; Urban and Suburban Transport Emissions; Monitoring and Laboratory Studies; Global Studies; Comparison of Modelling with Experiments; Indoor Pollution; Pollution Engineering; Fluid Mechanics for Environmental Problems; Chemistry of Air Pollution; Aerosols and Particles; and Health Problems.
Series: Advances in Air Pollution
ISBN: 1-85312-822-8   2000   apx 900pp apx £265.00/US$426.00

Water Pollution V

Modelling, Measuring and Prediction

Editors: P. ANAGNOSTOPOULOS, Aristotle University, Thessaloniki, Greece and C.A. BREBBIA, Wessex Institute of Technology, UK

Featuring the proceedings of the Fifth International Conference on Water Pollution, this book contains contributions from many different countries and covers a multitude of topics and techniques. Subjects covered include: Wastewater Treatment and Sewage Systems; Mathematical and Physical Modelling; Modelling as a Management Tool; Saltwater Intrusion; Biological Effects; Chemical and Nuclear Pollution; Pollution of Estuaries, Coastal Seas and Open Seas; Lakes, Rivers and Channels; Heavy Metals; Groundwater and Aquifer Contamination; Organic Contaminants; and Oil Spills.
Series: Progress in Water Resources, Vol 1
ISBN: 1-85312-685-3   1999   608pp
£195.00/US$299.00

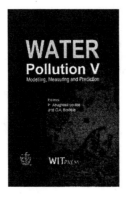

Management Information Systems

Editors: C.A. BREBBIA, Wessex Institute of Technology, UK and P. PASCOLO, Università degli Studi di Udine, Italy

Management Information Systems (MIS) are rapidly finding applications in many areas including environmental conservation, economic planning, resource integration, cartography, urban planning, risk assessment, pollution control, and transport management systems. Representing the state-of-the-art in MIS, the contributors to this book reflect on the ways in which this technology plays an active role in linking together economic development and environmental conservation planning through its own unique characteristics. They also review its growing use in various fields of application.
The papers featured were originally presented at the major international conference Management Information Systems 2000 and span the following broad range of topics: Environmental Management, Economic Development Management, Applications of GIS, Modelling Issues, MIS in Training and Education, The World Wide Web, Applications of MIS, Remote Sensing, Integrated Modelling and Management, and Hydro-Informatics and Geo-Informatics.
Series: Management Information Systems
ISBN: 1-85312-815-5   2000   516pp
£155.00/US$248.00

All prices correct at time of going to press but subject to change.
WIT Press books are available through your bookseller or direct from the publisher.